Applications of Functional Analysis in Engineering

Applications of Functional Analysis in Engineering

J. L. Nowinski

University of Delaware
Newark, Delaware

PLENUM PRESS · NEW YORK AND LONDON

Library of Congress Cataloging in Publication Data

Nowinski, J. L.
    Applications of functional analysis in engineering.

    (Mathematical concepts and methods in science and engineering; v. 22)
    Includes bibliographical references and index.
    1. Functional analysis. I. Title. II. Series.
TA347.F86N68   515.7′02462   81-5213
ISBN-13: 978-1-4684-3928-1   e-ISBN-13: 978-1-4684-3926-7   AACR2
DOI: 10.1007/978-1-4684-3926-7

© 1981 Plenum Press, New York
Softcover reprint of the hardcover 1st edition 1981

A Division of Plenum Publishing Corporation
233 Spring Street, New York, N.Y. 10013

To **Maria Kirchmayer Nowinski**,
granddaughter of **Jan Matejko**,
my dear wife and best friend.

Preface

Functional analysis owes its origins to the discovery of certain striking analogies between apparently distinct disciplines of mathematics such as analysis, algebra, and geometry. At the turn of the nineteenth century, a number of observations, made sporadically over the preceding years, began to inspire systematic investigations into the common features of these three disciplines, which have developed rather independently of each other for so long. It was found that many concepts of this triad—analysis, algebra, geometry—could be incorporated into a single, but considerably more abstract, new discipline which came to be called *functional analysis*. In this way, many aspects of analysis and algebra acquired unexpected and profound geometric meaning, while geometric methods inspired new lines of approach in analysis and algebra.

A first significant step toward the unification and generalization of algebra, analysis, and geometry was taken by Hilbert in 1906, who studied the collection, later called l_2, composed of infinite sequences $x = x_1, x_2, \ldots,$ $x_k, \ldots$, of numbers satisfying the condition that the sum $\sum_{k=1}^{\infty} x_k^2$ converges. The collection l_2 became a prototype of the class of collections known today as *Hilbert spaces*.

However great his contribution, Hilbert failed to interpret geometrically the terms of a sequence as the coordinates of a point, x, in some abstract space. This step was taken in 1907 by Schmidt and Fréchet, who boldly turned to geometric language and noted a complete analogy between the structure of Hilbert's collection and that of the family, named $\mathscr{L}_2$, of square-integrable functions.

After Riesz' observation that it is quite natural to define a *distance* between the members of $\mathscr{L}_2$, the idea of identifying collections (sets) with abstract spaces possessing specific geometric structures was finally born.

The underlying unity of algebra, geometry, and analysis so disclosed gave strong impetus to further studies continuing to this day. Especially important contributions in this direction have been made by Banach, Hahn, Moore, Sobolev, and Wiener, to mention only a few.

The theory of abstract spaces seems to have remained an academic construction, carried out for its own sake, until, beginning in 1923, studies on quantum mechanics indicated that the theory might provide mathematical equipment useful in applications to physics. This fact was clearly shown in two famous papers by von Neumann published in 1929–1930, in which a purely axiomatic approach to the theory of quantum mechanics was proposed. In the course of time, the interweaving between quantum mechanics and Hilbert space theory became so close that a physical reality has been ascribed to certain abstract postulates, for example, the statement that it is legitimate to represent dynamical variables by operators in Hilbert spaces, and the results of measurements on atomic scale by eigenvalues of these operators.

Another application of functional analysis was presented in 1940 by Weyl in a paper on the method of *orthogonal projections*. This paper was later to play a role in the discovery of similarities between the methods of Ritz and Trefftz and that of orthogonal projections. Applications of functional analysis have been made in optimization (e.g., Ref. 1), in the generalized moment problem, and, of course, in purely theoretical investigations such as existence and uniqueness of solutions to partial differential equations, calculus of variations, and approximate computations. Oddly enough, the impact of functional analysis on engineering sciences has been, up to the present time, relatively weak. A few exceptions† are the expositions, in book form, of Mikhlin, Gould, and, notably, of Synge. Among the research papers, many of these published by Diaz, Greenberg, McConnell, Payne, Prager, Washizu, and a few others are of a primarily applied character. More recently, Hilbert space ideas were applied by Eringen, Edelen, and others in the development of nonlocal continuum mechanics, and by Christensen in nonlinear viscoelasticity.[2–4]

The objective of the present book is to reduce the gap existing between the abundance of facts and methods available in abstract functional analysis, and their heretofore limited use in various areas of applied mechanics.‡ It is believed that these techniques can be employed in engineering problems in the same way as the standard methods of mathematical analysis, probability theory, and others have been utilized for years. This book is a brief review, of

---

† Bibliographical details are given later in the text.

‡ That this is not an isolated case brings to mind the warning, given long ago by the distinguished mathematician, Richard Courant, against "the tendency of many workers to lose sight of the roots of mathematical analysis in physics and geometric intuition and to concentrate their efforts on the refinement and the extreme generalization of existing concepts."[5]

a somewhat elementary character, of applications of functional analysis, principally to the theory of elasticity. It is not addressed to those who delight in pure logical deductive sequences of reasoning; these can be found in abundance in many excellent books on functional analysis. The presentation here is intended to appeal primarily to the geometric intuition of the reader, being deliberately shorn at each step of those refinements that distinguish arguments concerned with questions of rigor. In spite of its simplifications, it is nevertheless hoped that the exposition is correct or, at least, that the inaccuracies are few and venial.

The author would like to think that this book will introduce the novice to the subject and guide him to a point at which he is ready to broaden his knowledge. In addition, an attempt has been made to establish the applicability of the methods of abstract (specifically, function) spaces in solving practical problems as being on a par with the classical methods of Rayleigh–Ritz, Trefftz, and others. It is believed that with an increasing interest on the part of practitioners, the methods of functional analysis will be enriched and perfected.

In Chapter 1, the reader is reminded of the distinctions between physical and abstract spaces. Chapter 2 devotes a good deal of attention to the study of the (relatively) simple affine (linear) space of three dimensions. This chapter initiates a tour through a world of spaces with increasingly exotic structures: Euclidean three-dimensional (Chapter 3), Euclidean finite-dimensional (Chapters 4 and 5), and, finally, infinite-dimensional spaces (Chapter 6). Chapter 7 is devoted to establishing the axiomatic system of Hilbert space, an applicable representative of which—the function space—is studied in Chapters 8–10.

Chapter 10 completes the theoretical portion of the book; the next five chapters explore the possibilities for practical applications of the theory. The reader is there introduced to the derivation of bounds and inequalities for the estimation of quantities of physical interest, to the methods of hypercircle and orthogonal projections, and to the connections between the ideas of function space and Rayleigh–Ritz, Trefftz, and variational methods. Almost all of the chapters are followed by problems (122 in all) closely related to the text, and introducing worthwhile additional material. Their objective is to give the reader a better feeling for the purely theoretical questions discussed in the text. A detailed solution is given for each problem, all of which are collected in a single chapter at the end of the book. A few topics of a related character (such as quantum mechanics) are examined in comments following the appropriate chapters. The theory of distributions is discussed in Chapter 16. Many illustrative problems are explored directly in the text; six of them are carried through to numerical results.

This book owes much, both in style and in content, to the generous help of my friend and former student, Dr. Allan Dallas, from the Department of Mathematical Sciences of the University of Delaware, who read the entire text, improving and correcting a number of unclear points and errors in the first draft of the book. Of course, those that still remain are the responsibility of the author. My thanks also go to Mrs. Ruby L. Schaffer and Mrs. Alison Chandler for their expert typing. Last, but certainly not least, I express my sincere gratitude to the editor of this series, Professor Angelo Miele, for his exemplary cooperation, and to the anonymous reviewer whose constructive criticism and deep knowledge of the subject have contributed significantly to the clarity and correctness of the text. I also gratefully acknowledge the gracious cooperation of Mr. Robert Golden, Senior Production Editor, from the Plenum Publishing Corporation.

Contents

I

Physical Space. Abstract Spaces

Whereas the *Random House Dictionary of the English Language*[6] lists as many as nineteen connotations of the word "space," for the purposes of this exposition it is sufficient to consider only two, which we shall designate more specifically by the terms "physical" and "abstract" space.

The *physical space* is simply the unlimited expanse of the universe, in which all material objects are located and all phenomena occur.

An *abstract* or, more precisely, a *mathematical space*, is a conception, the result of a mental construction. It may denote different things: an ideal extent of any number of dimensions, a collection (set) of abstract objects (points, vectors, numbers, functions, or sequences, for example) or a collection of real objects (inert bodies, people, animals, and so on). Of the various examples of mathematical spaces, probably the most familiar is the Euclidean space—the subject of high-school geometry. Two others are the space of classical mechanics, assumed to be controlled by Newton's laws of motion, and the four-dimensional spatiotemporal world of the theory of relativity.

The mathematical spaces—and there are many of them—are, of course, more or less elaborate idealizations of the physical space. In order to serve the purposes for which they are conceived, they are described precisely, with full logical rigor. For instance, the definition of the high-school Euclidean space is given by a system of postulates, known as the axioms of Euclid, satisfying three basic requirements of independence, consistency, and completeness.

To each mathematical space one can ascribe a dimension, specified by a number which may be infinite. Examples of one- and two-dimensional spaces are a straight line and a plane, respectively.

We denote these particular spaces by $\mathscr{E}_1$ and $\mathscr{E}_2$, respectively, and imagine them as resting in a Euclidean space of three dimensions, $\mathscr{E}_3$, the

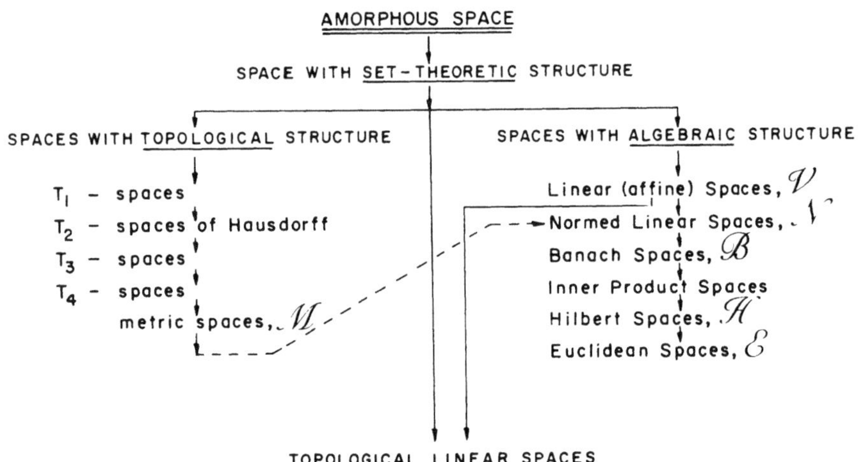

Figure 1.1. A family of mathematical spaces. The arrows indicate transition from a general structure to a more special one. Space T_1: each set consisting of a single point is postulated to be closed. Space T_2: for each pair P_1, P_2 of distinct points, there exists a pair S_1, S_2 of disjoint open sets such that $P_1 \in S_1$, $P_2 \in S_2$. Banach space: complete normed linear space. Hilbert space: complete inner product space. Euclidean space: the set of all n-tuples of real numbers equipped with the familiar distance $[\sum_{i=1}^{n} (x_i - y_i)^2]^{1/2}$. Symbols $P \in S$ ($P \notin S$) mean P belongs (does not belong) to S. Explanations of the terminology used here are given later in the text.

latter being an idealized image of the physical space, which is believed to have three dimensions. The space $\mathscr{E}_1$ is called a subspace of $\mathscr{E}_2$, while both $\mathscr{E}_1$ and $\mathscr{E}_2$ are subspaces of $\mathscr{E}_3$.†

A detailed examination of the various classes of mathematical spaces lies beyond the scope of this book; still, it is both interesting and informative to cast a glance at the "genealogical" tree of the most common abstract spaces (Figure 1.1).

Before actually doing this, it is good to recall that, from a purely mathematical standpoint, the term *space* designates, as we have noted, both a continuous extent and a set of separate objects. These objects are called *elements*, or *members*, of the space, and their number may be finite or infinite. As long as the elements are not clearly defined, there is no means for deciding whether a given object belongs to the space.‡ In a similar vein, if there

---

† See Comment 1.1 of this chapter, below.

‡ Basically, by a *set* is meant a *well-defined* collection of objects, so that it is always possible to decide whether an object belongs to the particular collection. Again, by a *space* we normally mean a set with some kind of mathematical *structure* (see below).

should be no defined operations on the elements of the space, the space has no structure; it is *amorphous*, and the question of carrying out analytical processes within the space is sterile (Figure 1.1).

The structure of a space, also called its *geometric structure*, has three aspects: set-theoretic, algebraic, and topological.[†]

As is well known, the *theory of sets* is concerned with the investigations of operations on sets and such fundamental mathematical concepts as those of function and relation. It serves as a groundwork on which rest the axiomatic systems of algebra and topology. Concerning the notions of *topology*, our interest in this book involves no more than those deriving from the idea of the closeness (distance between) the elements of a space. The *algebraic* concepts chiefly used are those of the addition of elements and the multiplication of the latter by scalars. Many specific items related to the geometric structure will be explained as they arise in the text. At this point, it will suffice to briefly examine only some of them.

A first step towards the establishment of a *geometric structure* for an amorphous space is the introduction of the set-theoretic axiomatics, which form the foundation of any branch of mathematics. This leads to the definitions of such concepts as the ordering of elements, equality, union, and intersection of sets.[‡]

Further development of the space structure is realized in the designation of a family of subsets as the "open sets."[§] This family, required to satisfy a certain set of axioms, defines a so-called *topology* of the space. Springing from this construction are such new concepts as a neighborhood of an element, continuity of a function, an accumulation point, convergence of a sequence, and compactness.[¶] Whenever a topology is identified for a space, that space with its topology is known as *topological space*. There is a hierarchy of types of topological spaces; some types are more specialized cases of others. A few of these types are indicated in Figure 1.1, denoted there by the symbols $\mathcal{T}_i$, where $i = 1, 2, 3,$ or 4. An analysis of the definitions of, and differences between, these $\mathcal{T}_i$-spaces (or any of the other types) would lead

---

[†] A detailed exposition of these topics would require a separate volume. Examples of excellent readable accounts include: Berberian,[7] Mikhlin,[8] Naylor and Sell,[9] and Sneddon.[10] One of the several fairly comprehensive texts is that by Taylor.[11] Popular presentations are those of Fletcher[12] and Hausner.[13] The more recent Oden's book[14] includes many interesting exercises and examples from classical and continuum mechanics. Very useful also is Kreyszig's book.[15]

[‡] A readable review of the theory of sets is found in, e.g., Lipschutz.[16]

[§] The concept of an "open" set need not be interpreted geometrically. The definition of "openness" depends entirely on the kind of topology. See also the second footnote following equation (4.5).

[¶] It is good to recall that the limiting process underlying most of these concepts is one of the most essential operations of mathematical analysis.

us too far afield; let it suffice to note here that the structure of a topological space becomes much more fruitful if the topology is generated by a measure of closeness, or nearness, of elements, known as the *distance function*. A space equipped with such a measure of distance between its elements is called a *metric space*; each metric space is an example of a topological space.

A next specialization in structure consists of two simultaneous acts: provision of the *algebraic* structure characteristic of the so-called *linear vector spaces*,† and generalization of the familiar notion of length through the concept of the *norm* (or *length*) of an element. A space so devised is called a *normed linear vector space* or, briefly, a normed vector space. A particular example of a normed space is a so-called *Banach* space, $\mathscr{B}$.

A welcome feature of normed vector spaces is the similarity of their structure to that of Euclidean spaces. This similarity becomes even more pronounced if the norm of each element is defined in terms of a binary operation (i.e., one performed on pairs of elements) called an *inner product*, which is formally analogous to the dot product of ordinary vectors. A space so enriched becomes an *inner product space*, well-equipped for both theoretical investigations and practical applications.

Particularly useful inner product spaces are those of the class named after the great mathematician David Hilbert. The geometric structure of a *Hilbert space*, $\mathscr{H}$, is more complete than that of the general inner product space, and, in spite of the presence of certain features not encountered in Euclidean spaces, is a direct progenitor of the latter. Thus, a finite-dimensional real Hilbert space coincides with a Euclidean space of the same dimension, except for the nature of, and the operations on, elements: in Hilbert space, these are defined in an abstract manner, whereas in a Euclidean space, the elements are the familiar points and vectors, with rules of operation which are quite concrete.

As an illustration of a geometric structure, we can examine that of the three-dimensional Euclidean space. Its set-theoretical structure is determined by specifying it as the set of all ordered triples of real numbers $x = x_1$, x_2, x_3, where x denotes a point of the space and x_1, x_2, x_3 its coordinates. The topological structure of $\mathscr{E}_3$ is established by introducing the concept of (Euclidean) *distance* between two points x and y: $d(x, y) = [(x_1 - y_1)^2 + (x_2 - y_2)^2 + (x_3 - y_3)^2]^{1/2}$. Finally, the algebraic structure is incorporated by defining the addition of elements: $x + y = (x_1 + y_1, x_2 + y_2, x_3 + y_3)$, and the scalar multiplication: $\alpha x = (\alpha x_1, \alpha x_2, \alpha x_3)$, where α is a real number.

Believing a transition from the concrete to the abstract to be more natural for a physically minded reader than an opposite course, we shall

---

† Many authors use the terms "vector space," "linear space," and "linear vector space" interchangeably.

begin the exposition proper with an examination of a relatively simple *affine* (or *linear*) *space*. It will be the familiar Euclidean space $\mathscr{E}_3$ stripped, however, of metric.

Comment 1.1. In the same sense in which a line and a plane are subspaces of $\mathscr{E}_3$ of dimension one and two, respectively, the origin is a subspace of dimension zero, $\mathscr{E}_0$, for it includes the single zero vector. This agrees with Definition 2.1 in Chapter 2 of a point as equivalent to the zero vector, as well as with the convention that the zero vector, θ, generates a *singleton* set, $\{\theta\}$, whose dimension is postulated as being zero. The conventional definition of a subspace implies that any subspace must include the zero vector (see the remarks following equation (4.1), for example). Thus, if we imagine vectors to remain bound and emanate from the space origin, O, then $\mathscr{E}_1$ and $\mathscr{E}_2$ represent lines and planes passing through O, respectively. In this understanding, the subspace $\mathscr{E}_0$ coincides with the point O. In our future studies, we shall meet geometric systems which have all the features of subspaces proper, but do not include the space origin.†

---

† See Chapter 9 of this book for a discussion of "translated" subspaces.

2

Basic Vector Algebra

As already stated, elementary geometry deals with properties of, and relations between, objects (figures) assumed to dwell in an ideal, Euclidean space of three dimensions. An important class of these objects is the collection of *directed line segments*, known as "ordinary" *vectors*. These quantities are endowed with both magnitude and direction, and are graphically represented by arrows. In the first four chapters of this book, a vector will be denoted by a boldface letter or by two letters with an arrow above them, such as $\overrightarrow{AB}$, indicating the origin, A, and the terminus, B, of the vector.

If one confines one's attention to vectors as the sole geometric elements of interest, it becomes natural to apply the terminology "space of vectors" or a *"vector space"* to a space occupied by vectors. By approaching the algebra of ordinary vectors from a point of view which is more dogmatic than that usually taken, we are afforded a means of gaining an appreciation for the issues involved in abstract vector spaces to be discussed later.

With this in mind, we designate "point" and "vector" as *primitive* concepts whose properties are intuitively clear. Indeed, the axiomatics to be established is considered as giving implicit definition to these concepts.†

We begin with the following three postulates:

Axiom 2.1. With each pair of points A and B, given in this order, there is associated a single vector, denoted by $\overrightarrow{AB}$ or, briefly, by **x**, say.

Axiom 2.2. To each point A and any vector **x** there corresponds a unique point B such that

$$\overrightarrow{AB} = \mathbf{x}. \qquad (2.1)$$

† The axiomatics follows that given by Rashevski.[17]

The actual meaning of Axiom 2.2 is better understood by thinking of a vector as a uniform translation of the entire space (together with its "contents"), rather than as a directed line segment. In this manner, any definite vector, say $\overrightarrow{AB}$, in fact generates a certain vector field, and equalities such as

$$\mathbf{x} = \overrightarrow{AB} = \overrightarrow{A'B'} = \overrightarrow{A''B''} = \cdots$$

define the same vector except at various locations A, A', A'', ... of its point of application. This fact enables us to shift vectors "*parallel*" to themselves from one point of space to another without impairing their identities, in effect treating them as "*free*." We note that, from this standpoint, the sign of equality of vectors $(=)$, used in equations such as (2.1), represents simply the sign of the identity of vectors.

Axiom 2.3. If

$$\overrightarrow{AB} = \overrightarrow{CD}, \qquad \text{then } \overrightarrow{AC} = \overrightarrow{BD}. \tag{2.2}$$

The pictorial sense of this axiom is† that if two opposite sides of a quadrilateral are equal and parallel, then the two remaining sides are also equal and parallel (Figure 2.1). This implies that the quadrilateral is a parallelogram.

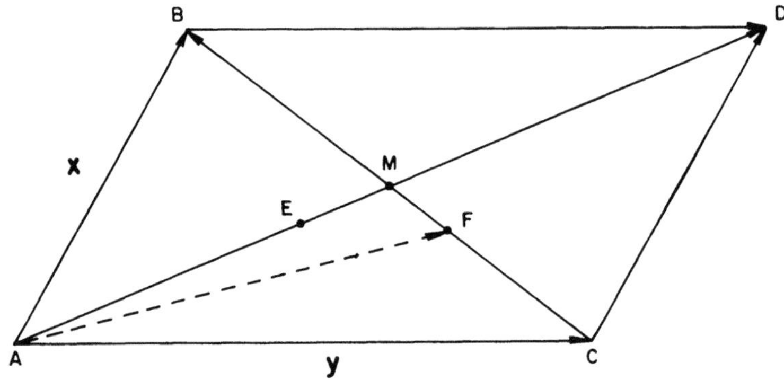

Figure 2.1. Illustration for Axiom 2.3.

---

† Note that Axioms 2.1, 2.2, and 2.3 implicitly define the notions of *parallelism* of vectors, and of lines carrying these vectors in such a way that the lines do not meet. For if they did, one could shift the concomitant vectors to the intersection point P and obtain two separate equal vectors at a single point P, contrary to Axiom 2.2.

At this stage, it is convenient to introduce two concepts: (1) of the *zero vector* by means of

Definition 2.1.

$$0 = \overrightarrow{AA}, \tag{2.3}$$

and (2) of the vector $-\mathbf{x}$ (negative of $\mathbf{x}$), of direction opposite to that of the given vector $\mathbf{x}$, by means of

Definition 2.2.

$$-\mathbf{x} = \overrightarrow{BA}, \qquad \text{where} \qquad \mathbf{x} = \overrightarrow{AB}. \tag{2.4}$$

With these preliminaries established, we introduce the fundamental operation of *addition of vectors* through

Definition 2.3. Let $\mathbf{x}$ and $\mathbf{y}$ be two arbitrary vectors, given in this order. Choose a point A as the origin of the vector $\overrightarrow{AB} = \mathbf{x}$ (Axiom 2.2), and construct the vector $\overrightarrow{BD} = \mathbf{y}$ at the point B. Points A and D determine the vector $\overrightarrow{AD} = \mathbf{z}$ (Axiom 2.1) called the *sum* of the vectors $\mathbf{x}$ and $\mathbf{y}$. Symbolically,

$$\mathbf{x} + \mathbf{y} = \mathbf{z}. \tag{2.5}$$

Thus far, there has been established a certain simple axiomatic system. It includes two undefined terms ("point" and "vector"), three primitive statements (axioms or postulates), and three definitions. It is interesting to note that the system provides a great deal of information about vectors, in spite of its simplicity. As illustrations, we offer the following conclusions.

Conclusion 2.1. Commutative Law of Addition.

$$\mathbf{x} + \mathbf{y} = \mathbf{y} + \mathbf{x} \tag{2.6}$$

or, verbally, the order of addition does not affect the sum of vectors. Indeed, using Axiom 2.2, we construct the vector $\overrightarrow{AB} = \mathbf{x}$ at a point A, and then the vector $\overrightarrow{BD} = \mathbf{y}$ at the point B, so that

$$\overrightarrow{AD} = \mathbf{x} + \mathbf{y} \tag{2.7}$$

by Definition 2.3. Next, at the same point A, we construct the vector $\overrightarrow{AC} = \mathbf{y}$. Since $\mathbf{y} = \overrightarrow{AC} = \overrightarrow{BD}$, we invoke Axiom 2.3, concluding first that $\overrightarrow{AB} = \overrightarrow{CD}$, and consequently that $\overrightarrow{CD} = \mathbf{x}$. If we now assume that, at the outset, we have erected the vector $\mathbf{y} = \overrightarrow{AC}$ at the point A and subsequently attached the vector $\overrightarrow{CD} = \mathbf{x}$ at the point C, we obtain by Definition 2.3,

$$\overrightarrow{AD} = \mathbf{y} + \mathbf{x}. \tag{2.8}$$

An inspection of relations (2.7) and (2.8) proves our conclusion.

In like manner, and just as simply, three further conclusions can be reached.

Conclusion 2.2.

$$(\mathbf{x} + \mathbf{y}) + \mathbf{z} = \mathbf{x} + (\mathbf{y} + \mathbf{z}), \tag{2.9}$$

that is, the *addition* of vectors is *associative*.

Conclusion 2.3.

$$\mathbf{x} + \mathbf{0} = \mathbf{x}, \tag{2.10}$$

that is, addition of the *zero vector* does not change a given vector.

Conclusion 2.4.

$$\mathbf{x} + (-\mathbf{x}) = \mathbf{0}. \tag{2.11}$$

With merely the law of vector addition, the vector calculus would provide no significant mathematical equipment. Still another binary operation is needed, that which forms a vector from a pair composed of a *scalar* (meaning a real number throughout this book) and a vector. This operation, called *scalar multiplication*, or the multiplication of a vector by a scalar, is defined by the following five axioms.

Axiom 2.4. To each vector $\mathbf{x}$ and to each scalar α, there corresponds a vector, denoted by $\alpha\mathbf{x}$, called the *scalar product* of α and $\mathbf{x}$.

Informally, the product $\alpha\mathbf{x}$ may be interpreted as a translation in the same "direction," but of different "magnitude" than that provided by $\mathbf{x}$; even so, the concepts of "direction" and "magnitude" remain undefined. What is really known is the "ratio" of vectors, "$\alpha = \alpha\mathbf{x}/\mathbf{x}$" (including negative ratios as well).

Axiom 2.5. There is

$$1\mathbf{x} = \mathbf{x}. \tag{2.12}$$

This axiom states that the multiplication of a vector by the scalar 1 does not change the vector.

The statement may seem vacuous, but is actually needed for completeness of the axiomatics (compare Berberian,[7] Theorem 1, p. 6).

Axiom 2.6.

$$(\alpha + \beta)\mathbf{x} = \alpha\mathbf{x} + \beta\mathbf{x}: \tag{2.13}$$

the *distributivity* of *scalar multiplication* with respect to the addition of scalars is established.

Axiom 2.7.

$$\alpha(\mathbf{x} + \mathbf{y}) = \alpha\mathbf{x} + \alpha\mathbf{y}: \tag{2.14}$$

the *distributivity* of *scalar multiplication* with respect to the addition of vectors is established.

Axiom 2.8.

$$\alpha(\beta\mathbf{x}) = (\alpha\beta)\mathbf{x}, \tag{2.15}$$

asserting the *associativity* of *multiplication* by scalars.

Supplemented by the preceding five postulates, the axiomatic system so far established is sufficiently reached to admit two basic operations on vectors—their addition and their multiplication by scalars. The system thus produced is known as the *affine (linear)* system, while the space of vectors satisfying Axioms 2.1–2.8 is called the *affine space*.

It is instructive to note certain implications of the last few axioms.

Conclusion 2.5. For an arbitrary vector $\mathbf{x}$,

$$0\mathbf{x} = \mathbf{0}. \tag{2.16}$$

In fact, by Axiom 2.6, we can write, for any number α,

$$0\mathbf{x} = (\alpha - \alpha)\mathbf{x}$$

$$= \alpha\mathbf{x} - \alpha\mathbf{x}$$

$$= \mathbf{0},$$

the last equality following by virtue of equation (2.11). This proves our assertion.

Conclusion 2.6. For an arbitrary scalar α,

$$\alpha 0 = 0. \tag{2.17}$$

To show this, we use Conclusion 2.5 and Axiom 2.8 to write, for an arbitrary vector,

$$\alpha 0 = \alpha(0\mathbf{x})$$
$$= (\alpha 0)\mathbf{x}$$
$$= 0\mathbf{x} = 0.$$

As another illustration, we provide the following theorem, with its proof.

Theorem 2.1. The diagonals of a parallelogram bisect each other (Figure 2.1).

It should be carefully noted that "bisection" here means nothing other than the equality of two segments of the severed diagonal, in the understanding of Axiom 2.2. With this in mind, let $ABCD$ be a parallelogram in the sense of Axiom 2.3, so that

$$\overrightarrow{AB} = \overrightarrow{CD} = \mathbf{x}, \quad \overrightarrow{AC} = \overrightarrow{BD} = \mathbf{y}. \tag{2.18}$$

By Definitions 2.2 and 2.3,

$$\overrightarrow{AD} = \mathbf{x} + \mathbf{y}, \quad \overrightarrow{CB} = \mathbf{x} - \mathbf{y}. \tag{2.19}$$

Suppose that M is the point of intersection of the diagonals, and select points E and F on the diagonals AD and CD, respectively, such that

$$\overrightarrow{AE} = \alpha\overrightarrow{AD} \quad \text{and} \quad \overrightarrow{CF} = \alpha\overrightarrow{CB}, \tag{2.20}$$

in the manner of Axiom 2.4. On account of equations (2.19), we have

$$\overrightarrow{AE} = \alpha(\mathbf{x} + \mathbf{y}), \quad \overrightarrow{CF} = \alpha(\mathbf{x} - \mathbf{y}), \tag{2.21}$$

as well as

$$\overrightarrow{AF} = \mathbf{y} + \overrightarrow{CF} = \alpha\mathbf{x} + (1 - \alpha)\mathbf{y}, \tag{2.22}$$

by virtue of the second equation (2.21). Now, let the points E and F coincide; then, they must also coincide with the point M, since M is the only point in which the diagonals meet. Thus, in this case,

$$\overrightarrow{AE} = \overrightarrow{AM} = \overrightarrow{AF},$$

and from the preceding equations,

$$\alpha = \tfrac{1}{2}. \tag{2.23}$$

This completes the proof, inasmuch as

$$\vec{AM} = \vec{AE} = \tfrac{1}{2}\vec{AD} \quad \text{and} \quad \vec{CF} = \vec{CM} = \tfrac{1}{2}\vec{CB}. \tag{2.24}$$

An important question arising at this stage concerns the absence from the established axiomatics of the affine space of any mention of such fundamental notions as *length* (of a line segment, for instance) or the *distance* between two points. Even the term "magnitude of a vector" is (not accidentally) only vaguely defined, signifying nothing other than the equality and ratio of parallel vectors in the sense of Axioms 2.3 and 2.4, respectively. We point out that, whereas Definition 2.3 enables us to sum nonparallel vectors, it has a rather descriptive character, and says nothing about the metrical relations between the vectors to be summed and their sum.

A peculiar feature of the affine axiomatics brought out in the above examples is the omission of the idea of *measure* in the strict sense of this word. Thus, not only are the concepts of length and distance lacking, but there is also no definition of the angle between two directions. The latter should come as no surprise, for an inclination is measured by a ratio of lengths, namely, those of the arc subtending the angle and the pertinent radius.

To clarify this matter, consider the following example. Imagine a triangle ABC, formed by three vectors, $\vec{AB}$, $\vec{BC}$, and $\vec{CA}$, resting in an affine space. We are permitted to translate the triangle parallel to itself to any location, for two identical and identically directed geometric figures at different locations are, in fact, precisely the same figure from the standpoint of affine geometry. However, a rotation of the triangle is not an identity-preserving transformation, since such an operation changes the triple of vectors into a different triple, the connections of which with the original triple—like any relations between nonparallel vectors—are nonexistent (at least in the axiomatics so far formulated†).

The foregoing example clearly shows the incompleteness of a geometric structure which fails to account for measure. In point of fact, the entire structure of modern civilization rests in part on association of numbers with things, only one instance of which is the measurement of extentions (lengths, areas, and volumes). It suffices to mention here, among the many other examples, the marking of sizes, weights, and prices, the numbering of phones, houses, and streets, and the measurement of times and speeds.

In order to fill the gaps in the affine structure and make it as complete as, say, high-school Euclidean geometry, we shall introduce a function which is said to be a *metric* for the space.

---

† This conclusion is perhaps surprising to those taking for granted the mobility of material objects in our physical space and of geometric objects in high-school geometry.

Before examining the latter concept in some detail in Chapter 3, it is worth mentioning that the study of nonmetric spaces is not as barren as it may seem. Actually, much attention is devoted in the modern literature to topological spaces devoid of a metric,† although some of these possess properties which may generalize or be reminiscent of certain characteristics of the structure endowed by a metric. As examples, we cite the topological spaces of the type $\mathscr{T}_i$, $i = 1, 2, 3, 4$, which fit into the hierarchy of spaces as shown in the diagram of Figure 1.1.

Problems

1. Can the affine geometry distinguish between an ellipse and a circle? Between a rhombus and a square?

2. Show that if $\mathbf{x} + \mathbf{x} = \mathbf{x}$, then $\mathbf{x} = \mathbf{0}$.

3. In a triangle ABC there hold $\overrightarrow{AB} = \mathbf{y}$, $\overrightarrow{AC} = \mathbf{x}$, and $\overrightarrow{CB} = \overrightarrow{CM} + \overrightarrow{MB}$, where $\overrightarrow{CM} = \overrightarrow{MB}$. Express $\overrightarrow{AM} = \mathbf{z}$ in terms of $\mathbf{x}$ and $\mathbf{y}$.

4. Show that the quadrilateral formed by the lines joining the midpoints of consecutive sides of an arbitrary quadrilateral is a parallelogram (Fig. 2.2).

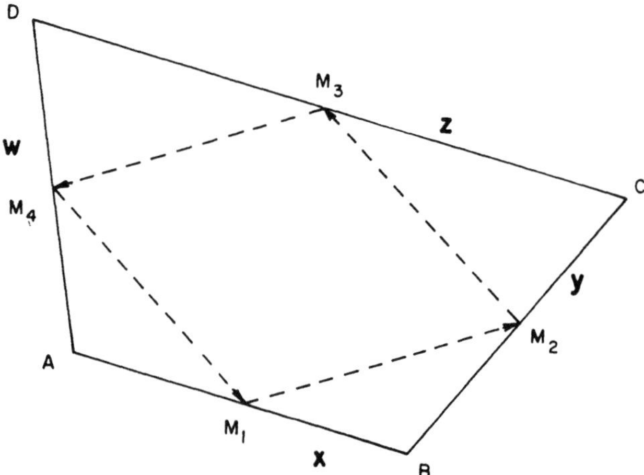

Figure 2.2. Illustration for Problem 4.

---

† These are constructed by specifying a system of open sets with suitable properties, in place of introducing a metric.

5. Show that the line joining the midpoints of two sides of a triangle is parallel to the third side of the triangle.

6. Show that, in a parallelogram, a median intersecting a diagonal divides the latter into segments whose lengths are in the ratio $2 : 1$.

7. Assume that the concepts of a plane and a line lying on a plane have been established. Under these assumptions, let a line l lying in the plane of two lines l_1 and l_2 carrying the vectors $\overrightarrow{A_1 B_1}$ and $\overrightarrow{A_2 B_2} = \overrightarrow{A_1 B_1}$, respectively, meet l_1 at the point A_1. Show that l must also meet the line l_2.

8. Define $\mathbf{x} = (x_1, x_2, x_3)$, $\mathbf{y} = (y_1, y_2, y_3)$, $\mathbf{0} = (0, 0, 0)$, where x_i, y_i, $i = 1, 2, 3$, are scalars. Define also $\mathbf{x} + \mathbf{y} = (x_1 + y_1, x_2 + y_2, x_3 + y_3)$ and $\alpha\mathbf{x} = (\alpha x_1, \alpha x_2, \alpha x_3)$, where α is a scalar. Consider that $\mathbf{x} = \mathbf{y}$ if $x_i = y_i$, $i = 1, 2, 3$. Accepting the usual algebra of scalars, show that Conclusions 2.1, 2.2–2.4, and Axioms 2.5–2.8 are verified.

9. Accepting the definitions of the preceding problem, show that the equality $(1, 0, 0) = \alpha(0, 1, 0)$ cannot hold, whatever the value of α.

10. Show that the medians of a triangle intersect at a point which divides each of them in the ratio $2 : 1$ (reckoning from each vertex).

It is only with the multiplication of vectors by one another that the geometry becomes rich—J. L. SYNGE

3

Inner Product of Vectors. Norm

The incorporation of a metric into the axiomatic system of affine spaces depends on the method chosen for evaluating the magnitude of arbitrarily directed vectors. The procedure universally adopted was probably suggested by Grassman and Gibbs. It consists of deriving the notion of *length*, or *norm*, of vectors from the concept of the *inner* (also called scalar, or dot) *product* of vectors. We shall subsequently denote the inner product of two vectors, $\mathbf{x}$ and $\mathbf{y}$ say, by the symbol $(\mathbf{x}, \mathbf{y})$ instead of the usual notation $\mathbf{x} \cdot \mathbf{y}$. As we shall see later, this convention makes the transition from Euclidean to abstract spaces simpler and more natural.

It is almost self-evident that, since the postulational basis so far established has avoided the concept of *length*, the latter must either be introduced in an *a priori* manner through a special definition or recognized as primitive and *intuitively* clear. For the time being, we select the second alternative as less abstract. Later, we shall demonstrate how, from a system of axioms (inspired by the present heuristic approach), one can derive both the notion of the *length* of a vector and the *angle* between two vectors.

We first define the *length* of a vector to be the distance between the terminal points of the arrow representing the vector. We denote the length of a vector $\mathbf{x}$ by $\|\mathbf{x}\|$ and call it the *norm* of $\mathbf{x}$. The unit of length is taken the same for all vectors, independent of their direction, and it is assumed that the rule for effecting the measurement of the length of arrows is known. By its very nature, the norm always represents a real non-negative number.

In order to enrich the axiomatics of affine spaces by the concepts of length and angle, it is convenient to supplement the operations of addition and scalar multiplication of vectors by a third operation, known as *inner multiplication*. Such an operation, theoretically at least, may, of course, be defined in a variety of ways. A particularly fruitful form, however, has

17

proved to be that suggested by Gibbs and endowed with the following rather attractive characteristic properties:

Symmetry

$$(\mathbf{x}, \mathbf{y}) = (\mathbf{y}, \mathbf{x}). \tag{3.1}$$

Distributivity (with respect to the second vector)

$$(\mathbf{x}, \mathbf{y} + \mathbf{z}) = (\mathbf{x}, \mathbf{y}) + (\mathbf{x}, \mathbf{z}). \tag{3.2}$$

Homogeneity (with respect to the first vector)

$$\alpha(\mathbf{x}, \mathbf{y}) = (\alpha\mathbf{x}, \mathbf{y}). \tag{3.3}$$

Here $\mathbf{x}$, $\mathbf{y}$, and $\mathbf{z}$ are vectors and α is a real scalar. These three properties, when combined, define the so-called *linearity* (or rather bilinearity) of the product. A final property, of fundamental importance when extended to abstract spaces, involves the positive definiteness of the inner (or, rather, of the "self-inner") product, defined by the following conditions:

$$\begin{aligned}
(\mathbf{x}, \mathbf{x}) &> 0 \qquad \text{for } \mathbf{x} \neq \mathbf{0}, \\
(\mathbf{x}, \mathbf{x}) &= 0 \qquad \text{for } \mathbf{x} = \mathbf{0}.
\end{aligned} \tag{3.4}$$

Explicitly, the Gibbs inner product, accepted as standard for elementary vector algebra, is defined by the formula

$$(\mathbf{x}, \mathbf{y}) = \|\mathbf{x}\| \, \|\mathbf{y}\| \cos(\mathbf{x}, \mathbf{y}). \tag{3.5}$$

The preceding simple definition not only meets the conditions (3.1)–(3.4), but also seems rather natural, inasmuch as it: (a) involves the product of the magnitudes of vectors (and this is reminiscent of the product of scalars), (b) includes a function of their mutual inclination—a matter of special importance, and (c) implies that the inner product may take any real value, since the value of the cosine lies between -1 and 1 inclusive.

Surely, instead of the cosine, some other function of the inclination angle α, such as α itself, α^2, $\sin\alpha$, or $\tan\alpha$, might be selected, but would most likely be found deficient. For example, the linearity of the inner product is in fact realized by the use of the cosine of the angle (the latter being represented by a line segment in a unit circle), rather than the length of the circular arc subtending the angle. Again, replacement of the cosine in (3.5) by the sine would result in a failure to fulfill the requirement of distributivity. In fact, assuming, for simplicity, the perpendicularity† of three vectors $\mathbf{x}$, $\mathbf{y}$, and $\mathbf{z}$ (Figure 3.1a), we notice immediately that the area of the rectangle $OAA'''B'''$

---

† We define the perpendicularity of two intersecting lines as the equality of adjacent angles. It is important to note here that, temporarily, our argument is principally intuitive.

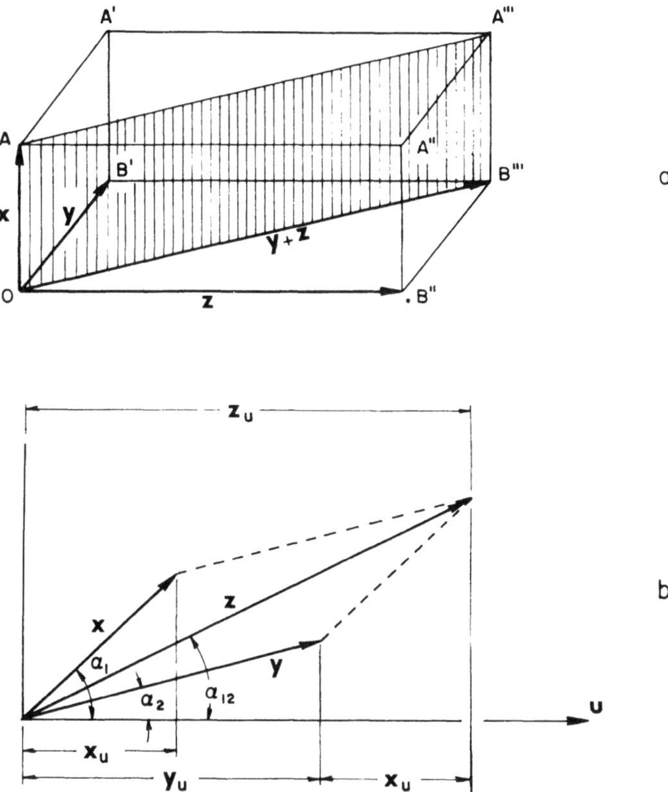

Figure 3.1. Illustration for the inner product.

constructed over the sum $\mathbf{y} + \mathbf{z}$ is, in general, smaller than the sum of the areas of the rectangles $OAA'B'$ and $OAA''B''$; this, of course, violates the relation (3.2).

Returning to the definition (3.5), we first note the obvious conformity with the symmetry requirement (3.1). In order to show that it satisfies (3.2), for example, we see that it is permissible to use the concept of the *orthogonal projection* (of a point on a line as the foot of the perpendicular from the point to the line, and of a vector on a line, as the directed line segment joining the projections of its end points), upon noting that it derives from our definition of perpendicularity. Assume, then, for the sake of argument, that three vectors $\mathbf{x}$, $\mathbf{y}$, and $\mathbf{u}$ lie in one plane,† and consider projecting $\mathbf{x}$, $\mathbf{y}$, and $\mathbf{z}$ orthogonally on $\mathbf{u}$ (Figure 3.1b), where $\mathbf{z} = \mathbf{x} + \mathbf{y}$. Denote by x_u the signed length

---

† It is not difficult to relax this requirement.

of the orthogonal projection of $\mathbf{x}$ on $\mathbf{u}$, so that $x_u = \|\mathbf{x}\|\cos(\mathbf{x}, \mathbf{u})$. A glance at Figure 3.1b reveals that

$$z_u = x_u + y_u \tag{3.6}$$

or

$$\|\mathbf{x} + \mathbf{y}\|\cos \alpha_{12} = \|\mathbf{x}\|\cos \alpha_1 + \|\mathbf{y}\|\cos \alpha_2. \tag{3.7}$$

Multiplying both sides of the preceding equation by $\|\mathbf{u}\|$ and taking account of (3.5), we arrive at the desired result [compare with (3.2)].

An important next step is the formation of the inner product of a vector with itself,

$$(\mathbf{x}, \mathbf{x}) = \|\mathbf{x}\|^2, \tag{3.8}$$

from which we observe that, in terms of the self-inner product, the *length* or *norm* of a vector is represented by

$$\|\mathbf{x}\| = \sqrt{(\mathbf{x}, \mathbf{x})}, \tag{3.9}$$

the square root taken, as always, to be the positive one.

We shall repeatedly find evidence in this book that the relation (3.9), perceived from a general point of view, is one of the central equations of functional analysis. In the meantime, we return to the definition (3.5) and write it in the form

$$\cos \phi = \frac{(\mathbf{x}, \mathbf{y})}{\|\mathbf{x}\|\ \|\mathbf{y}\|}, \tag{3.10}$$

assuming that neither of the vectors $\mathbf{x}$ and $\mathbf{y}$ is the zero vector. We conclude that if two nonzero vectors are mutually perpendicular, or *orthogonal*, that is, if $\phi = \pi/2$, then

$$(\mathbf{x}, \mathbf{y}) = 0. \tag{3.11}$$

Clearly, (3.11) holds if either $\mathbf{x}$ or $\mathbf{y}$, or both, are zero. We note that the converse of this statement is also true, that is, if (3.11) holds, then $\mathbf{x}$ and $\mathbf{y}$ are mutually orthogonal. This conclusion employs the convention that the *zero vector*, $\|\mathbf{0}\| = 0$, is *perpendicular* to *every* vector (including itself).

The reader has already been warned that, in our attempt to explain the concept of length, we have violated the rules of logic by basing our argument on circuitous reasoning. In fact, the course of our analysis ran as follows: *we first* acknowledged the concept of length to be a primitive concept, *next* introduced the definition of an inner product [equation (3.5)] in a form enabling us to arrive at the relation (3.9), and finally recognized the self-inner product as a motivation for the concept of length, returning us to the point of our departure.

While such a line of approach may be pedagogically helpful, the reverse of this procedure, that is, declaring equation (3.5) to be the definition of the inner product, would have been far less cavalier and considerably more far-reaching. The explicit meaning of the length of a vector would then automatically follow from the metric equation (3.9). With this in mind, we enunciate the following definitions.

Definition 3.1. The *length*, or *norm*, of a vector $\mathbf{x}$, denoted by $\|\mathbf{x}\|$, is given by

$$\|\mathbf{x}\| = \sqrt{(\mathbf{x}, \mathbf{x})}, \tag{3.12}$$

where the inner product satisfies conditions (3.4).†

Definition 3.2. The *distance* AB between two points A and B is defined as the length of the vector $\overrightarrow{AB}$. Thus,

$$AB = (\mathbf{x}, \mathbf{x})^{1/2}, \quad \text{where } \mathbf{x} = \overrightarrow{AB}. \tag{3.13}$$

With the concept of length thus established, we are now in a position to determine the length of any rectifiable curve as well as the angle between any two directions. Likewise, the concepts of a circle (as the locus of points equidistant from a given point, its center), of an isosceles or equilateral triangle, of perpendicularity, congruence, and others, become meaningful.

A more detailed examination of the concept of the inner product (in its general sense) must be deferred to subsequent chapters. At this point, however, it is important to emphasize that the incorporation of the notion of length into the affine axiomatics constitutes a significant enrichment of the latter: the affine space is transformed into the classical Euclidean space—the last member of the family of spaces in Figure 1.1 (note the dashed line in this figure).

To illustrate, consider the following simple examples.

(α) Let $\mathbf{a}$ and $\mathbf{b}$ be two orthogonal vectors whose sum is $\mathbf{c}$:

$$\mathbf{a} \perp \mathbf{b}, \quad \mathbf{c} = \mathbf{a} + \mathbf{b}, \tag{3.14}$$

and whose norms are a, b, and c, respectively. We form the inner product of each side of the preceding equation with itself and write

$$(\mathbf{c}, \mathbf{c}) = (\mathbf{a} + \mathbf{b}, \mathbf{a} + \mathbf{b}). \tag{3.15}$$

---

† That is, positive definite. An important instance in which it is useful to allow the product $(\mathbf{x}, \mathbf{x})$ to be *indefinite*, that is, positive, zero, or negative, depending on the choice of $\mathbf{x}$, is mentioned at the end of Chapter 5.

By equations (3.3) and (3.8), we have

$$\|\mathbf{c}\|^2 = \|\mathbf{a}\|^2 + 2(\mathbf{a}, \mathbf{b}) + \|\mathbf{b}\|^2,$$

and, by the first relation (3.14),

$$c^2 = a^2 + b^2. \tag{3.16}$$

This yields the familiar *Pythagorean theorem* for a right triangle.

(β) Let the vectors $\mathbf{a} \equiv \overrightarrow{AB}$, $\mathbf{b} \equiv \overrightarrow{BC}$, and $\mathbf{c} \equiv \overrightarrow{AC}$ form a triangle ABC in which the angle $\measuredangle\, BAC = \alpha$. It is required to prove the law of cosines for the triangle.

We write $\mathbf{b} = \mathbf{c} - \mathbf{a}$ and form the inner product

$$(\mathbf{b},\, \mathbf{b}) = (\mathbf{c} - \mathbf{a},\, \mathbf{c} - \mathbf{a})$$

$$= (\mathbf{c},\, \mathbf{c}) + (\mathbf{a},\, \mathbf{a}) - 2(\mathbf{a},\, \mathbf{b}),$$

$$= \|\mathbf{c}\|^2 + \|\mathbf{a}\|^2 - 2\|\mathbf{a}\|\,\|\mathbf{b}\|\cos\alpha.$$

By reverting to the conventional notation, we write

$$b^2 = a^2 + c^2 - 2ac\,\cos\alpha, \tag{3.17}$$

the required result.

(γ) Just as simply, it can be shown that the diagonals of a rhombus are mutually perpendicular (see Figure 3.2).

By definition of a rhombus, we have $\|\mathbf{a}\| = \|\mathbf{b}\|$ or

$$(\mathbf{a},\, \mathbf{a}) - (\mathbf{b},\, \mathbf{b}) = 0. \tag{3.18}$$

But the last equation is easily transformed into

$$(\mathbf{a} - \mathbf{b},\, \mathbf{a} + \mathbf{b}) = 0, \tag{3.19}$$

and this proves our assertion [compare the definition of the norm, equation (3.8)].

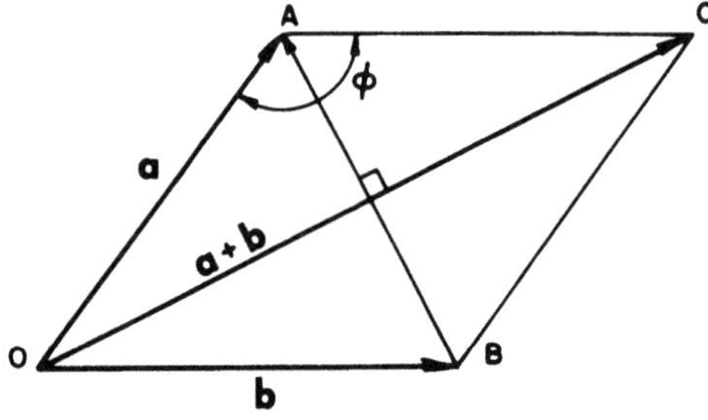

Figure 3.2. Perpendicularity of the diagonals of a rhombus.

(δ) We now prove an inequality of considerable importance in applications, known as the *triangle* (or Minkowski) *inequality*. It states that the length of a side of a triangle: (1) is no greater than the sum of the lengths of its two other sides; (2) is no less than the absolute value of the difference of the lengths of the other sides. To prove the first part of this assertion, we write (see triangle OAC in Figure 3.2)

$$\|\mathbf{a} + \mathbf{b}\|^2 = (\mathbf{a} + \mathbf{b}, \mathbf{a} + \mathbf{b})$$
$$= (\mathbf{a}, \mathbf{a}) + 2(\mathbf{a}, \mathbf{b}) + (\mathbf{b}, \mathbf{b}).$$

By equation (3.6), and in view of the fact that $|\cos \phi| \le 1$, we have

$$\|\mathbf{a} + \mathbf{b}\|^2 \le \|\mathbf{a}\|^2 + 2\|\mathbf{a}\| \, \|\mathbf{b}\| + \|\mathbf{b}\|^2 = (\|\mathbf{a}\| + \|\mathbf{b}\|)^2. \tag{3.20}$$

By taking the square root of each side of the preceding inequality, we confirm the first part of our claim.

To verify the second part of the assertion, we note that $\cos \phi \ge -1$, implying that

$$(\mathbf{a}, \mathbf{b}) \ge -\|\mathbf{a}\| \, \|\mathbf{b}\| \tag{3.21}$$

and

$$(\mathbf{a}, \mathbf{a}) + 2(\mathbf{a}, \mathbf{b}) + (\mathbf{b}, \mathbf{b}) \ge (\mathbf{a}, \mathbf{a}) - 2\|\mathbf{a}\| \, \|\mathbf{b}\| + (\mathbf{b}, \mathbf{b})$$

or

$$\|\mathbf{a} + \mathbf{b}\|^2 \ge (\|\mathbf{a}\| - \|\mathbf{b}\|)^2. \tag{3.21a}$$

This completes the proof.

(ε) We now write the inequality (3.21), augmented by $(\mathbf{a}, \mathbf{b}) \le \|\mathbf{a}\| \, \|\mathbf{b}\|$, in terms of absolute values and obtain

$$|(\mathbf{a}, \mathbf{b})| \le \|\mathbf{a}\| \, \|\mathbf{b}\|. \tag{3.22}$$

The foregoing inequality is known as the *Cauchy–Schwarz inequality*.† It is one of the *central* inequalities of functional analysis. Its great generality will become evident later in this book when we take up abstract spaces whose elements need not be conventional vectors.

The Cauchy–Schwarz inequality assumes a somewhat different meaning if it is written in the form

$$\|\mathbf{a}\| \, \|\mathbf{b}\| \, |\cos \phi| \le \|\mathbf{a}\| \, \|\mathbf{b}\|, \tag{3.23}$$

and either $\|\mathbf{a}\|$ or $\|\mathbf{b}\|$ is canceled.‡ The resulting inequality expresses the

---

† The name of Bunyakovsky is also associated with this inequality.
‡ Assuming, of course, that neither **a** nor **b** is the zero vector.

well-known fact that the length of the orthogonal projection of a line segment on a line is no greater than the length of the segment itself. This is, of course, but another way of saying that a leg of a right triangle is no greater than its hypotenuse.

To introduce the basic ideas of Euclidean spaces of not necessarily three, but any finite (and, later, an infinite) number of dimensions, it is essential to examine certain features of Euclidean geometry in establishing certain preliminary facts. This is done in the next chapter.

Problems

1. The length of the sum of two perpendicular vectors is 25. The length of one of these vectors is 3. Find the length of the second vector and the angle between this vector and the sum.

2. Let a vector $\mathbf{v}$ be perpendicular to two sides AB and AC of a triangle ABC. Show that $\mathbf{v}$ is also perpendicular to the third side of this triangle.

3. Show that, in a circle, the chords joining the end points of a diameter with a point on the circumference are perpendicular to each other.

4. Find the equation of a line passing through two given points.

5. Do the following "definitions" of inner products satisfy the distributive law (3.2)? If not, why not?
   (a) $(\mathbf{x}, \mathbf{y}) = \|\mathbf{x}\| \, \|\mathbf{y}\| \cos^2(\mathbf{x}, \mathbf{y})$
   (b) $(\mathbf{x}, \mathbf{y}) = \|\mathbf{x}\| \, \|\mathbf{y}\| \sin(\mathbf{x}, \mathbf{y})$.

6. Verify: (a) the so-called parallelogram rule,
$$\|\mathbf{x} + \mathbf{y}\|^2 + \|\mathbf{x} - \mathbf{y}\|^2 = 2\|\mathbf{x}\|^2 + 2\|\mathbf{y}\|^2,$$
   where the norm derives from an inner product; (b) the identity
$$(\mathbf{x}, \mathbf{y}) = \tfrac{1}{2}[\|\mathbf{x} + \mathbf{y}\|^2 - \|\mathbf{x}\|^2 - \|\mathbf{y}\|^2].$$

7. For any vectors $\mathbf{x}$ and $\mathbf{y}$, prove that
$$|\, \|\mathbf{x}\| - \|\mathbf{y}\| \,| \leq \|\mathbf{x} - \mathbf{y}\|$$
   and give the geometric interpretation of this inequality.

8. Show that the relation
$$\|\mathbf{x} - \mathbf{y}\|^2 = \|\mathbf{x} - \mathbf{z}\|^2 + \|\mathbf{y} - \mathbf{z}\|^2 - 2\|\mathbf{x} - \mathbf{z}\| \, \|\mathbf{y} - \mathbf{z}\| \cdot \cos(\mathbf{x} - \mathbf{z}, \mathbf{y} - \mathbf{z}),$$
   between any three vectors $\mathbf{x}$, $\mathbf{y}$, and $\mathbf{z}$, is another form of the cosine law (3.17).

4

Linear Independence. Vector Components. Space Dimension

In developing the axiomatic system in the preceding chapters, we had in mind Euclidean spaces of three or fewer dimensions. The idea of space dimension, however, did not explicitly enter into our axiomatics and has been so far devoid of any specific meaning. In order to correct this omission, it is necessary to adjoin a new postulate to those already adopted. Before doing this in the next chapter, we have first to examine the closely related concept of *linear independence of vectors*.

At the start, let us imagine a representative of the simplest, that is, one-dimensional, Euclidean space, $\mathscr{E}_1$, realized by a straight line—a carrier of collinear vectors, $\mathbf{x}$, say. Suppose that we shift the vectors along the line to make their initial points coincide with a point O selected as the origin. Consider then a vector $\mathbf{x}_1$ from the set $\{\mathbf{x}\}$. By Axiom 2.4, every vector of the set can be represented in the form

$$\mathbf{x} = \alpha \mathbf{x}_1, \tag{4.1}$$

where α is a scalar, $-\infty < \alpha < \infty$. Since the foregoing relation is linear in α, it is natural to say that the vectors of the set depend *linearly* on $\mathbf{x}_1$. The vector $\mathbf{x}_1$ itself can be thought of as an "*independent*" vector.

We note that, in view of Axiom 2.5 and Conclusion 2.3, equation (4.1) generates the zero vector, $\mathbf{0}$, for $\alpha = 0$, so that the zero vector belongs to the set $\{\mathbf{x}\}$. This, of course, confirms the fact that the origin O lies on the given line.

Concentrating, in turn, on a Euclidean space of two dimensions, $\mathscr{E}_2$, represented by a plane, we select two coplanar, nonparallel vectors, $\mathbf{x}_1$ and

$\mathbf{x}_2$, with their initial points at some origin O. By Definition 2.3, any other vector of the set $\{\mathbf{x}\}$ lying in the plane can be represented by the two selected vectors

$$\mathbf{x} = \alpha_1 \mathbf{x}_1 + \alpha_2 \mathbf{x}_2, \tag{4.2}$$

where $-\infty < \alpha_1, \alpha_2 < \infty$. Vectors $\mathbf{x}_1$ and $\mathbf{x}_2$ can now be regarded as *"independent"* in the same sense as vector $\mathbf{x}_1$ in the preceding example. This statement should mean that every vector of the given set can be represented as a linear combination of $\mathbf{x}_1$ and $\mathbf{x}_2$ (viz., is *linearly dependent* on $\mathbf{x}_1$ and $\mathbf{x}_2$), but neither of the latter two can be expressed solely in terms of the other (inasmuch as, by hypothesis, they are not parallel). If $\alpha_1 = \alpha_2 = 0$, then $\mathbf{x}$ becomes the zero vector, and this verifies the fact that the plane passes through the origin O.

A similar argument implies that the number of linearly independent vectors in the space $\mathscr{E}_3$ equals three, $\mathbf{x}_1$, $\mathbf{x}_2$, and $\mathbf{x}_3$, say, provided the vectors are neither coplanar nor any two of them are parallel.† As before, every other vector in the space *depends linearly* on $\mathbf{x}_1$, $\mathbf{x}_2$, and $\mathbf{x}_3$.

With the preceding examples in mind, the concept of linear dependence (or independence) of an arbitrary number of vectors suggests itself automatically, and we adopt the following.

Definition 4.1. Given a set of vectors, $\mathbf{x}_1$, $\mathbf{x}_2$, ..., $\mathbf{x}_n$, it is said that the vectors are *linearly dependent* if there exist scalars α_1, α_2, ..., α_n, *not all zero*, such that the relation

$$\alpha_1 \mathbf{x}_1 + \alpha_2 \mathbf{x}_2 + \cdots + \alpha_n \mathbf{x}_n = \mathbf{0} \tag{4.3}$$

holds.

Now, if all coefficients α_i, $i = 1, 2, \ldots, n$, in the preceding equation are zero, the result is trivial. If all coefficients except a single one, say α_k, are zero, then $\alpha_k \mathbf{x}_k = \mathbf{0}$, and therefore $\mathbf{x}_k = \mathbf{0}$. This being so, we accept the convention that the zero vector depends linearly on any other vector.

If the relation (4.3) holds only when all coefficients α_i equal zero, we say that the vectors $\mathbf{x}_i$ are *linearly independent*. Such is, for instance, the triple of coordinate vectors of any Cartesian (rectangular or oblique) coordinate system in a Euclidean space $\mathscr{E}_3$.

Suppose now that there is given a set of vectors $\{\mathbf{x}_i\}$ such that the coefficient α_k in the equation (4.3) is different from zero. We divide (4.3) through by α_k to obtain

$$\mathbf{x}_k = \beta_1 \mathbf{x}_1 + \beta_2 \mathbf{x}_2, + \cdots + \beta_{k-1} \mathbf{x}_{k-1} + \beta_{k+1} \mathbf{x}_{k+1} + \cdots + \beta_n \mathbf{x}_n, \tag{4.4}$$

---

† The reader with a geometric turn of mind will find it natural to treat $\mathscr{E}_1$ and $\mathscr{E}_2$ as *subspaces* of $\mathscr{E}_3$ of some sort. This matter will be examined in some detail in Chapter 6.

where $\beta_i = -\alpha_i/\alpha_k$, $i = 1, 2, \ldots, n$. A glance at the preceding equation convinces us that, if a set of vectors is *linearly dependent*, then some vector of the set can be represented by a *linear combination* of the remaining vectors of the set.

By way of illustration, imagine that we select two linearly independent (that is, two nonparallel) vectors, $\mathbf{x}_1$ and $\mathbf{x}_2$, in $\mathscr{E}_2$. We wish to show that the vectors $\mathbf{x}_1$ and $3\mathbf{x}_1 + 2\mathbf{x}_2$ are also linearly independent.

Assume that, for some scalars α_1 and α_2,

$$\alpha_1 \mathbf{x}_1 + \alpha_2(3\mathbf{x}_1 + 2\mathbf{x}_2) = \mathbf{0}. \tag{4.5}$$

We must demonstrate that $\alpha_1 = \alpha_2 = 0$. Upon rearranging, we have

$$(\alpha_1 + 3\alpha_2)\mathbf{x}_1 + 2\alpha_2 \mathbf{x}_2 = \mathbf{0}.$$

By hypothesis, however, $\mathbf{x}_1$ and $\mathbf{x}_2$ are linearly independent; therefore,

$$\alpha_1 + 3\alpha_2 = 0, \qquad \alpha_2 = 0,$$

or $\alpha_1 = \alpha_2 = 0$, as required.

A conclusion which can be drawn from the preceding example is that, inasmuch as linear combinations of linearly independent vectors may provide linearly independent vectors, there is always an infinite number of equivalent sets of linearly independent vectors. These sets include, however, the same number of vectors, in spite of the fact that the vectors themselves are different.

It is customary to say that a set of vectors *spans* (or determines, or generates) a space† if every vector in the space can be represented as a linear combination of the vectors of the set.

It is interesting to note that the vectors of a *spanning* set need not be linearly independent, and some may be "redundant." However, if they are linearly independent, then it is said that they form a *basis* for the space.‡

So far we have verified that, for $n \leq 3$, the number of base vectors is the same as the "dimension" of the space. We shall show later that this statement is also true for $n > 3$, and that, conversely, if a basis for a space consists of n vectors, then the *dimension* of the space is also n.

---

† Throughout this chapter, we assume that the dimension of any space is not greater than three ($n \leq 3$ in $\mathscr{E}_n$), but most of our conclusions remain valid for $n > 3$. This is discussed in subsequent chapters. [As regards spanning, see the text preceding equation (5.17) in Chapter 5.]

‡ A basis of this type is often called a *Hamel* basis. The existence of such bases expresses an algebraic property of the vector space. These are but one of the many types of bases introduced in analysis. In particular, bases of topological spaces are suitably defined collections of "open sets." Models of open sets are an open interval of the real axis ("x axis"), the interior of a circle, and the set of all points on one side of a straight line in a plane. Models of closed sets are a closed interval of the real axis and the set of points on and within a circle.

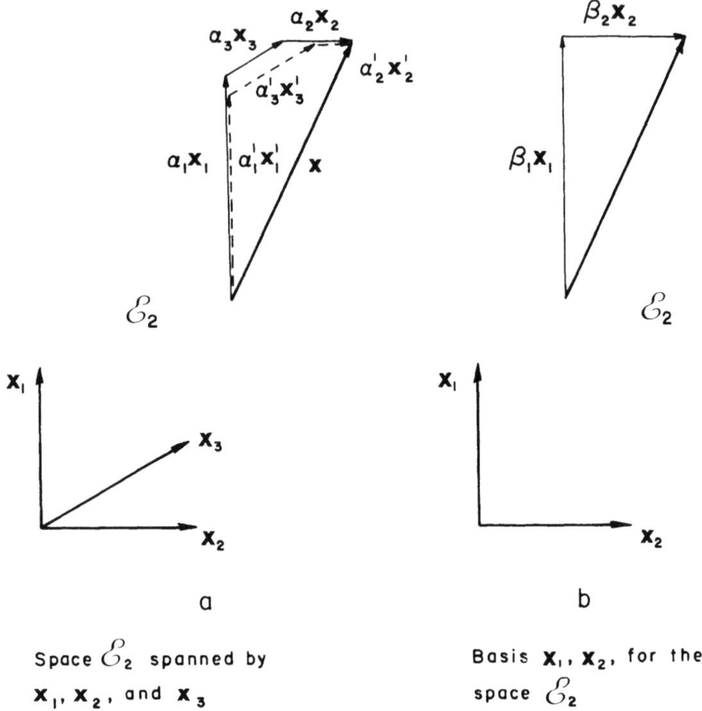

Figure 4.1. A spanning set and a basis.

As an illustration of the difference between a *spanning set* and a *basis*, consider a two-dimensional space, represented by a plane in Figure 4.1. Inasmuch as the spanning vectors are not necessarily linearly independent, the representation of a vector x in terms of these vectors is not unique (cf. Figure 4.1a, in which $x = \alpha_1 x_1 + \alpha_2 x_2 + \alpha_3 x_3$ and $x = \alpha_1' x_1 + \alpha_2' x_2 + \alpha_3' x_3$).

On the other hand, if a set constitutes a basis (Figure 4.1b), the representation of any vector x is unique. Thus, if a spanning set is to be converted into a basis, some of the spanning vectors may have to be discarded.

It is apparent that in any space there is an infinite number of bases (including always the same number of vectors). This follows from the fact that in any space the number of sets of linearly independent vectors is infinite. Among the possible bases, those composed of *orthogonal* (in particular, *orthonormal*) vectors are most useful.† In $\mathcal{E}_3$, an *orthonormal basis*, that

---

† Orthonormal vectors are mutually orthogonal unit vectors.

is, a basis composed of orthonormal vectors, is formed by the well-known unit coordinate vectors (sometimes called *versors*), $\mathbf{e}_i$, $i = 1, 2, 3$, $\|\mathbf{e}_i\| = 1$, defining a rectangular Cartesian frame.[†] Obviously,

$$(\mathbf{e}_i, \mathbf{e}_k) = \delta_{ik}, \qquad i, k = 1, 2, 3, \tag{4.6}$$

where δ_{ik} is the Kronecker delta, equal to 1 for $i = k$ and otherwise zero. It is intuitively clear that the vectors $\mathbf{e}_i$ are linearly independent, yet it may be of interest to verify this analytically.

Assume then that the scalars α_1, α_2, and α_3 are such that the relation

$$\alpha_1 \mathbf{e}_1 + \alpha_2 \mathbf{e}_2 + \alpha_3 \mathbf{e}_3 = \mathbf{0}$$

holds. We form the inner product of each side of this equation with $\mathbf{e}_1$. By appeal to equation (4.6), we conclude that $\alpha_1 = 0$. In a similar manner, we find that $\alpha_2 = \alpha_3 = 0$, which completes our verification.

Inasmuch as every vector $\mathbf{x}$ in a space is represented in terms of the base vectors, we can write, for three-space, for example,

$$\mathbf{x} = x_1 \mathbf{g}_1 + x_2 \mathbf{g}_2 + x_3 \mathbf{g}_3, \tag{4.7a}$$

where the $\mathbf{g}_i$'s are base vectors and the x_i's are *components* of $\mathbf{x}$.[‡] The preceding equation may be thought of an an analytic representation of the given space.

Similarly, in terms of an *orthonormal* basis $\{\mathbf{e}_i\}$, a vector $\mathbf{x}$ in $\mathscr{E}_3$ is given by

$$\mathbf{x} = x_1 \mathbf{e}_1 + x_2 \mathbf{e}_2 + x_3 \mathbf{e}_3, \tag{4.7b}$$

where the values of the components in (4.7a) and (4.7b) are different.

By forming inner products of both members of the preceding equation successively with the vectors of the basis, we find

$$x_i = (\mathbf{x}, \mathbf{e}_i), \qquad i = 1, 2, 3. \tag{4.8}$$

This shows that the components x_i of $\mathbf{x}$ are the *orthogonal projections* of $\mathbf{x}$ on the base vectors of an orthonormal basis.

The components of the versors $\mathbf{e}_i$ are, of course, $(1, 0, 0)$, $(0, 1, 0)$, and $(0, 0, 1)$, respectively. By (4.8), the components of the zero vector $\mathbf{0}$ are $(0, 0, 0)$, since the zero vector was assumed orthogonal to any other vector (including itself).

The identification of vectors with *sets* of their *components*, initiated in

---

[†] A more familiar notation is, of course, $\mathbf{i}, \mathbf{j}, \mathbf{k}$.

[‡] By components (or coordinates) of a vector we mean here signed lengths of component vectors $\mathbf{x}_i$ whose resultant is the given vector. Thus, $\mathbf{x} = \sum_i \mathbf{x}_i$, $\mathbf{x}_i = x_i \mathbf{e}_i$, for example.

this chapter, is a radical departure from our earlier treatment of vectors as directed line segments. In the present context, a vector ceases to be a *single* entity *invariant* with respect to changes of coordinate frames, but is split into components and represented by a *set of scalars*. The values of the latter clearly depend on the selected frame of reference.

Summing up matters, we can state that, by changing our line of treatment, we have abandoned the approach of synthetic geometry and replaced it by that of analytic geometry, more convenient for our purposes.

In order to examine briefly the operations on vectors represented as sets of components, let

$$\mathbf{z} = \gamma_1 \mathbf{e}_1 + \cdots + \gamma_n \mathbf{e}_n \tag{4.9}$$

be the sum of two vectors,

$$\mathbf{x} = \alpha_1 \mathbf{e}_1 + \cdots + \alpha_n \mathbf{e}_n,$$
$$\mathbf{y} = \beta_1 \mathbf{e}_1 + \cdots + \beta_n \mathbf{e}_n, \tag{4.10}$$

where $\{\mathbf{e}_k\}$ is a set of n orthonormal vectors. Such vectors obey condition (4.6) for $i, k = 1, 2, \ldots, n$, and in the next chapter they are shown to be linearly independent.† Adding (4.10) and comparing term by term with (4.9) implies that the components of a sum of two (or more) vectors are equal to the sums of the corresponding components of the vectors. Similarly,

$$\alpha \mathbf{x} = \alpha\alpha_1 \mathbf{e}_1 + \cdots + \alpha\alpha_n \mathbf{e}_n, \tag{4.11}$$

so that multiplication of a vector by a scalar is equivalent to multiplication of its components by the scalar.

Let us now consider two vectors in three-space $\mathscr{E}_3$:

$$\mathbf{x} = x_1 \mathbf{e}_1 + x_2 \mathbf{e}_2 + x_3 \mathbf{e}_3,$$
$$\mathbf{y} = y_1 \mathbf{e}_1 + y_2 \mathbf{e}_2 + y_3 \mathbf{e}_3. \tag{4.12}$$

By forming their inner product, we have

$$(\mathbf{x}, \mathbf{y}) = x_1 y_1 + x_2 y_2 + x_3 y_3. \tag{4.13}$$

For $\mathbf{x} = \mathbf{y}$, this gives

$$\|\mathbf{x}\|^2 = (x_1)^2 + (x_2)^2 + (x_3)^2. \tag{4.14}$$

---

† See the text following equation (5.16). Actually, the orthogonality of vectors alone suffices for their linear independence.

If the coordinates of the end points of a vector $\mathbf{x} = \overrightarrow{AB}$ are x_i and x_i', $i = 1, 2, 3$, respectively, then the last equation is

$$AB^2 = (x_1' - x_1)^2 + (x_2' - x_2)^2 + (x_3' - x_3)^2. \qquad (4.15)$$

Either of the two preceding equations expresses the familiar *Pythagorean theorem* extended to three dimensions. They determine the *length* of a vector and the *distance* between two points, provided the components of the vector and the coordinates of the points are known, respectively.

It is important to emphasize that one of the main objectives in constructing the axiomatic system in the preceding chapters was to arrive at metric formulas such as (4.14) and (4.15). That the latter turned out to be those of Euclidean geometry is not a matter of chance, but is due to the fact that the axiomatics was deliberately patterned after that of Euclid.

The Pythagorean theorem, and specifically its generalization to spaces of many (and even infinitely many) dimensions, is one of the central theorems of mathematics.†

With this, we end our review of Euclidean spaces of dimension no greater than three, and in the next chapter confine our attention to spaces of many dimensions.‡ A first idea that comes here to mind is a simple extension of the basic concepts of three-dimensional geometry to spaces of many dimensions. In doing this, we shall be forced to relinquish representation of elements and relationships between elements in such spaces, inasmuch as our geometric imagination is too limited for this purpose. Even so, many-dimensional spaces do not involve anything mysterious or metaphysical. For a number of years, they have been successfully employed in various branches of physics under the name of *configuration spaces*.

In theoretical mechanics, for instance, motions of a system composed of n free particles (possessing $3n$ degrees of freedom) are often projected into a $3n$-space in which the system is represented by a single point with $3n$ coordinates. Likewise, in thermodynamics, states of systems undergoing changes are depicted by the so-called Gibbs' diagrams in spaces of many dimensions. An even more dramatic example is provided by motions of deformable continuous bodies (e.g., vibrations of beams or plates), the representation of which requires configuration spaces of an infinite number of dimensions.

Similar circumstances are met in our everyday experience. As an example, we can cite the samplings of opinions by means of polls, the results of which are analyzed in so-called sample spaces, often of many dimensions.

---

† Compare, e.g., Friedrichs.[18] Lanczos, in his book,[19] goes so far as to state that from the Pythagorean theorem "we can deduce all the laws of Euclidean geometry."

‡ For brevity, we subsequently refer to spaces of dimension greater than three as *many*- or *multi*dimensional spaces.

Problems

1. Let e_i, $i = 1, 2, 3$, be unit vectors along of the axes of a Cartesian rectangular frame. Are the vectors $x = 2e_1 + e_2 - e_3$, $y = e_1 - 3e_2 + 2e_3$, and $z = -2e_1 - e_2 + 3e_3$ linearly independent?

2. Show that the vector $v = 3e_1 - 2e_2 - 5e_3$ depends linearly on the x, y, and z defined in the preceding problem.

3. Let x and y be linearly independent. Are x and $x + 10y$ also independent?

4. Given $x = 2e_1 + 3e_2 + 4e_3$, $y = e_1 + e_2 + e_3$, and $z = 5e_1 + e_2 + 2e_3$, find the norm of $v = x + y + z$.

5. Find the projection of the vector $x = 4e_1 + 6e_2 - 2e_3$ on the direction of the vector $y = -9e_1 - 3e_2 + 3e_3$.

6. Find the equation of a plane passing through the tip A of the position vector $\overrightarrow{OA} = 2e_1 + 10e_2 + 6e_3$ and perpendicular to the vector $n = 4e_1 + 6e_2 + 12e_3$.

7. Show that the vectors $f_1 = (3e_1 + 6e_2 + 6e_3)/9$, $f_2 = (6e_1 + 3e_2 - 6e_3)/9$ and $f_3 = (6e_1 - 6e_2 + 3e_3)/9$ form a triad of mutually orthogonal unit vectors.

8. Which of the three vectors $x = (6, 2, 6)$, $y = (8, 16, 4)$, $z = (2, -6, 4)$, and $v = (3, 4, 5)$ may serve as a basis for Euclidean three-space? ($x = (6, 2, 6)$ means $x = 6e_1 + 2e_2 + 6e_3$.)

9. Let two noncollinear vectors x_1 and x_2 be given in $\mathscr{E}_2$, so that any vector in this space (plane) can be represented by $z = xx_1 + yx_2$. Find the relation between $z = \|z\|$ and x and y if: (a) x_1, x_2 form an orthogonal basis, (b) an orthonormal basis.

10. Find a linearly independent subset of the set $\{x = (2, 4, -3)$, $y = (4, 8, -6)$, $z = (2, 1, 3)$, $v = (8, 13, -6)\}$ which spans the same space as the given set.

5

Euclidean Spaces of Many Dimensions

In the preceding chapters, we have studied the main concepts concerning vectors in the familiar three-dimensional space. It was pointed out that the maximal number of linearly independent vectors was characteristic of those sets which could be identified pictorially with either a line, plane, or the entire space. Such a close association between the number of base vectors and the dimension of a space proves to be so fundamental that the generalization of the very idea of space proceeds virtually parallel to the extension of the idea of vector. Clearly, the simplest extension which comes to mind consists of imagining a vector in a space whose dimension is greater than three. It turns out that this can be done on various levels of abstraction, and we shall learn about such constructions in subsequent chapters. At this stage, we wish to give some thought to the notion of *components*, which was so characteristic of a vector in Euclidean three-space. With this in mind, we adopt the following definition.

Definition 5.1. Let n be a fixed positive integer. A vector, $\mathbf{x}$, *is* an ordered set of n real scalars, $x_1, x_2, \ldots, x_n$. We denote this by writing

$$\mathbf{x} = (x_1, x_2, \ldots, x_n) \tag{5.1}$$

and call the scalars the *components* of $\mathbf{x}$. Guided by the axiomatics adopted in Chapter 2, we shall set up the operation on the vectors so defined through the definitions given below.

Before actually doing this, it is important to note that the just-given definition requires discarding the pictorial image of a vector as a directed line segment, but relieves the difficulty in dealing with vectors in spaces of many dimensions.

In order to underline the abstract nature of n-dimensional vectors, we drop—from now on—their boldface designation and write, in *lightface* type,[†]

$$x = (x_1, x_2, \ldots, x_n), \tag{5.2}$$

where the x_i's are the components of x. Again, to avoid confusion, we shall mark the components of a vector by a subscript (e.g., x_i) and the vectors themselves by a superscript (e.g., x^i), except that the base vectors shall be denoted temporarily by e_i.[‡]

Relation (5.2) implies that each vector is identified with an ordered n-tuple of scalars and, conversely, each ordered n-tuple of scalars represents a vector. This relation can be considered to be a counterpart of Axiom 2.1. The definition of equality of vectors, corresponding to Axiom 2.2, now reads

Definition 5.2. Two vectors $x = (x_1, x_2, \ldots, x_n)$ and $y = (y_1, y_2, \ldots, y_n)$, of the same dimension, are equal, written $x = y$, if for each $k = 1, 2, \ldots, n$, there is $x_k = y_k$.

Definition 5.3. That vector of which each component is zero is the *zero* (or null) vector, denoted by θ.[§] In the notation of equation (5.2),

$$\theta = (0, 0, \ldots, 0). \tag{5.3}$$

The negative (inverse) of a given vector x is introduced by

Definition 5.4.

$$-x = (-x_1, -x_2, \ldots, -x_n). \tag{5.4}$$

The operations of addition of vectors and scalar multiplication are established by the following two definitions.

Definition 5.5.

$$x + y = (x_1 + y_1, x_2 + y_2, \ldots, x_n + y_n), \tag{5.5}$$

where $x = (x_1, x_2, \ldots, x_n)$ and $y = (y_1, y_2, \ldots, y_n)$.

[†] This symbolism agrees with the traditional notation in functional analysis. The notation (5.2) is often called a *n-tuple* notation of vectors.

[‡] To match our previous notation e_i. In the future, orthonormal vectors will, for the most part, be denoted by i^v. As the need arises, however, different notation for orthogonal and orthonormal sets will be used (provided no confusion is likely to arise).

[§] Also, occasionally, by 0 if it is clear from the context whether the matter concerns the zero vector or the number zero.

Definition 5.6.

$$\alpha x = (\alpha x_1, \alpha x_2, \ldots, \alpha x_n), \tag{5.6}$$

where $x = (x_1, x_2, \ldots, x_n)$ and α is a scalar.

It is not difficult to verify that the operations on vectors obey the following laws.

(a) Commutativity and associativity of addition:

$$x + y = y + x, \qquad (x + y) + z = x + (y + z). \tag{5.7}$$

These equations correspond to equations (2.6) and (2.9).

(b) Distributivity and associativity of multiplication:

$$\alpha(x + y) = \alpha x + \alpha y, \qquad (\alpha + \beta)x = \alpha x + \beta x, \qquad \alpha(\beta x) = (\alpha\beta)x. \tag{5.8}$$

These equations correspond to equations (2.14), (2.13), and (2.15).

(c) Multiplication of a vector by zero gives the zero vector, and by 1, the vector itself:

$$0x = \theta, \qquad 1x = x; \tag{5.9}$$

compare equations (2.16) and (2.12).

To illustrate, we can verify the last of relations (5.8) by writing its left-hand side as $\alpha(\beta x_1, \beta x_2, \ldots, \beta x_n)$, by virtue of Definition 5.6, and $(\alpha\beta x_1, \alpha\beta x_2, \ldots, \alpha\beta x_n)$, again by the same definition. Now, reading equation (5.6) from right to left, we convince ourselves (with a trivial change of notation) that $(\alpha\beta x_1, \alpha\beta x_2, \ldots, \alpha\beta x_n) = \alpha\beta(x_1, x_2, \ldots, x_n)$. By equation (5.2), the last result reads simply $(\alpha\beta)x$, which completes the verification.

Likewise, if $\alpha = 0$, Definitions 5.3 and 5.6 imply directly the first of relations (5.9).

The set of all vectors described by Definitions 5.1–5.6 constitutes, for a given value of n, what may be called an *n-dimensional Euclidean space devoid of metric*. We denote this space by $\mathscr{R}_n$ in order to indicate that it is composed of *n*-tuples of real numbers. It is an *affine* space, in the sense given to that term in Chapter 2, and is often called an *arithmetic* or *coordinate* space.

In order to produce a "true Euclidean *n*-space," it is, in addition, required to provide a *norm* for each (*n*-dimensional) vector x by the definition

$$\|x\| = [(x_1)^2 + (x_2)^2 + \cdots + (x_n)^2]^{1/2}, \tag{5.10}$$

thus endowing the *n*-space considered with a metric analogous to that of Euclidean three-space.† Furthermore, since the entire geometric structure of

---

† It is important to note that there are many other definitions giving specific norms for $\mathscr{R}_n$ (examples of various metrics are given at the end of this chapter), but only the definition (5.10) converts $\mathscr{R}_n$ into $\mathscr{E}_n$. Some authors are more liberal, calling any real inner product space Euclidean. See, e.g., Halmos.[20]

the n-space is modeled on that of $\mathscr{E}_3$, it is quite natural to call the n-space a *Euclidean space of dimension n, $\mathscr{E}_n$*. For $n = 3$ this space becomes the more common one. Definition (5.10) complies with our general requirement that the norm (and the self-inner product defined below) be positive definite:

$$\|x\| > 0 \qquad \text{unless} \quad x = \theta \quad \text{and then} \quad \|x\| = 0. \tag{5.11}$$

It is easily shown that (5.10) satisfies all other properties required of a norm, for example,

$$\|\alpha x\| = |\alpha| \, \|x\|, \qquad \text{where } |\alpha| \text{ is the absolute value of } \alpha. \tag{5.12}$$

It is now straightforward to take another step and generalize the concept of the inner product [cf. equation (4.13)] by means of the following.

Definition 5.7. The inner product (x, y) of two vectors, x and y, of the same dimension n is given by the formula in $\mathscr{E}_n$:

$$(x, y) = x_1 y_1 + x_2 y_2 + \cdots + x_n y_n, \tag{5.13}$$

where the x_i's and y_i's are the components of x and y, respectively.

In view of this definition and definition (5.10) of the norm, the relation between the latter and the self-inner product is

$$\|x\| = \sqrt{(x, x)}. \tag{5.14}$$

An inspection of equation (5.13) shows that the inner product is symmetric and obeys the rules of associativity and distributivity,

$$(\alpha x, y) = \alpha(x, y), \qquad (x, y + z) = (x, y) + (x, z). \tag{5.15}$$

A generalization of the notion of orthogonality of vectors [cf. (3.11)] is achieved by means of

Definition 5.8. Two vectors, x and y, are said to be orthogonal if their inner product vanishes. Symbolically, $x \perp y$ if $(x, y) = 0$.

An example of a set of orthogonal vectors is provided by the n vectors

$$e_1 = (1, 0, \ldots, 0), \quad e_2 = (0, 1, \ldots, 0), \ldots, \quad e_n = (0, 0, \ldots, 1). \tag{5.16}$$

In fact, in view of Definition 5.7, we find immediately that

$$(e_i, e_j) = \delta_{ik}, \qquad i, k = 1, 2, \ldots, n, \tag{5.16a}$$

where δ_{ik} is the already-mentioned Kronecker delta. By (5.14), of course, $\|e_i\| = 1, i = 1, 2, \ldots, n$, so that the set (5.16) is, in fact, an orthonormal set. It may be considered as an analog, in Euclidean spaces of dimension greater than three, of the triple of versors (4.6), contained in $\mathscr{E}_3$.[†]

By virtue of their mutual orthogonality, the vectors (5.16) are *lineraly* independent in the sense of Definition 4.1,[‡] and every vector $x = (x_1, x_2, \ldots, x_n)$ in $\mathscr{E}_n$ can be represented (uniquely) as a linear combination of the vectors of the set (5.16). Indeed, by Definitions 5.5 and 5.6,

$$x = (x_1, x_2, \ldots, x_n) = \sum_{i=1}^{n} x_i e_i = x_1(1, 0, \ldots, 0) + x_2(0, 1, \ldots, 0)$$

$$+ \cdots + x_n(0, 0, \ldots, 1), \tag{5.16b}$$

where $x_1, x_2, \ldots, x_n$ are, of course, scalars.

Consequently, if we agree to extend to $\mathscr{E}_n$, for $n > 3$, the concepts of spanning set (a set of vectors spans $\mathscr{E}_n$ if every vector in $\mathscr{E}_n$ can be expressed as a linear combination of the vectors of the set) and linear independence,[§] defined previously for $\mathscr{E}_3$, then the set (5.16) provides a *basis* for $\mathscr{E}_n, n > 3$, in the same sense in which this term was understood in Chapter 4 [cf. the text following equation (4.5)]. In order to avoid repetition, we shall not pursue this question further, postponing to the next chapter a more precise clarification of the concepts of spanning set and basis in finite- and infinite-dimensional Euclidean spaces. At this point, it is worth emphasizing, however, that any basis for (what we have defined as) $\mathscr{E}_n$ is composed of exactly n vectors.

A pictorial representation of an orthonormal basis of dimension n, using the conventional graphical symbol for perpendicularity of vectors, is given in Figure 5.1a. While not particularly graceful, it may be of some assistance in forming a mental image of n-space.

In analogy to equation (4.8), we deduce from equation (5.16b)

$$x_i = (x, e_i), \qquad i = 1, 2, \ldots, n, \tag{5.17}$$

and identify the components x_i with the *orthogonal projections*[¶] of the vector x on the axes of the orthonormal *basis* $\{e_i\}$.

[†] Some authors give to the set (5.16) the pictorial name "coordinate vectors" oriented along the "axes of the n-dimensional orthogonal Cartesian reference frame." Compare, e.g., Sokolnikoff (Ref. 21, p. 13).

[‡] Actually, an *arbitrary* collection of nonzero mutually orthogonal vectors, $x^1, x^2, \ldots, x^n$, is linearly independent. Suppose that $\sum_{k=1}^{n} a_k x^k = \theta$, and form the self-inner product of the sum, which must vanish: $0 = \|\theta\|^2 = \sum_{k=1}^{n} \|\alpha_k x^k\|^2 = \sum_{k=1}^{n} |\alpha_k|^2 \|x^k\|^2$. Since $\|x_k\| > 0$ for all k, necessarily $\alpha_k = 0$ for all k.

[§] The obvious extension of the notion of linear independence to collections of vectors $\mathscr{E}_n, n > 3$ was, in fact, tacitly carried out and employed above during the discussion of orthogonal sets.

[¶] An orthogonal projection is defined here as the *norm* of the vector, whose end points are the projections of the end points of the given vector, prefixed with the appropriate sign.

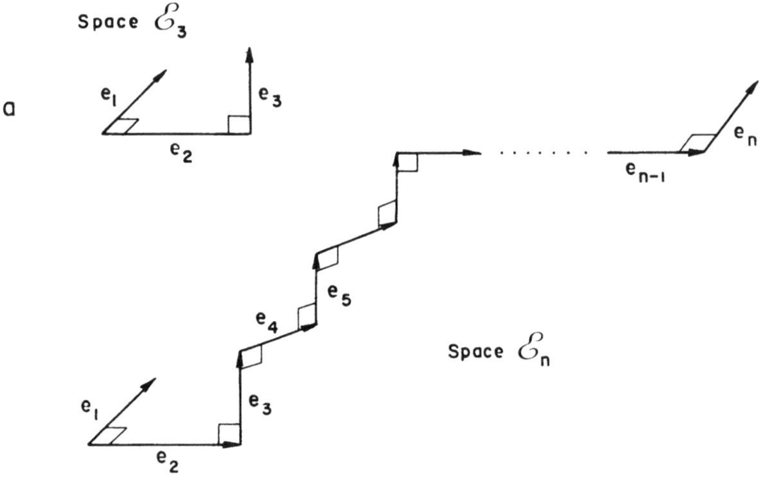

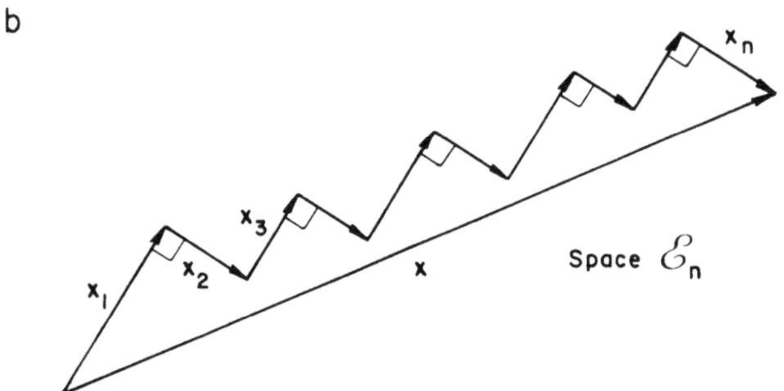

Figure 5.1. (a) Symbolic representation of an orthonormal basis. (b) Resolution of a vector in the space $\mathcal{E}_n$.

This being so, we arrive at the decomposition

$$x = (x, e_1)e_1 + (x, e_2)e_2 + \cdots + (x, e_n)e_n \tag{5.18}$$

of a vector x referred to an n-dimensional orthonormal basis (Figure 5.1b).

Denote now by $\bar{\mathbf{x}}_i$ the ith component vector of a vector $\mathbf{x}$, that is, $\bar{\mathbf{x}}^i = x_i\mathbf{e}_i$, $i = 1, 2, \ldots, n$ (do not sum). We take the self-inner product of equation (5.18) and, with equation (5.16a) in mind, conclude that (Figure 5.1b).

$$\|\mathbf{x}\|^2 = \|\bar{\mathbf{x}}^1\|^2 + \|\bar{\mathbf{x}}^2\|^2 + \cdots + \|\bar{\mathbf{x}}^n\|^2. \tag{5.19}$$

Thus, referred to an orthonormal basis, the square of the norm of a vector is equal to the sum of the squares of the norms of its vectorial components. This brings us back to the *Pythagorean* theorem, generalized to the setting of n-space.

Just as easily, we can extend the validity of the two fundamental inequalities of Cauchy–Schwarz and Minkowski, discussed in Chapter 3. First, we note that, by the positive definiteness of the self-inner product, the form

$$(x, x) = x_1{}^2 + x_2{}^2 + \cdots + x_n{}^2 \tag{5.20}$$

is positive whenever at least one x_k is nonzero and zero if each x_k is equal to zero. Accordingly, for any scalar r,

$$(x + ry, x + ry) \geq 0 \tag{5.21}$$

or, expanding,

$$R(r) \equiv (y, y)r^2 + 2(x, y)r + (x, x) \geq 0. \tag{5.22}$$

The quadratic equation $R(r) = 0$ represents a parabola which is, by (5.22), supposed to touch the r-axis in at most one point. Hence, the equation $R(r) = 0$ can have at most a single real root. This implies that the discriminant

$$(x, y)^2 - (x, x)(y, y) \leq 0. \tag{5.23}$$

Taking the positive roots and rewriting, we have

$$|(x, y)| \leq \|x\| \, \|y\|, \tag{5.24}$$

which is the *Cauchy–Schwarz inequality.*

The *triangle inequality* follows from the identity

$$\|x + y\|^2 = (x + y, x + y)$$
$$= (x, x) + 2(x, y) + (y, y)$$

upon application of the Cauchy–Schwarz inequality to the middle term. We find

$$(x, x) + 2(x, y) + (y, y) \leq \|x\|^2 + 2\|x\| \, \|y\| + \|y\|^2$$

or, after taking square roots,

$$\|x + y\| \leq \|x\| + \|y\|, \tag{5.25}$$

as asserted.

In component representation, the just-mentioned inequalities take the self-explanatory forms

$$(x_1 y_1 + x_2 y_2 + \cdots + x_n y_n)^2$$
$$\leq (x_1{}^2 + x_2{}^2 + \cdots + x_n{}^2)(y_1{}^2 + y_2{}^2 + \cdots + y_n)^2 \tag{5.26a}$$

and

$$[(x_1 + y_1)^2 + (x_2 + y_2)^2 + \cdots + (x_n + y_n)^2]^{1/2}$$
$$\leq (x_1{}^2 + x_2{}^2 + \cdots + x_n{}^2)^{1/2} + (y_1{}^2 + y_2{}^2 + \cdots + y_n{}^2)^{1/2}. \quad (5.26b)$$

An orthonormal basis for n-space can be obtained from an arbitrary set of n linearly independent vectors by means of the so-called *Gram–Schmidt orthogonalization process*.

Assume then that the set $g_1, g_2, \ldots, g_n$ constitutes a basis for $\mathscr{E}_n$, i.e., is linearly independent. We first construct a unit vector (note that $g_1 \neq \theta$)

$$e_1 = \frac{g_1}{\|g_1\|} \quad (5.27)$$

and require that the second vector

$$e_2 = c_{21}e_1 + c_{22}g_2 \quad (5.28)$$

be orthogonal to e_1. This gives

$$(e_2, e_1) = c_{21} + c_{22}(g_2, e_1) = 0 \quad (5.29)$$

so that

$$e_2 = c_{22}[g_2 - (g_2, e_1)e_1]. \quad (5.30)$$

The just-written equation has a simple geometric meaning (Figure 5.2). It states that—to within the scaling factor c_{22}—the vector e_2 is a difference between two vectors: the vector g_2 (subject to orthogonalization) and the vector $(g_2, e_1)e_1$, which is the orthogonal projection of g_2 on e_1 (or g_1).

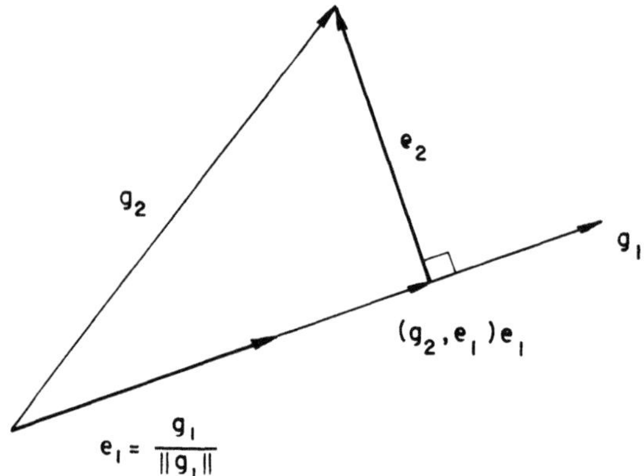

Figure 5.2. Illustration of the Gram–Schmidt process.

Consequently, e_2 gives the component of g_2 perpendicular to e_1, as required. In order that e_2 be a unit vector, the scaling factor should be chosen as

$$c_{22} = [(g_2, g_2) - (g_2, e_1)^2]^{-1/2}. \tag{5.31}$$

Substitution of this into equation (5.30) yields the second member of the orthonormalized basis $\{g_i\}$.

The next step is to assume that

$$e_3 = c_{31}e_1 + c_{32}e_2 + c_{33}g_3 \tag{5.32}$$

and to determine the coefficients c_{ij} from the orthonormality conditions,

$$(e_3, e_1) = 0, \qquad (e_3, e_2) = 0, \qquad (e_3, e_3) = 1. \tag{5.32a}$$

We continue this process until the g_i's are exhausted, the general recurrence formula being

$$e_k = c_{kk}\left[g_k - \sum_{i=1}^{k-1}(g_k, e_i)e_i\right],$$

$$c_{kk} = \left[(g_k, g_k) - \sum_{i=1}^{k-1}(g_k, e_i)^2\right]^{-1/2}, \tag{5.32b}$$

$$k = 1, 2, \ldots, n.$$

The norm (5.10), which imparts the features of a metric space to the partially unorganized space, was, as already noted, intentionally selected in such a form as to convert the space into a Euclidean space. Generally speaking, the decision on the form of the metric is left to one's own judgment, and there are many special expressions producing norms, each of which may be appropriate to adopt in a particular circumstance. Of course, each metric so induced leads to a different metric space, even though the rest of the axiomatic system remains the same.

By way of example, consider the norm

$$\|x\| = \max(|x_1|, |x_2|, \ldots, |x_n|), \tag{5.33}$$

equal to the greatest of the absolute values of the components of each vector of the space. It is easily verified that this definition satisfies all requirements imposed earlier on a norm: it is positive definite, linear (in the sense that $\|\alpha x\| = |\alpha|\|x\|$), and obeys the triangle inequality, $\|x + y\| \le \|x\| + \|y\|$.

Another example is provided by the norm associated with the so-called $l_p(n)$ spaces,

$$\|x\|_p = \left[\sum_{i=1}^{n}|x_i|^p\right]^{1/p} \tag{5.34}$$

for $1 \le p < \infty$. Evidently, for $p = 2$, $\|x\|_2$ transforms into the Euclidean norm (5.10) and, for $p = \infty$, $\|x\|_\infty$ coincides with the norm (5.33).[†]
Some authors make use of the so-called "taxicab" norm

$$\|x\| = \sum_{i=1}^{n} |x_i|. \tag{5.35}$$

Euclidean spaces in which the inner product $(x, x) > 0$ for all vectors except the zero vector and has the form $(x, x) = \sum_{i=1}^{n} (x_i)^2$ are often called *real proper* Euclidean space. It is, of course, possible to imagine spaces in which the product (x, x) is real, but takes both positive and negative values. Such spaces are called real *pseudo-Euclidean* spaces, and play an important role in the spatiotemporal world of relativity. Odd as it may seem, in such spaces the length of a vector may become imaginary. This happens for the Minkowski metric $ds^2 = -(dx_1)^2 - (dx_2)^2 - (dx_3)^2 + c^2(dx_4)^2$, where c is the light velocity and the dx_i's are coordinate elements.

Problems

1. Using the Gram–Schmidt orthogonalization process, convert the vector triad: $\{g_1 = (1, 1, 0), g_2 = (0, 1, 1), g_3 = (1, 1, 1)\}$ into an orthonormal triad.

2. In an n-dimensional space, there are given $m(\le n)$ orthogonal vectors $f_1, f_2, \ldots, f_m$. Show that for any vector x there is

$$\|x\| \ge \frac{(x, f_1)^2}{\|f_1\|^2} + \cdots + \frac{(x, f_m)^2}{\|f_m\|^2}.$$

3. Is any set $\{f_1, f_2, \ldots, f_n\}$ of mutually orthogonal nonzero vectors linearly independent?

4. Let $e_1, e_2, \ldots, e_n$ be an orthonormal basis in an n-dimensional space, and let $f_1, f_2, \ldots, f_n$ be another basis. Find the conditions for the set $\{f_i\}$ to be orthonormal.

5. What is the dimension of the space spanned by the vectors $x_1 = (\frac{1}{2}, -1, 0, \frac{3}{2})$, $x_2 = (1, \frac{3}{2}, 0, -\frac{1}{2})$, $x_3 = (1, -\frac{1}{2}, 1, \frac{1}{2})$, $x_4 = (6, 18, -8, -4)^2$?

6. Show that the vectors $(2, 0, 0, 0)$, $(0, 3, 0, 0)$, $(0, 0, 4, 0)$, $(0, 0, 0, 5)$ span a four-dimensional space. Are they linearly independent?

[†] Since $\|x\|_\infty = \lim_{p \to \infty} [\max |x_i| (\sum_{i=1}^{n} |x_i/\max x_i|^p)^{1/p}] = \max |x_i|$. The associated space is often denoted by $l_\infty(n)$.

7. Represent the vector $x = (2, 3, 4)$ as a sum of the vectors $e_1 = (1, 0, 0)$, $e_2 = (0, 1, 0)$, and a vector h perpendicular to both e_1 and e_2. Give a geometrical interpretation and find the orthogonal projections of x on h and $e_1 + e_2$.

8. Represent a vector f in the space $\mathscr{E}_{n+1}$ by a vector g belonging to a subspace $\mathscr{E}_n$ of $\mathscr{E}_{n+1}$ and a vector h in $\mathscr{E}_{n+1}$ perpendicular to $\mathscr{E}_n$. (Note: a vector is perpendicular to a subspace if it is perpendicular to every vector in the subspace.)

6

Infinite-Dimensional Euclidean Spaces

The passage from a finite to an infinite number of space dimensions does not require fundamental changes in the axiomatic system developed in the preceding chapters. It must, however, be carried out with caution, just as in the case of any other operation in mathematics involving the infinite: evaluation of improper integrals, summation of infinite series, and other limit processes.

As a first step, we note that, in the same manner as in a finite-dimensional Euclidean space $\mathscr{E}_n$, where a vector is identified with a finite ordered set of scalars [cf. equation (5.2)], so in what will be called an *infinite dimensional Euclidean space*, devoid of metric, $\mathscr{R}_\infty$, a vector is identified with an infinite ordered set of real scalars,

$$x = (x_1, x_2, \ldots, x_k, \ldots). \tag{6.1}$$

Again, the scalars are called the *components* of x; the ordered set in (6.1) is assumed to be *countable* or *denumerable*. It is recalled that a set is countable if it is finite or if its members may be put in a one-to-one correspondence with the set of positive integers.†

The just-established definition of the collection of objects comprising $\mathscr{R}_\infty$ as the set of *all* (infinite) sequences of real numbers is perhaps the most natural infinite-dimensional generalization of the finite-dimensional space $\mathscr{E}_n$: instead of considering all n-tuples of real numbers, we should apparently now consider all infinite sequences as the most appropriate extension. From another point of view, however, this is too naive, the space $\mathscr{R}_\infty$ being "too

---

† For example, the sets of all integers and of all rational numbers are countable, but the set of all real numbers is not.

big" to be of use in most cases. To see what is meant by this, recall that each element $x = (x_1, \ldots, x_n)$ of $\mathscr{E}_n$ has a magnitude (norm) given by $(\sum_{i=1}^{n} x_i^2)^{1/2}$. Now, if we wish our infinite-dimensional space to have a similar structure, we should apparently provide each of its elements with a norm which is the obvious extension of the norm used in $\mathscr{E}_n$. Thus, we come to consider the infinite series $\sum_{i=1}^{\infty} x_i^2$ as the square of the norm of $x = (x_1, x_2, \ldots, x_n, \ldots)$. However, since this series need not converge for many elements in $\mathscr{R}_\infty$, it becomes clear that $\mathscr{R}_\infty$ has too many elements for it to carry a (topological) structure directly analogous to that of $\mathscr{E}_n$. Rather, the best generalization of $\mathscr{E}_n$ is the much smaller *space* l_2 (or $\mathscr{E}_\infty$) consisting of all those $x = (x_1, x_2, \ldots, x_n, \ldots)$ in $\mathscr{R}_\infty$ for which the sum $\sum_{i=1}^{n} x_i^2$ converges, i.e., $\sum_{i=1}^{\infty} x_i^2 < \infty$. Thus, l_2 may be dubbed "infinite-dimensional Euclidean space,"† if for its norm we take this generalized Euclidean norm, $\|x\|^2 = \sum_{i=1}^{\infty} x_i^2$. In pictorial language, the imposed restriction implies that, while $\mathscr{R}_\infty$ is a space of vectors of which some may have "infinite length," l_2 is a space of vectors of finite lengths.

The basic algebraic operations, established earlier for vectors in $\mathscr{R}_n$, carry over to the case when n becomes infinite. These are the equality of vectors, their addition, and multiplication by scalars; the definition of the zero vector is likewise a straightforward extension. As an illustration, if

$$x = (x_1, x_2, \ldots, x_i, \ldots) \quad \text{and} \quad y = (y_1, y_2, \ldots, y_i, \ldots)$$

are two vectors in $\mathscr{R}_\infty$, then their sum is

$$x + y = (x_1 + y_1, x_2 + y_2, \ldots, x_i + y_i, \ldots). \tag{6.1a}$$

Evidently, the space endowed with these latter algebraic properties alone still remains an affine space, in the sense of Chapter 2. A subspace of $\mathscr{R}_\infty$ is made an actual Euclidean space, $\mathscr{E}_\infty \equiv l_2$, if one introduces appropriately the concept of the norm of a vector. This is, of course, done in the form already noted above, that is,

$$\|x\|^2 = \sum_{i=1}^{\infty} x_i^2, \tag{6.1b}$$

where the series is assumed to converge. Naturally, the distance between two vectors x and y is now given by

$$\|x - y\|^2 = \sum_{i=1}^{\infty} (x_i - y_i)^2; \tag{6.1c}$$

the fact that $\|x - y\| < \infty$, provided that $\sum_{i=1}^{\infty} x_i^2 < \infty$ and $\sum_{i=1}^{\infty} y_i^2 < \infty$, will follow once it has been shown that the Cauchy–Schwarz inequality remains valid in l_2. We shall see that this is indeed the case.

---

† Speaking later of $\mathscr{E}_\infty$, we shall use the symbols "$\mathscr{E}_\infty$" and "l_2" interchangeably.

6.1. Convergence of a Sequence of Vectors in $\mathscr{E}_\infty$

It should be now clear that in the analysis of a vector space possessing a topological structure, specifically that of spaces of an infinite number of dimensions, an important part is played by the concept of *convergence* of infinite sequences (and series) of vectors.

As is well known from calculus, a sequence of scalars $\{\alpha_n\}$ *converges* to a limit α if, loosely speaking, with increasing index n the terms of the sequence approximate α better and better (within a prescribed accuracy). A necessary and sufficient condition for this to happen is the so-called *Cauchy condition*, requiring that for every $\varepsilon > 0$, there exists an N_ε such that

$$|\alpha_n - \alpha_m| < \varepsilon \qquad \text{if } m, n \geq N_\varepsilon. \tag{6.2}$$

Quite similar definitions of *limit* and *convergence* are set up for sequences of vectors, emanating from a point in a Euclidean space of *dimension three*. Imagine then a sequence,[†] $\{x^k\}$, for which the directions and norms of its members approximate more and more closely a given vector, x (Figure 6.1a). Let us agree to evaluate the deviation of a generic member of the sequence, x^k, from the vector x by the value of the norm $\|x - x^k\|$. It is then said that the sequence $\{x^k\}$ *tends* to x if

$$\lim \|x - x^k\| = 0 \qquad \text{for } k \to \infty; \tag{6.3}$$

in the usual notation, this is written[‡]

$$x^k \to x \quad \text{or} \quad x = \lim x^k \qquad \text{for } k \to \infty. \tag{6.3a}$$

To clarify the geometric meaning of the foregoing definition, let us recall that in the space $\mathscr{E}_3$, a δ-*neighborhood* of a point is envisaged as the interior of a sphere of radius δ, centered at the point. We note parenthetically that in spaces of many dimensions, a neighborhood of a point is the interior of an n-sphere or a hypersphere with center at that point, depending on whether the space dimension is finite or infinite, respectively.[§] In the case under discussion, any neighborhood, however small, of the limit vector x always includes all but a finite number of the terminal points of the vectors x^k. We describe this situation by calling the end point of x an

---

[†] We recall that, according to our earlier convention, a superscript distinguishes a vector from a component of a vector (with a subscript).

[‡] Convergence in the norm is often called "strong convergence," in contrast to "weak convergence," requiring $\lim_{n \to \infty} L[x^n] = L[x]$ for every continuous linear functional L [see the second footnote following equation (8.20)].

[§] Compare the closing paragraphs in Chapter 9.

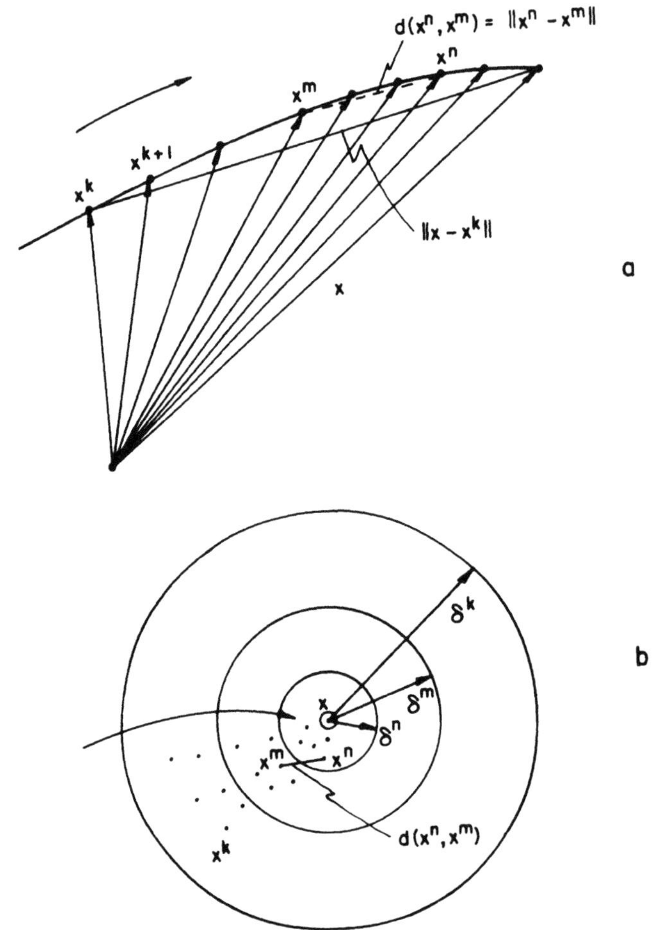

Figure 6.1. Convergence of a sequence of vectors.

accumulation (or *cluster*, or *limit*) point of the end points of the vector x^k (Figure 6.1b).

It is worth noting that, in view of the relation (4.14), $\lim\|x - x^k\| \to 0$ for $k \to \infty$ implies (in $\mathscr{E}_3$)

$$\lim[(x_1^k - x_1)^2 + (x_2^k - x_2)^2 + (x_3^k - x_3)^2] \to 0 \qquad \text{for } k \to \infty, \quad (6.3b)$$

where $x = (x_1, x_2, x_3)$ and $x^k = (x_1^k, x_2^k, x_3^k)$. Consequently, if $x^k \to x$ for $k \to \infty$, then $x_i^k \to x_i$ as $k \to \infty$, for $i = 1, 2, 3$. Also, conversely, if $x_i^k \to x_i$ as $k \to \infty$, for each i, then the sequence $\{x^k\}$ converges to x.

Using the generalized triangle inequality (5.25), it is not difficult to show that in $\mathscr{E}_3$, condition (6.3) implies

$$\|x^l - x^k\| \to 0 \tag{6.4}$$

for $k, l \to \infty$.

Thus, (6.4) is a *necessary* condition[†] for the convergence of the sequence because it follows from the assumption of convergence in the sense of (6.3). On account of its similarity with the Cauchy criterion (6.2), a sequence of vectors obeying condition (6.4) is said to converge in the sense of Cauchy or to be a *Cauchy* (or *fundamental*) sequence. It can be easily demonstrated that, in $\mathscr{E}_3$, (6.4) is also a *sufficient* condition for convergence, that is, if (6.4) holds, then (6.3) follows. The same is true of the finite-dimensional Euclidean spaces examined in the preceding chapter, in which convergence in the norm is a simple generalization of the condition (6.3b). Actually, it can be demonstrated (Ref. 22, p. 9) that in any finite-dimensional abstract space (to be discussed later), in which a norm is defined, (6.4) is both a necessary and a sufficient condition for (6.3) to be true (more precisely, for there to exist an x in the space for which (6.3) is true).

Turning our attention now to the infinite-dimensional space $\mathscr{E}_\infty$, including vectors with an infinite number of components and satisfying conditions such as (6.1)–(6.1b), we extend the concept of *convergence*, established thus far only for Euclidean spaces of finite dimension. Consequently, with the distance between two vectors, $x = (x_1, x_2, \ldots, x_i, \ldots)$ and $x^k = (x_1{}^k, x_2{}^k, \ldots, x_i{}^k, \ldots)$, defined as $\|x^k - x\| = [\sum_{i=1}^{\infty} (x_i{}^k - x_i)^2]^{1/2}$, we say that a sequence of vectors x^k, $k = 1, 2, \ldots$, in $\mathscr{E}_\infty$ converges to the vector x in $\mathscr{E}_\infty$ if

$$\|x^k - x\| \to 0 \qquad \text{for } k \to \infty. \tag{6.5}$$

In view of the definition of $\|x^k - x\|$, this, of course, implies that $x_i{}^k \to x_i$ as $k \to \infty$ for each $i = 1, 2, \ldots$. Thus, the convergence of a sequence of vectors in $\mathscr{E}_\infty$ implies the convergence of each sequence of respective components of these vectors, under the assumption that x and all the x^k's are in $\mathscr{E}_\infty$ (that is, l_2).

It is demonstrated (Ref. 23, p. 49), but we shall refrain from doing so, that for vectors in l_2, a necessary and sufficient condition for convergence of a sequence $\{x^k\}$ to some x in l_2 is

$$\|x^n - x^m\| \to 0 \qquad \text{for } m, n \to \infty, \tag{6.6}$$

which is just the Cauchy condition. Consequently, in l_2 every Cauchy se-

---

[†] We recall that a necessary condition is a logical consequence of a given statement, that is, if (6.3) holds, then (6.4) follows.

quence converges. In the geometric language of Figure 6.1, this means that the distances $\|x - x^k\|$ go to zero simultaneously with the distances $\|x^n - x^m\|$, and conversely.

6.2. Linear Independence. Span. Basis

The concepts of linear independence, span, and basis in $\mathscr{E}_\infty$ are slightly more complicated than those defined earlier for Euclidean spaces of finite dimension. This will be apparent in the definitions which we now intend to set up.

We first note that, for an *infinite* set of vectors, there are two possible definitions of *linear independence*. We can state that (Ref. 15, p. 53)

(1) An infinite set of vectors, $\{x^1, x^2, \ldots\}$, in $\mathscr{E}_\infty$ is *linearly independent* if each of its finite subsets is linearly independent, that is, if

$$\alpha_1 x^1 + \alpha_2 x^2 + \cdots + \alpha_n x^n = 0 \tag{6.7}$$

implies that all scalars

$$\alpha_1 = \alpha_2 = \cdots = \alpha_n = 0 \tag{6.8}$$

for each positive integer n. Naturally, if this is not the case, the set is said to be *linearly dependent*.

A characteristic feature of this definition is that, no matter what the number of vectors (whether finite or infinite) and the dimension of the space, only finite combinations of vectors are considered [recall equation (4.3), where the set $\{x^i\}$ is finite].

The second possible definition is contained in Ref. 22, p. 13 and Ref. 24, p. 228.

(2) A *linear combination* of an infinite set of vectors $x^1, x^2, \ldots$ is a formal expression

$$\alpha_1 x^1 + \alpha_2 x^2 + \cdots + \alpha_i x^i + \cdots, \tag{6.9}$$

where the α_i's are scalars. We wish to state that, if

$$\alpha_1 x^1 + \alpha_2 x^2 + \cdots + \alpha_i x^i + \cdots = 0 \tag{6.10}$$

implies that $\alpha_1 = \alpha_2 = \cdots = 0$, then the vectors of the set $\{x^1, x^2, \ldots\}$ are *linearly independent*. However, in order for the equality (6.10) to have a meaning, an explanation of the concept of convergence of an *infinite series* of vectors must first be provided, since (6.9), in fact, requires that the given series have a limit (here zero). Quite like the corresponding definition of elementary analysis, we say that if the sequence of partial sums $\sigma_n = \sum_{i=1}^n x^i$

of a sequence of vectors $x^1, x^2, \ldots$ where $n = 1, 2, \ldots$, converges to a vector σ, that is,

$$\sigma_n \to \sigma \quad \text{or} \quad \|\sigma_n - \sigma\| \to 0 \tag{6.10a}$$

for $n \to \infty$, then the *infinite series* $\sum_{i=1}^{\infty} x^i$ *converges*, and the vector σ is called the *sum* of the series,

$$\sigma = \sum_{i=1}^{\infty} x^i. \tag{6.11}$$

Clearly, with only slight changes, the preceding can serve as a definition of a *limit* of *finite linear combinations* of vectors of an infinite set $\{x^1, x^2, \ldots\}$.

With these preliminaries out of the way, we shall agree to accept, from now on, *definition* (2) of the linear independence of an infinite number of vectors in an infinite dimensional space, and proceed to introduce the notions of span and basis as in Chapter 5. We thus state that an infinite set of vectors $\{x^1, x^2, \ldots\}$ *spans* an infinite dimensional space if every vector in the space can be represented either as a linear combination of a finite number of the vectors of the set or as a limit of such linear combinations. The set is a *basis* for the space if every vector in the space can be represented in a *unique* manner as a linear combination of a finite or infinite number of the vectors of the set.

Unlike the case for a finite-dimensional manifold, an infinite set of *linearly independent* vectors that *span* a space need not form a basis for the space. As an illustration, consider the set of vectors†

$$\overset{\displaystyle \downarrow k\text{th component}}{a^k = (1, 0, 0, \ldots, 0, 1, 0, \ldots),} \tag{6.12}$$

the second nonzero component of which is the kth component, $k = 2, 3, \ldots$. It can be proved (Ref. 22, pp. 13 and 16), but we shall omit the demonstration, that the set (6.12) is linearly independent and actually *spans* the Euclidean space $\mathscr{E}_\infty$ (recall our convention $\mathscr{E}_\infty \equiv l_2$).

Consider now the vector

$$e_1 = (1, 0, 0, \ldots), \tag{6.13}$$

having all its components equal to zero except the first. Clearly, the sequence of sums, with the a^k's defined by (6.12),

$$\frac{1}{n}(a^2 + a^3 + \cdots + a^{n+1}), \tag{6.14}$$

---

† Each element of the sequence (6.12), $a^k = (a_1{}^k, a_2{}^k, \ldots, a_i{}^k, \ldots)$, obviously satisfies the convergence criterion $\sum_{i=1}^{\infty} (a_i{}^k)^2 < \infty$ and thus belongs to l_2.

where $n = 1, 2, \ldots$, or, explicitly, the sequence

$$\frac{a^2}{1} = (1, 1, 0, \ldots), \qquad \frac{a^2 + a^3}{2} = (1, \tfrac{1}{2}, \tfrac{1}{2}, 0, \ldots),$$

$$\frac{a^2 + a^3 + a^4}{3} = (1, \tfrac{1}{3}, \tfrac{1}{3}, \tfrac{1}{3}, 0, \ldots), \ldots, \tag{6.15}$$

converges to e_1 in the sense of the norm,

$$\left\| e_1 - \frac{1}{n}(a^2 + a^3 + \cdots a^{n+1}) \right\| \to 0 \qquad \text{for } n \to \infty, \tag{6.16}$$

inasmuch as the differences

$$\left\| e_1 - \frac{a^2}{1} \right\|^2 = (1 - 1)^2 + (0 - 1)^2 = 1,$$

$$\left\| e_1 - \frac{a^2 + a^3}{2} \right\|^2 = (1 - 1)^2 + 2(0 - \tfrac{1}{2})^2 = \tfrac{1}{2},$$

$$\left\| e_1 - \frac{a^2 + a^3 + a^4}{3} \right\|^2 = (1 - 1)^2 + 3(0 - \tfrac{1}{3})^2 = \tfrac{1}{3}, \ldots$$

go monotonically to zero. Thus, e_1 is in the span of the vectors (6.12). However, these vectors do *not* form a basis, since—contrary to the definition of the latter—there exists no finite or infinite representation of e_1 in the form†

$$e_1 = \alpha_2 a^2 + \alpha_3 a^3 + \cdots, \tag{6.17}$$

or, explicitly,

$$(1, 0, 0, \ldots) = \alpha_2(1, 1, 0, \ldots) + \alpha_3(1, 0, 1, 0, \ldots) + \cdots \tag{6.17a}$$

whatever the values of the scalars $\alpha_2, \alpha_3, \ldots$. It follows that, in spite of the fact that the vectors (6.12) are linearly independent and span the given manifold, they do not provide a basis for $\mathscr{E}_\infty$ (or rather l_2).

6.3. Linear Manifold

So far, mostly for simplicity's sake, we have spoken of a space as a whole. But from the very definition of the space as a set of vectors, it is intuitively clear that one can conceive of "portions" of the space if one

---

† It is important to note that $e_1 = \lim_{n \to \infty}((1/n)a^2 + (1/n)a^3 + \cdots + (1/n)a^{n+1})$ is *not* in the form of (6.17) since the coefficients $1/n$ depend on n.

identifies these portions with (sub)sets of vectors belonging to the given set. The idea of "linear manifold" arises by considering certain distinguished portions of a linear space. It is said that a set of vectors, $\mathcal{M}$, constitutes a *linear manifold* in a space if, for all scalars α and β, $\mathcal{M}$ contains the linear combination $\alpha x + \beta y$ whenever it contains the vectors x and y.[†]

A typical example of a linear manifold is a plane in $\mathcal{E}_3$ passing through the origin (taking $\alpha = \beta = 0$ above, we see that a linear manifold must contain the zero vector). Still simpler is the line (through the origin) formed by the set of vectors αx, where $x \neq 0$, and α runs through all real numbers.

Bearing in mind the definitions of a span and a basis in Chapter 4, we now extend these concepts to multidimensional spaces.

(1) We first consider *finite-dimensional* linear manifolds.

We say that a finite set of vectors, x^1, x^2, ..., x^n, forms a *basis* for a *finite-dimensional linear manifold*, $\mathcal{M}_n$, if it *spans* the manifold and the vectors are *linearly independent*. The number of vectors of the basis determines the *dimension* of the manifold.

(2) *Infinite-dimensional* linear manifolds, $\mathcal{M}_\infty$. If no finite set of vectors spans a given manifold, the dimension of the latter is said to be *infinite*. The situation in this case becomes more complicated.

A linear manifold is said to be *closed* if the limit, x, of every convergent sequence of vectors, x^1, x^2, ..., contained in the manifold *belongs* to the manifold. The convergence of a sequence is understood here in the sense established previously.

We are now in a position to introduce the important concept of a *subspace*, $\mathcal{S}$, as a *closed linear manifold*. We note that the definition of this term is not universally established, some authors making no distinction between a linear manifold and a subspace, and allowing the latter to be nonclosed (Ref. 11, p. 84). In the informal spirit of this book, the terms linear manifold and subspace (and, often, the space itself) are used interchangeably where no confusion is likely to arise.

It is shown (Ref. 7, p. 61; Ref. 22, p. 15) that every *finite-dimensional* linear manifold is *closed*, and therefore is a subspace. This is *not* the case with *infinite-dimensional* manifolds. To illustrate this point, consider the space composed of all vectors of the form (Ref. 7, p. 35; Ref. 22, p. 14)

$$x = (a_1, a_2, ..., a_k, 0, ...),\qquad\qquad(6.18)$$

that is, of all vectors possessing only a *finite number of nonzero* components. Since any linear combination of these vectors generates a vector belonging to the set, the set represents a linear manifold.

---

† Some authors, e.g., Synge,[40] introduce systems slightly more general than linear manifolds, namely, the *convex systems*. A *convex set* is one which contains $\alpha x + \beta y$ whenever it contains x and y, and $\alpha + \beta = 1$, α, $\beta \geq 0$.

Let us now consider the sequence of vectors

$$y^1 = (1, 0, 0, \ldots),$$

$$y^2 = (1, \tfrac{1}{2}, 0, \ldots), \tag{6.19}$$

$$y^k = (1, \tfrac{1}{2}, \tfrac{1}{3}, \ldots, 1/k, 0, \ldots),$$

selected from the given manifold. We claim that the limit of this sequence is the vector (since $\sum_{n=1}^{\infty} 1/n^2 < \infty$, y is, in fact, in l_2)

$$y = (1, \tfrac{1}{2}, \tfrac{1}{3}, \ldots, 1/k, \ldots). \tag{6.20}$$

Indeed, we have

$$\|y - y^n\| = \left\| \left(0, \ldots, 0, \frac{1}{n+1}, \frac{1}{n+2}, \ldots\right) \right\|$$

$$= \left(\sum_{k=n+1}^{\infty} \frac{1}{k^2} \right)^{1/2}, \tag{6.21}$$

so that $\|y - y^n\|$ tends to zero as $n \to \infty$. Note also that the infinite series $\sum_{k=1}^{\infty} (1/k^2)$ converges. This proves that our claim is correct, by definition of convergence of a sequence in l_2. However, all components of y are not zero, showing that the limit of the sequence $\{y^k\}$ is not in the original (6.18) manifold. It follows that the latter is not a subspace, i.e., is not closed.

In applications, we often choose as a basis for an infinite-dimensional space those vectors that belong to one of a number of well-known families. A general proof is then given that the vectors of the family do, in fact, serve as a basis. We shall examine this point in more detail in Chapter 10.

Let us now proceed with the infinite-dimensional generalizations of other definitions, established before for finite-dimensional Euclidean spaces. We first extend the concept of the inner product [equation (5.13)] by putting

$$(x, y) = \sum_{i=1}^{\infty} x_i y_i$$

$$= x_i y_i, \qquad i = 1, 2, \ldots, \tag{6.22}$$

the last form by appeal to Einstein's convention.† We shall use this abbreviation repeatedly in the subsequent text (unless the contrary is stated). Clearly,

---

† We recall that any term in which the same index appears twice stands for the sum of all such terms, the index being given its range of values.

the infinite series (6.22) should be convergent in order for the inner product and the *norm* of vectors to remain finite. That is, we actually work within the space l_2 contained in $\mathscr{E}_\infty$ (as remarked above). For the norm we have

$$\|x\| = (x_1{}^2 + x_2{}^2 + \cdots + x_k{}^2 + \cdots)^{1/2}, \qquad (6.23)$$

identical with the Pythagorean formula (5.10) except for the number of terms, which is now infinite.

Continuing our extension process, we define the *distance* between two vectors, x and y, by the formula

$$d(x, y) \equiv \|x - y\| = (x - y, x - y)^{1/2} = \left[\sum_{i=1}^{\infty} (x_i - y_i)^2 \right]^{1/2}, \qquad (6.24)$$

reminiscent of the definition (4.15) for a space of finite dimension.

As regards the orthonormalization of a basis for l_2 (or of any infinite set of linearly independent vectors), the process does not essentially differ from that of Gram–Schmidt described in Chapter 5, the only distinction being that (from a theoretical standpoint) the construction never ends.

It is important to emphasize that, although a basis in an infinite-dimensional space consists of an infinite number of linearly independent vectors, not every infinite set of linearly independent vectors constitutes a basis. For example, if from an infinite set of base vectors one would eliminate a single vector, the remaining set would still be infinite, but would have ceased to be a basis because the vectors would no longer possess the property of spanning the entire space. In practice, it is often difficult (and sometimes impossible) to decide whether a given infinite set of linearly independent vectors constitutes a basis for a given space or is simply a "mutilated basis," with one or more vectors missing. As a simple illustration of such mutilation, one can take the finite basis of three unit coordinate vectors $e_1 = (1, 0, 0)$, $e_2 = (0, 1, 0)$, and $e_3 = (0, 0, 1)$, in ordinary three-space, and delete one. Clearly, any effort to represent every vector in the space in terms of the remaining two coordinate vectors would be futile.

It is a fortunate circumstance that, if an *orthonormal set* of vectors spans an infinite-dimensional space, then the set gives a *basis* for the space (Ref. 22, p. 17). Having such a basis, call it $\{f_i\}$, we can represent each vector in the space by the series

$$x = \sum_{k=1}^{\infty} x_k f_k, \qquad (6.25)$$

where

$$x_k = (x, f_k). \qquad (6.26)$$

It can be shown that if the condition of convergence,

$$\sum_{i=1}^{\infty} x_i^2 < \infty, \tag{6.27}$$

considered repeatedly above, is satisfied for every vector in the space, then also the infinite series giving the scalar product of any two vectors, (6.22), converges. In addition, the expansions given by (6.25) will converge.

We now cite—without proof—the generalization of the *Cauchy-Schwarz* inequality, derived earlier for finite-dimensional spaces [equation (5.24)],

$$\left(\sum_{i=1}^{\infty} x_i y_i\right)^2 \le \sum_{i=1}^{\infty} x_i^2 \sum_{i=1}^{\infty} y_i^2 \tag{6.28}$$

or

$$|(x, y)| \le \|x\| \, \|y\|, \tag{6.29}$$

where $x = (x_1, x_2, \ldots)$ and $y = (y_1, y_2, \ldots)$.

The triangle inequality (5.25) now takes the more general form

$$\left[\sum_{i=1}^{\infty} (x_i + y_i)^2\right]^{1/2} \le \left[\sum_{i=1}^{\infty} x_i^2\right]^{1/2} + \left[\sum_{i=1}^{\infty} y_i^2\right]^{1/2}. \tag{6.30}$$

Inasmuch as the relationship between a scalar product and the corresponding norm does not depend on the dimension of the space, the equation

$$(x, y) = 0 \tag{6.31}$$

also expresses the perpendicularity of two vectors, x and y, in an infinite-dimensional space.

Remark 6.1. To enable the reader to feel more at ease, we have established in this chapter several concepts in the context of the fairly easily visualized space $\mathscr{E}_\infty$. Actually, most conclusions drawn here also apply in infinite-dimensional abstract spaces with norm, examined later in this book.

Problems

1. Show that every convergent sequence is Cauchy.

2. Show that in the subset (open interval) (1, 3), the sequence $3 - (1/n)$, $n = 1, 2, 3, \ldots$, is Cauchy, but does not converge in the sense of the definition (6.3).

3. Make a list of all subspaces of the space $\mathscr{E}_3$, all elements of which are of the form $(\alpha_1, \alpha_2, \alpha_3)$, where the α_i's are real scalars and at least one of the coordinates is zero. Verify that $\mathscr{E}_3$ itself as well as the zero vector are subspaces of $\mathscr{E}_3$.

4. Let x^1, x^2, ..., x^m and y^1, y^2, ..., y^n be elements of a space $\mathscr{V}$. Show that the subspace $\mathscr{S}_1$ spanned by $\{x^i\}$ is the same as the subspace $\mathscr{S}_2$ spanned by $\{x^i\}$ and $\{y^j\}$ if each vector y^j is linearly dependent on the vectors $\{x^i\}$.

5. If $\{x^1, x^2, \ldots\}$ is an orthonormal set of vectors in an inner product space and $y^n = \sum_{k=1}^{n} \alpha_k x^k$, where the α_k's are scalars, show that the sequence y^n is Cauchy if $\sum_{k=1}^{\infty} \alpha_k{}^2 < \infty$.

6. Show that a closed interval $[\alpha, \beta]$ of the real axis is a closed subset of the space of real numbers, represented by the real axis.

7. The *orthogonal complement* of a subspace $\mathscr{S}$ of an inner product space $\mathscr{V}$ is the set $\mathscr{S}^{\perp}$ of all vectors orthogonal to all vectors x in $\mathscr{S}$. It is to be shown that $\mathscr{S}^{\perp}$ is closed.

8. Let the set $\{x^k\}$ be an infinite orthogonal basis for an infinite-dimensional inner product space. Why, after removal of merely a single element of this set, does the new set fail to be a basis for the space?

9. Show that in an infinite-dimensional space, there exists an infinite number of linearly independent vectors.

7

Abstract Spaces. Hilbert Space

In this chapter we take a decisive step and strip the concepts of "space" and "vector" of any concrete meaning; a space then becomes a set endowed with a certain structure described by axioms, and a vector an unspecified element of the set.

To illustrate the abstract nature of the corresponding axiomatics, consider a function $f(x)$, of a single variable x, defined in the interval $-1 \leq x \leq 1$. We regard the function as representative of the (perhaps infinite) collection of values of the function, taken for the infinitely many values of its argument in the interval $[-1, 1]$. From this standpoint, the function becomes a set of scalars, and, if the latter are looked upon as components of a vector, it is natural to identify the function with a vector in a space of infinite dimension.

A similar, although less direct, approach is employed for functions of several variables, systems of functions, and so on. For instance, a stress tensor represented by six functions of three or four variables (three spatial coordinates and time) can be symbolized by a single vector. The decision to consider space elements—independent of their nature—as vectors in some such manner places at our disposal the arsenal of implements developed in the preceding chapters of our exposition. In this vein, the sum of two functions, $f(P)$ and $g(P)$, is understood as the resultant of two vectors; the scalar α in the product $\alpha f(P)$ becomes a scaling factor for the vector $f(P)$; the scalar α itself represents a constant function, hence also a vector (with zero playing the role of the null vector). The analogy can be extended further: it can appear that we can treat a certain set of functions, call it $\{\phi_i(P)\}$ say, as a basis for a given space, with the expansion

$$f(P) = \sum_{i=1}^{\infty} \alpha_i \phi_i(P) \tag{7.1}$$

being understood as a *resolution* of the vector $f(P)$ in terms of the base vectors. Here, the coefficients α_i play the part of *components* of the vector $f(x)$ referred to a *coordinate frame* defined by the coordinate vectors $\phi_i(P)$.

On this level of abstraction, it is natural to recast our previous axiomatics into a form equally abstract. The geometric structure at which we shall arrive is called a *Hilbert space*, the progenitor of Euclidean spaces (cf. Figure 1.1).

We begin the construction of the axiomatic system of Hilbert space with the following:

Definition 7.1. A *vector*, or *linear*, space, $\mathcal{V}$, is a collection of objects, x, y, z, ..., (of an unspecified nature), possessing the first two of the six groups of properties listed below. The objects are called *vectors* (or elements) of $\mathcal{V}$.†

Group A. For every pair of elements, x and y, in $\mathcal{V}$, there exists an element $x + y$ in $\mathcal{V}$, called their sum, such that
(a) $x + y = y + x$, or addition is commutative,
(b) $x + (y + z) = (x + y) + z$, or addition is associative.
(c) There exists in $\mathcal{V}$ a unique element θ, called the *zero vector*, with the property that $x + \theta = x$ for every x.
(d) To every element x in $\mathcal{V}$, there corresponds a unique element $-x$, called the negative of x, with the property $x + (-x) \equiv x - x = \theta$.

Group B. For each scalar α (here and hereafter, always a *real* number) and each x in $\mathcal{V}$, there exists an element αx, called the *scalar product* of α and x, such that whenever y is an element and β is a scalar,
(a) $\alpha(x + y) = \alpha x + \alpha y$, or multiplication by a scalar is distributive with respect to element addition,
(b) $(\alpha + \beta)x = \alpha x + \beta x$, or multiplication by elements is distributive with respect to scalar addition,
(c) $(\alpha\beta)x = \alpha(\beta x)$, or multiplication by scalars is associative,
(d) $1x = x$ for every element x,
(e) $0x = \theta$, or multiplication of every element by zero gives the zero element.

The alternate name "linear" for a vector space, defined by the preceding two groups of properties, arises from the fact that it is precisely spaces of this sort on which familiar linear operations (such as differentiation and integra-

---

† See any treatise on functional analysis such as, e.g., Gould (Ref. 25, Chap. 1, Sec. 4 and Chap. 4, Sec. 2), Berberian (Ref. 7, p. 39), Sneddon (Ref. 10, Sec. 4.6), Vulikh (Ref. 23, Chap. 6).

tion, for example) are considered as acting (recall that, in general, a linear operation L is one such that $L(\alpha x + \beta y) = \alpha L x + \beta L y$).

We note that property (e) of Group B is listed merely for convenience, inasmuch as it follows from properties (c) of Group A and (b) of Group B. In fact, from (b) of Group B, for $\alpha = 1$, $\beta = 0$, we have $x = 1x = [(0 + 1)x] = 1x + 0x = x + 0x$. By (c) of Group A, then $0x = \theta$.

It is often said that the axioms of Group A define the *additive* fabric of the vector space; axioms (c), (d), and (e) of Group B define its *multiplicative* fabric; and the remaining axioms [(a) and (b) of Group B] define the relationships between the two groups of operations. In establishing the linear character of $\mathscr{V}$, we were, of course, guided by the rules established in the preceding chapters for ordinary vectors.

The fact that to every pair of elements, x and y, and to every pair consisting of a vector and a scalar, x and α, correspond elements $x + y$ and αx in the space, respectively, is expressed by saying that the space is *closed* under addition and scalar multiplication.†

The third group of properties introduces a *metric* space character for $\mathscr{V}$ through the definition of an inner product of its elements.

Group C. For every pair of elements in $\mathscr{V}$, x and y, there exists a scalar, denoted by (x, y) and called the *inner product* of x and y, such that for every scalar α and every element z,

(a) $(x, y) = (y, x)$, or the inner product is symmetric,
(b) $(\alpha x, y) = \alpha(x, y)$,
(c) $(x + y, z) = (x, z) + (y, z)$,
(d) $(x, x) > 0$ if $x \neq \theta$ and $(x, x) = 0$ if $x = \theta$.

The (non-negative) square root of the inner product of an element x with itself, $(x, x)^{1/2}$, is called the *norm* of x and is denoted by $\|x\|$. Hence,

$$\|x\| = (x, x)^{1/2}. \tag{7.2}$$

We note that a linear space $\mathscr{V}$ provided with a metric derived from an inner product, that is, a space satisfying the groups of axioms A–C, is called a *pre-Hilbert* space by some authors.‡

The remaining three groups of axioms establish the concepts of *dimension, completeness,* and *separability* for a pre-Hilbert space.

---

† This term should not be confused with that introduced in Chapter 6 [cf. the text preceding equation (6.18)] for linear manifolds.
‡ For examples, Berberian (Ref. 7, p. 25) or Taylor (Ref. 11, p. 115). It is important to observe that an inner product space is automatically a normed linear space, since the definition (7.2) provides an admissible norm, that is, a norm satisfying conditions (7.13) below.

Group D. A pre-Hilbert space is *n-dimensional* if there exist n linearly *independent* vectors, x^1, x^2, ..., x^n, i.e., such that the equation

$$\alpha_1 x^1 + \alpha_2 x^2 + \cdots + \alpha_n x^n = 0 \tag{7.3}$$

implies that all coefficients α_k are zero, but every set consisting of $n + 1$ vectors chosen from the space is linearly *dependent*. If n linearly independent elements can be found for any $n = 1, 2, \ldots$, the vector space is said to be *infinite-dimensional*.

Henceforth, we shall be primarily concerned with infinite-dimensional spaces.

Group E. The pre-Hilbert space is *complete*. This means that every sequence consisting of elements of $\mathscr{V}$ and converging in the *Cauchy* sense converges to an element *in* $\mathscr{V}$. Symbolically, from

$$\lim \|x^m - x^n\| = 0 \qquad \text{for } m, n \to \infty \tag{7.4}$$

there must follow

$$\lim \|x - x^m\| = 0 \qquad \text{for } m \to \infty \tag{7.5}$$

for some x in $\mathscr{V}$.

There is an inherent connection between the concepts of *completeness* and *closedness*† of a linear manifold (mentioned earlier). More precisely, in a complete metric linear space (in particular, in a Hilbert space), a complete linear manifold is closed and vice versa: a closed linear manifold is complete.

Most authors call a *complete* inner-product space (that is, a complete pre-Hilbert space) a *Hilbert space*. To the already-listed properties, the requirement is often added that the Hilbert space be *separable*; this imparts to the space a simpler structure. In fact, most examples of Hilbert spaces encountered in applications possess this property of separability. It is said that a Hilbert space is *separable* if it contains a countable set S of points which is *dense* in the space, that is, which is such that any neighborhood of any point of the space contains a point of S. Examples of dense sets are the set of all rational points in the real line $\mathscr{R}$, and the set of all points $x = (x_1, x_2, \ldots, x_n)$ with rational coordinates in the space $\mathscr{R}_n$. Since these sets are countable, the spaces $\mathscr{R}$ and $\mathscr{R}_n$ are separable.

The Hilbert space defined by the preceding axioms is sometimes called a *real* Hilbert space, since the scalars involved are assumed to be real; we

---

† Compare the text preceding equation (6.18) in Chapter 6.

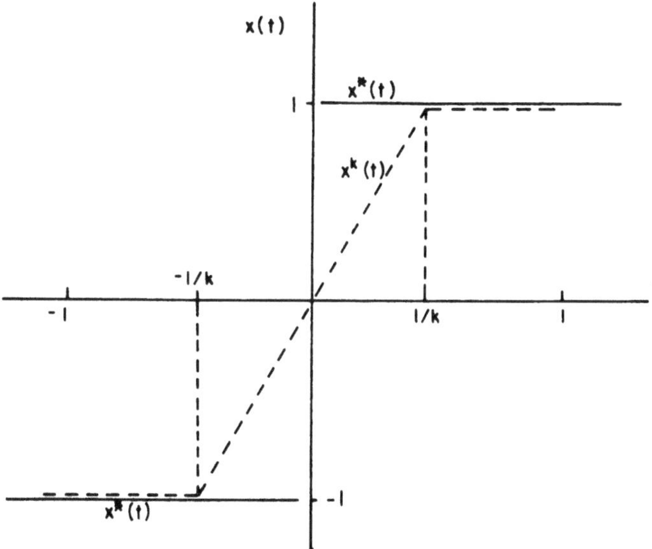

Figure 7.1. Illustration for space completeness.

denote it by $\mathscr{H}$. Axiom (d) of Group C implies that the metric of $\mathscr{H}$ is positive definite.

It may be of interest to note that completeness is not an "absolute" property of a space, but depends on the selected norm. As an illustration, consider the space $C_{-1 \leq t \leq 1}$ of functions $x(t)$ of a single variable t, *continuous* in the interval $-1 \leq t \leq 1$. Select in this space the sequence of functions $\{x^k\}$ defined by (Ref. 26, p. 59)

$$x^k(t) = \begin{cases} -1 & \text{for } -1 \leq t \leq -\dfrac{1}{k}, \\[2ex] kt & \text{for } -\dfrac{1}{k} \leq t \leq \dfrac{1}{k}, \\[2ex] 1 & \text{for } \dfrac{1}{k} \leq t \leq 1, \end{cases} \tag{7.6}$$

where $k = 1, 2, \ldots$, (Figure 7.1). Define the norm by the rule $\|x(t)\| = \{\int_{-1}^{1} [x(t)]^2 \, dt\}^{1/2}$. Then the sequence $\{x^k\}$ is a Cauchy sequence in the selected norm:

$$\int_{-1}^{1} [x^m(t) - x^n(t)]^2 \, dt \leq \frac{2}{\min(m, n)} \to 0 \qquad \text{as } m, n \to \infty.$$

However, $\{x^k\}$ does not converge in this norm to a function belonging to the space $C_{-1 \leq t \leq 1}$. Indeed, select the discontinuous function

$$x^*(t) = \begin{cases} -1 & \text{for } -1 \leq t < 0, \\ 0 & \text{for } \quad\; t = 0, \\ 1 & \text{for } \quad 0 \leq t \leq 1. \end{cases} \tag{7.7}$$

From the known Minkowski inequality for integrals,[†] we have

$$\left| \int_{-1}^{1} [\phi(t) - x^*(t)]^2 \, dt \right|^{1/2} \leq \left| \int_{-1}^{1} [\phi(t) - x^k(t)]^2 \, dt \right|^{1/2}$$

$$+ \left| \int_{-1}^{1} [x^k(t) - x^*(t)]^2 \, dt \right|^{1/2}, \quad (7.8)$$

where $\phi(t)$ is a continuous function. By the continuity of $\phi(t)$, the left-hand member in (7.8) is different from zero; furthermore, as seen from Figure 7.1,

$$\int_{-1}^{1} [x^k(t) - x^*(t)]^2 \, dt \to 0 \qquad \text{as } k \to \infty. \tag{7.9}$$

Consequently, the first integral on the right side of (7.8) cannot converge to zero as $k \to \infty$, whence the sequence $\{x^k\}$ cannot converge to any function ϕ in the space $C_{-1 \leq t \leq 1}$ endowed with the norm defined above. We conclude that the space is incomplete. It can be *completed*, by adjoining appropriate elements, to produce the space $\mathscr{L}_2(-1, 1)$ of functions square-integrable in the Lebesgue sense.[‡]

On the other hand, we shall now demonstrate that $C_{-1 \leq t \leq 1}$ can be endowed with another norm, with respect to which completeness is realized. Specifically, let the norm on $C_{-1 \leq t \leq 1}$ now be given by $\|x\| = \max_{-1 \leq t \leq 1} |x(t)|$. Supposing that $\{x^k\}$ is a Cauchy sequence in this new space, it follows that, for any $\varepsilon > 0$, there exists a positive integer N_ε such that

$$\|x^m - x^n\| = \max_{-1 \leq t \leq 1} |x^m(t) - x^n(t)| < \varepsilon \tag{7.10}$$

whenever $m, n > N_\varepsilon$. Now, (7.10) says, in particular, that the sequence of $\{x^k(t)\}$ is a Cauchy sequence of *real* numbers for any fixed t, $-1 \leq t \leq 1$, and so must converge to a real number, denoted by $x^0(t)$; here, we have appealed to the completeness of the real numbers. As noted, this reasoning holds for each t, $-1 \leq t \leq 1$, whence we have defined a function x^0 on $[-1, 1]$. We

---

[†] Compare inequality (5.25) in Chapter 5.

[‡] A square-integrable function is also integrable here, since we are working on the interval $[-1, 1]$. For completeness of the space $\mathscr{L}_2$ and Lebesgue integration, compare Oden (Ref. 14, pp. 219 and 201).

shall complete the proof of our claim by showing that x^0 is in $C_{-1 \leq t \leq 1}$ (that is, x^0 is continuous) and $x^k \to x^0$ in the present norm. Again using (7.10), it can easily be shown that the sequence $\{x^k\}$ converges *uniformly*[†] to x^0—which is precisely the statement that $\{x^k\}$ converges to x^0 in the sense of the present norm. Finally, it is well known that the limit of a uniformly convergent sequence of continuous functions in $[-1, 1]$ is also continuous (Ref. 27, p. 135). Thus, x^0 is continuous and, as we have indicated, $x^k \to x^0$ or

$$\lim_{k \to \infty} \|x^k - x^0\| = 0. \tag{7.11}$$

With this new norm, then $C_{-1 \leq t \leq 1}$ is complete.

In addition to Hilbert space, more general spaces frequently studied are the metric spaces, already mentioned in Chapter 1.

A *metric space* is a set for which the *distance* between any two of its elements has been defined.[‡] This distance, $d(x, y)$, mentioned somewhat informally in previous chapters,[§] is now required to satisfy the three properties laid down in the following axiom:

Axiom M.

   (a) $d(x, y) > 0$ if $x \neq y$ and $d(x, y) = 0$ if $x = y$,

   (b) $d(x, y) = d(y, x)$,   (7.12)

   (c) $d(x, z) \leq d(x, y) + d(y, z)$.

The preceding properties are the salient features usually connected with the concept of the distance between two points in the familiar three-space, $\mathscr{E}_3$. They express the positiveness of the distance (or its vanishing if the points coincide), its symmetry, and its compliance with the triangle inequality.

There are many examples of metric spaces, of which it is worthwhile to mention the following two.

(1) The space $C_{a \leq t \leq b}$ of all continuous functions $x(t)$ defined on an interval $a \leq t \leq b$.[¶] The distance between any two elements in this space is defined as the maximum of the absolute values of the *deviations* of one function from the other in the given interval $[a, b]$, as in (7.10) (cf. Ref. 23, p. 75)

$$d(x, y) = \max_{a \leq t \leq b} |x(t) - y(t)|. \tag{7.12a}$$

---

† See equation (10.14) in Chapter 10.

‡ Almost all of the metric spaces which we usually consider are also linear spaces.

§ Compare, e.g., equation (6.24).

¶ Clearly, this is a generalization of the space $C_{-1 \leq t \leq 1}$ just discussed.

The metric (7.12a) is sometimes called the Chebyshev metric. It is easily verified that the definition above is consistent with Axiom M for a metric. Furthermore, continuous functions satisfy Axioms A and B, so that $C_{a \leq t \leq b}$ is a linear vector space, as well as a metric space.

(2) The space l_1 consisting of vectors having an infinite number of components, $x = (x_1, x_2, \ldots)$, and satisfying the condition $\sum_{i=1}^{\infty} |x_i| < \infty$. This is a linear space, and the metric axiom is satisfied if the distance is determined by [compare the taxicab norm (5.35)]

$$d(x, y) = \sum_{i=1}^{\infty} |x_i - y_i|. \tag{7.12b}$$

A second important class of spaces consists of the *normed spaces*. Being linear vector spaces, they satisfy Groups A and B of the axioms. In contrast to metric spaces, each element, x, of a normed space is associated with a unique scalar, $\|x\|$, called the *norm* of x and defined by the following axiom:

Axiom N.

(a) $\|x\| > 0$ if $x \neq 0$, $\|x\| = 0$ if $x = 0$,

(b) $\|\alpha x\| = |\alpha|\, \|x\|$ for every scalar α, $\tag{7.13}$

(c) $\|x + y\| \leq \|x\| + \|y\|$ (the triangle inequality).

The distance in a normed space is determined by the previously used relation [compare, e.g., (6.24)]

$$d(x, y) = \|x - y\|. \tag{7.13a}$$

Since, evidently, a Hilbert space is a normed space, the same applies in Hilbert spaces. Such a definition of the metric is rather natural, since (7.12) follows from (7.13) and (7.13a).

Ordinarily, a normed space is not required to be complete;† if it is, it becomes a so-called *Banach* space. Analysis in such spaces is of importance in connection with many problems, both linear and nonlinear. An example of a *Banach space* is the space $\mathscr{E}_n$ with the usual representation of the norm [see equation (5.10)]; another is the space of continuous functions $C_{a \leq t \leq b}$ mentioned before, with norm given by $\|x\| = \max_{a \leq t \leq b} |x(t)|$ $[= d(x, \theta)$ where d is as in (7.12a)]. Both of these spaces are linear spaces and are complete; it can be shown that their norms satisfy the conditions (7.13) (see Ref. 28, pp. 44 and 46).

We now return to the Hilbert space and note that an inspection of the proofs of the Cauchy–Schwarz and triangle inequalities given in Chapter 5

---

† Compare Group E of properties, above.

for a Euclidean n-space indicates that, except for some nonessential details, they remain valid for the Hilbert space as well. We are thus permitted to transcribe the equations (5.24) and (5.25) verbatim,

$$|(x, y)| \leq \|x\| \, \|y\| \tag{7.14}$$

and

$$\|x + y\| \leq \|x\| + \|y\|, \tag{7.15}$$

and consider them proven for the Hilbert space.

It is interesting to note that, while being a Banach space, there is no choice of inner product which makes $C_{a \leq t \leq b}$ into a Hilbert space. In fact, Group C of the axioms and the equality (7.2) imply that, for any two elements x and y in a Hilbert space,

$$\|x + y\| = (x + y, x + y)$$
$$= \|x\|^2 + 2(x, y) + \|y\|^2$$

and

$$\|x - y\| = \|x\|^2 - 2(x, y) + \|y\|^2,$$

so that

$$\|x + y\|^2 + \|x - y\|^2 = 2(\|x\|^2 + \|y\|^2). \tag{7.15a}$$

This is, of course (an abstract version of) the *parallelogram rule* discussed in Problem 6 of Chapter 3.

Now set $b = 3a$ and define

$$x = \begin{cases} 0 & \text{for } 0 \leq t \leq a, \\ \text{linear} & \text{for } a \leq t \leq 2a, \\ 1 & \text{for } 2a \leq t \leq 3a, \end{cases}$$

$$y = \begin{cases} 1 & \text{for } 0 \leq t \leq a, \\ \text{linear} & \text{for } a \leq t \leq 2a, \\ 0 & \text{for } 2a \leq t \leq 3a, \end{cases}$$

both x and y being continuous in the closed interval $[0, 3a]$. Now, since

$$\|x - \theta\| = \|x\| = \max_{a \leq t \leq 3a} |x(t)|$$

in accordance with (7.6) and (7.10), we have, in the case considered,

$$\|x\| = \|y\| = \|x + y\| = \|x - y\| = 1. \tag{7.15b}$$

This violates condition (7.15a) and proves our assertion. Also, we have justified, as indicated in Figure 1.1, the statement that the class of Hilbert spaces is a proper subclass of the family of Banach spaces.

Two other important results, besides (7.14) and (7.15), are *Bessel's inequality and Parseval's equality.*

Before examining these, suppose that we have an *orthonormal basis, S* of vectors x^k, $k = 1, 2, \ldots$, in a Hilbert space, so that [compare text preceding equation (6.25)]

$$(x^i, x^j) = \delta_{ij}, \qquad i, j = 1, 2, \ldots. \tag{7.16}$$

Assume, for the time being, that one or more of the vectors of S are excluded from the set, so that the remaining set, call it S_{inc}, represents something which was called earlier a "mutilated basis."

We wish to approximate an arbitrary vector in the space by a linear combination $\sum_{k=1}^{\infty} \alpha_k x^k$ of the vectors of the set S_{inc}, where the α_k's are coefficients to be determined. The approximation should consist of the minimization of the distance

$$d^2\left(x, \sum_{k=1}^{\infty} \alpha_k x^k\right) = \left\|x - \sum_{k=1}^{\infty} \alpha_k x^k\right\|^2$$

$$= \left(x - \sum_{k=1}^{\infty} \alpha_k x^k, \; x - \sum_{k=1}^{\infty} \alpha_k x^k\right) \tag{7.17}$$

with respect to the coefficients α_k. Expanding the scalar product above in accordance with the rules C(b) and (c), we have

$$d^2\left(x, \sum_{k=1}^{\infty} \alpha_k x^k\right) = (x, x) - 2 \sum_{k=1}^{\infty} \alpha_k(x, x^k) + \sum_{k=1}^{\infty} \alpha_k^2. \tag{7.18}$$

The last expression reaches a minimum for

$$\alpha_k = (x, x^k), \qquad k = 1, 2, \ldots, \tag{7.19}$$

and the coefficients so determined are termed the *Fourier coefficients* of the vector x with respect to the system $\{x^k\}$.

It is interesting to note that the values (7.19) of the coefficients are closely connected with those obtained from an approximate representation

$$x \approx \sum_{k=1}^{\infty} \alpha_k x^k, \tag{7.19a}$$

upon using the orthogonality conditions (7.16). In fact, inner multiplication of (7.19a) with x_k leads directly to approximate relations coinciding with (7.19).

The actual minimum of (7.18) is

$$\min d^2\left(x, \sum_{k=1}^{\infty} \alpha_k x^k\right) = (x, x) - \sum_{k=1}^{\infty} \alpha_k^2, \tag{7.20}$$

so that, in view of the positive definiteness of the inner product [property C(d)],

$$\|x\|^2 = (x, x) \geq \sum_{k=1}^{\infty} \alpha_k^2. \tag{7.21}$$

This is the *Bessel inequality*, an objective mentioned previously. Its geometric sense becomes clear if one identifies the set $\{x^k\}$ with the set of coordinate vectors of an infinite-dimensional Cartesian frame, and the coefficients $\{\alpha_k\}$ with the scalar components of the vector x along the axes of the frame [compare equations (7.19)]. In this interpretation, the inequality (7.21) expresses the fact that, if the system of coordinates is "incomplete"—in the sense that some of the coordinate axes are disregarded—then the "incomplete" sum of the squares of the components of a vector is no greater than the square of the norm of the vector. This result should be intuitively obvious, inasmuch as the right-hand side of inequality (7.21) represents (in the present case) a "truncated" Pythagorean formula for the vector x.

Imagine now that the "mutilated basis," S_{inc}, is replaced by the basis S. In this case, the approximate equality (7.19a) becomes an exact one, and Bessel's inequality transforms into the *Parseval equality*. The geometric sense of the latter coincides now with the Pythagorean theorem generalized to infinitely many dimensions:

$$\|x\|^2 = \sum_{k=1}^{\infty} \alpha_k^2. \tag{7.22}$$

Figure 7.2, illustrating the resolution of a vector x in $\mathscr{E}_3$ with respect to a Cartesian rectangular frame $\{e_i\}$, $i = 1, 2, 3$, provides a crude visualization of

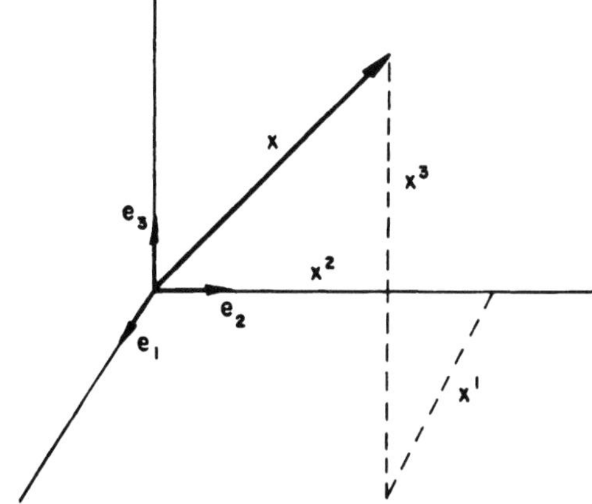

Figure 7.2. Illustration for the formulas of Bessel and Parseval.

both Bessel's and Parseval's formulas. Parseval's equality, referred to the actual basis $\{e_i\}$, $i = 1, 2, 3$, reads here

$$\|x\|^2 = x_1{}^2 + x_2{}^2 + x_3{}^2. \tag{7.23}$$

On the other hand, any of the "mutilated bases," $\{e_1, e_2\}$, $\{e_1, e_3\}$, or $\{e_2, e_3\}$, leads to a Bessel inequality,

$$\|x\|^2 \geq x_1{}^2 + x_2{}^2, \quad \|x\|^2 \geq x_1{}^2 + x_3{}^2, \quad \text{or} \quad \|x\|^2 \geq x_2{}^2 + x_3{}^2, \tag{7.23a}$$

respectively.

As one more example, let us consider Figure 7.3, illustrating the resolution of a vector x into its components along the axes of an orthogonal multidimensional frame, identified by orthonormal coordinate vectors i^k, $k = 1, 2, \ldots, n$. In order for the figure to have visual appeal, the graphical representation of the frame is done in the spirit of Figure 5.1.

If the orthonormal set $\{i^k\}$ is a basis and the space of finite dimension, then the end point of the train of component vectors, such as $\overrightarrow{A^{m-1}A^m}$, coincides with the extremity of the vector $x \equiv \overrightarrow{OA}$ under decomposition. On the other hand, if the set $\{i^k\}$ is a basis, but the space is infinite-dimensional, then the tip A of the vector x becomes the limit point of the set of terminal points of the trains of component vectors as the number n of the latter increases infinitely. The corresponding polygon of vectors, $OA^1A^2 \cdots A^n$, either becomes closed (provided the limit is actually reached) or the gap $\overline{A^nA}$

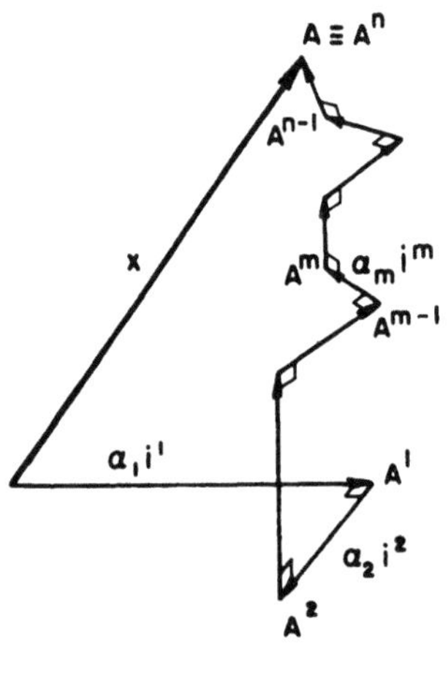

Equation (7.21): $\sum_k a_k{}^2 \leq (x,x)$

Figure 7.3. Illustration for Bessel's inequality.

tends to zero as n tends to infinity. The latter case is reflected in the Parseval equality. If the set $\{i^k\}$ is "deficient," the train of component vectors ends at some point A^m and a gap of finite width, perhaps small, remains. The corresponding situation is described by the Bessel inequality.

We conclude this chapter with the remark that the *separability* of a Hilbert space, defined earlier, implies the existence of a countable orthonormal basis. In fact, the statement that a Hilbert space is separable is equivalent to the statement that the space possesses a countable complete orthonormal basis (Ref. 7, p. 53). We recall that a set is countable if its members can be put into one-to-one correspondence with positive integers.

7.1. Contraction Mapping

The far-reaching implications of the completeness of a space are nicely illustrated by Banach's *principle of contraction mappings*.

For the setting, consider a metric space, with a distance function $d(x, y)$, where x, y are elements of the space, and let $f(x)$ be a function† mapping the space into itself. The function $f(x)$ is called a *contraction* (or contracting) mapping if there is a constant α, $0 \leq \alpha < 1$, such that

$$d(f(x), f(y)) \leq \alpha d(x, y) \qquad (7.24)$$

for all x, y in the space. Since $\alpha d(x, y) < d(x, y)$, in a contraction mapping the distance between any two elements is greater than that between the images of the elements (as if the former were "contracted"). If $x_n \to x$ for $n \to \infty$, then $d(f(x), f(x_n)) \to 0$, and this implies that every contracting mapping function is *continuous*.

The *principle of contraction mappings* says that if $f(\cdot)$ is a contraction mapping on a complete metric space, then there exists one, and only one, fixed point of the mapping, i.e., one, and only one, x in the space such that

$$f(x) = x. \qquad (7.25)$$

In case the metric is also a vector space, we can state the result by saying that the equation $f(y) - y = 0$ has one, and only one, solution, x. The restriction that the mapping be a contraction constitutes a sufficient condition for the (existence and) uniqueness of a solution, although equation (7.25) may have

---

† For applications of the theoretical foundations of functional analysis expounded so far, we shall find it often convenient to depart from our earlier general designation of (unspecified) space elements by the letters such as x, y, or z, and to use for *functions* the more familiar notation such as f and ϕ or u and v (the last two symbols mostly for functions of several variables).

a unique solution also in cases in which the mapping fails to be a contraction.

A proof of the above principle is straightforward.

Let x_0 be an arbitrary element in the space considered. Introduce the notation $x_1 = f(x_0)$, $x_2 = f(x_1) = f^2(x_0)$, ..., $x_k = f(x_{k-1}) = f^k(x_0)$, and evaluate the distance $d(x_n, x_m)$, where $m > n$, say. Since

$$d(x_n, x_m) = d(f(x_{n-1}), f(x_{m-1})) \leq \alpha d(x_{n-1}, x_{m-1})$$
$$= \alpha d(f(x_{n-2}), f(x_{m-2})) \leq \alpha^2 d(x_{n-2}, x_{m-2}), \text{ etc.,}$$

then (using the triangle inequality)

$$d(x_n, x_m) \leq \alpha^n d(x_0, x_{m-n})$$
$$\leq \alpha^n [d(x_0, x_1) + d(x_1, x_2) + \cdots + d(x_{m-n-1}, x_{m-n})]$$
$$\leq \alpha^n d(x_0, x_1)[1 + \alpha + \alpha^2 + \cdots + \alpha^{m-n-1}]$$
$$\leq \alpha^n d(x_0, x_1)\frac{1 - \alpha^n}{1 - \alpha}$$
$$\leq \alpha^n d(x_0, x_1)\frac{1}{1 - \alpha}. \tag{7.26}$$

But $\alpha < 1$ so that, for n sufficiently large, the last expression may be made as small as we wish. This implies that the sequence $\{x^k\}$ is Cauchy and, since the space is complete, that the limit x of $\{x_n\}$ is in the space,

$$x = \lim_{k \to \infty} x_k. \tag{7.27}$$

In view of the continuity of the mapping $f(x)$, however,

$$f(x) = f\left(\lim_{k \to \infty} x_k\right) = \lim_{k \to \infty} f(x_k) = x, \tag{7.28}$$

and this evidences the existence of a fixed point [equation (7.25)].

It is now easy to conclude that the latter is unique. In fact, let there be two such points, x and y. Then

$$f(x) = x, \qquad f(y) = y, \tag{7.29}$$

and the inequality (7.24) gives

$$d(x, y) \leq \alpha d(x, y). \tag{7.30}$$

But $\alpha < 1$, so that, to avoid contradiction, $d(x, y) = 0$ or $x = y$ [cf. (7.12a)].

Example 7.1. Problems of applied mechanics often reduce to solving systems of linear nonhomogeneous algebraic equations such as

$$\sum_{j=1}^{n} \alpha_{ij} x_j + \beta_i = y_i, \qquad i = 1, 2, \ldots, n, \tag{7.31}$$

and it is of interest to know beforehand whether the system has, or has not, a unique solution.[†] The principle of contraction mappings may occasionally be put to use here.

Assume then that $x = \{x_i\}$ and $y = \{y_i\}$ are n-tuples of real numbers and take the distance in the form

$$d(x, y) = \max_i \, |x_i - y_i|, \tag{7.32}$$

reminiscent of (7.12a).

We have for two pairs of elements x^1, x^2 and their respective images y^1, y^2, so $y^1 = f(x^1)$, $y^2 = f(x^2)$, where $f(\cdot)$ stands for the operation represented by the left-hand member of equation (7.31):

$$\begin{aligned}
d(y^1, y^2) &= d(f(x^1), f(x^2)) \\[4pt]
&= \max_i \, |y_i{}^1 - y_i{}^2| \\[4pt]
&= \max_i \, \left| \sum_{j=1}^{n} \alpha_{ij}(x_j{}^1 - x_j{}^2) \right| \\[4pt]
&\le \max_i \, \sum_{j=1}^{n} |\alpha_{ij}| \, |x_j{}^1 - x_j{}^2| \\[4pt]
&\le \max_j \, |x_j{}^1 - x_j{}^2| \, \max_i \, \sum_{j=1}^{n} |\alpha_{ij}| \\[4pt]
&= \max_i \, \sum_{j=1}^{n} |\alpha_{ij}| \, d(x^1, x^2).
\end{aligned} \tag{7.33}$$

If now

$$\sum_{j=1}^{n} |\alpha_{ij}| \le \alpha < 1 \tag{7.34}$$

for each $i = 1, 2, \ldots, n$, then the contraction principle can be applied,[‡] and

---

[†] As is known, such a system may be undetermined (when two or more equations coincide), thus having an infinite number of solutions.

[‡] We must note that the set of n-tuples of real numbers with the metric defined by (7.32) is, in fact, a complete metric space.

consequently the operator $f(\cdot)$ possesses exactly one (fixed) point x^* such that $f(x^*) = x^*$. Since this is so, the system of equations

$$x_i - \sum_{j=1}^{n} \alpha_{ij} x_j = \beta_i, \qquad i = 1, 2, \ldots, n, \tag{7.35}$$

has a unique solution $x^* = \{x_i^*\}$ for arbitrary $\{\beta_i\}$ provided that $\sum_{j=1}^{n} |\alpha_{ij}| \leq \alpha < 1$, $i = 1, 2, \ldots, n$. As a numerical illustration, consider the following simple system:

$$7x + 2y + 2z = 1,$$
$$x + 3y + z = 1, \tag{7.36}$$
$$x + 2y + 4z = 1.$$

We divide the system through by 10, say, and find easily,

$$\alpha_{11} = 0.3, \qquad \alpha_{12} = 0.2, \qquad \alpha_{13} = 0.2,$$
$$\alpha_{21} = 0.1, \qquad \alpha_{22} = 0.7, \qquad \alpha_{23} = 0.1,$$
$$\alpha_{31} = 0.1, \qquad \alpha_{32} = 0.2, \qquad \alpha_{33} = 0.6.$$

Inequality (7.34) being satisfied for $i = 1, 2, 3$, the system has a unique solution. Indeed, Cramer's rule yields immediately $x^* = 1/31$, $y^* = 9/31$, $z^* = 3/31$.

Problems

1. Show that, in a Hilbert space, if the sequences $\{x^k\}$ and $\{y^k\}$ converge to x and y, respectively [see equation (6.3)], then the sequence of inner products $\{(x^k, y^k)\}$ converges to (x, y).

2. Show that the norm (5.10) in a Euclidean space $\mathscr{E}_n$, $\|x\| = [\sum_{i=1}^{n} x_i^2]^{1/2}$, $x(x_1, x_2, \ldots, x_n)$, satisfies the requirements of Axiom N.

3. Let m vectors $x^1, x^2, \ldots, x^m$ span a subspace $\mathscr{S}$. Show that the dimension n of the subspace is less than or equal to m (i.e., $n \leq m$).

4. By applying the Gram–Schmidt procedure in the interval $[-1, 1]$, convert the first three terms of the sequence $1, t, t^2, t^3, \ldots$, of linearly independent vectors into the orthonormalized Legendre polynomials $P_1(t)$, $P_2(t)$, and $P_3(t)$. Verify the orthonormality of the vectors obtained.

5. Show that in a metric space $|d(x, z) - d(y, z)| \leq d(x, y)$. Give a geometric interpretation.

6. Prove the completeness of the space of continuous functions $C_{a \leq t \leq b}$ with respect to the Chebyshev metric (7.12a).

7. A set in a vector space is convex [see the footnote preceding (6.18)] if it contains an entire line segment $[x = \beta_1 x^1 + \beta_2 x^2, \beta_1 + \beta_2 = 1$; in Chapter 9 see the text concerning (9.14)] if it contains the two end points of the line (x^1 and x^2). Show that a closed ball (a "solid" hypersphere, $\|x - c\| \leq R$, $c =$ radius; refer to the end of Chapter 9) in a normed space is convex. Give a geometrical interpretation.

8. Show that for $k = 1$, Bessel's inequality (7.21) reduces to the Cauchy–Schwarz inequality (5.24).

9. Show that the space $C_{0 \leq t \leq 1}$ is separable defining the distance between two functions according to (7.12a).

8

Function Space

While the course of reasoning in the preceding chapters led us "from the particular to the general"—from a directed line segment, for example, to an abstract space element—at this stage we choose to change our approach and concentrate on the particular class of abstract spaces known as function spaces. A *function space* is an abstract space, the elements of which are functions or sets of functions defined in an appropriate domain.

To emphasize the fact that functions are visualized as vectors, we shall call the elements of a function space *function vectors*[†] and denote the latter temporarily by the letters f, g, or h, reminiscent of the traditional notation for functions. The letters x, y, and z will meanwhile be assigned to independent variables, and the symbol i^k, $k = 1, 2, \ldots$, to orthonormal systems of functions.

Assume temporarily that the function space has all of the properties of a Hilbert space[‡] except that the requirement C(d) is relaxed, in the sense that the self-inner product is supposed to be positive semi-definite, that is,

$$(f, f) = 0 \tag{8.1}$$

not only for $f = \theta$, but perhaps for some other $f \neq \theta$ as well. This liberalization of the definition of the metric turns out to be convenient in certain cases [compare Diaz (Ref. 29, p. 7)].

The explicit definition of the inner product in a given function space will depend considerably on the nature of the problem at hand, the form of the

---

[†] The name "vector functions" shall be reserved for vector-valued functions.

[‡] There are many advantages to be gained from working with Hilbert spaces, since they are the natural extension of Euclidean spaces; their structure is usually compatible with our spatial intuition. In the sequel, a function space will most frequently appear as a Hilbert space (thus enjoying *all* of the properties of the latter, unless stated otherwise).

product being either imposed by the problem itself or selected so as to make the solution of the problem as simple as possible.

A useful form for the inner product is the so-called *Hilbert product*,

$$(f, g) = \int_{\Omega} f(P)g(P) \, d\Omega, \tag{8.2}$$

where $f(P)$ and $g(P)$ are functions of position in the domain Ω. The function vectors here stand for single functions. It is apparent that the corresponding metric is positive-definite (at least on any space of functions of interest to us), and there is no difficulty in verifying that the Hilbert product—like all products listed below—satisfies the requirements imposed on an inner product by Group C of the axioms in Chapter 7. The origin of the form (8.2) can be traced to the familiar form (4.13) of the dot product of two ordinary vectors in Euclidean three-space,

$$\mathbf{a} \cdot \mathbf{b} = a_i b_i \tag{8.3}$$

if the summation over the subscripts is extended to infinity [as indicated by the formula (6.22) for l_2] and replaced by an integration. In fact, suppose that a function $f(x)$ of a single variable x ranging over the interval $[x_a, x_b]$ is viewed as the infinite collection of its values, $f(x_i)$, at points x_i, $x_a \leq x_i \leq x_b$. In the spirit of our remarks at the opening of Chapter 7, it is natural to identify the set of numbers $f(x_i)$ with the infinite set of *components* of the function vector $f(x)$. Thus, using the notation

$$f(x_i)(dx)^{1/2} \equiv f_i, \qquad g(x_i)(dx)^{1/2} \equiv g_i, \tag{8.4}$$

the integral (8.2) can be thought of as the limit of the sum $\sum_{i=1}^{n} f_i g_i$ for n tending to infinity,

$$\int_{x_a}^{x_b} f(x)g(x) \, dx = \lim_{n \to \infty} \sum_{i=1}^{n} f_i g_i. \tag{8.5}$$

A comparison of the right-hand sides of equations (8.3) and (8.5) makes apparent their identical structure, and verifies the claim that the Hilbert product is a descendent of the dot product.

From the form (8.2), the transition to functions of several variables is immediate; it suffices to simply treat the symbol Ω as a designation of a domain of two or more dimensions.

It is interesting to note that, if use is made of the Hilbert metric for *square integrable* functions, the condition of convergence (7.5) becomes precisely that of *convergence in the mean* (discussed later in Chapter 10). This follows from the fact that by combining equations (7.5) and (8.2) we have, for example,

$$\lim_{k \to \infty} \| f - f^k \| = \lim_{k \to \infty} \left[\int_a^b [f_k(t) - f(t)]^2 \, dt \right]^{1/2} = 0, \tag{8.5a}$$

where $\{f_k\}$ is a sequence of functions approximating (converging to) the given function f.

A somewhat different form for the inner product is provided by the *Dirichlet* inner product,

$$(f, g) = \int_\Omega (f_1 g_1 + f_2 g_2)\, d\Omega. \tag{8.6}$$

This product involves function vectors representing ordered pairs of functions, in the present case $f(P) \equiv [f_1(P), f_2(P)]$ and $g(P) \equiv [g_1(P), g_2(P)]$. Function vectors of this type may be called function vectors of the *second kind*. More generally, a function vector representing an n-tuple of functions may be called a function vector of the *nth kind*. The corresponding inner product has the form

$$(f, g) = \int_\Omega \sum_{i=1}^{n} f_i g_i\, d\Omega. \tag{8.6a}$$

Returning to equation (8.6), we observe that if $(f, f) = 0$, i.e., $\int_\Omega (f_1{}^2 + f_2{}^2)\, d\Omega = 0$, then $f_1 = f_2 = 0$. In spite of this conclusion, the form of the inner product (8.6) is often used to generate a *positive semi-definite metric*. In fact, select, for example, $f_1 = f_{,x}$ and $f_2 = f_{,y}$, a comma denoting differentiation with respect to the succeeding variable. Then the equality $(f, f) = 0$ holds in a two-dimensional domain, not only for f identically zero, but for $f = \text{const}$ as well.

A third realization for the product is the *Minkowski* inner product, involving triples of functions, $f = (f_1, f_2, f_3)$, i.e., function vectors of the *third kind*. On account of its definition,

$$(f, g) = \int_\Omega (f_1 g_1 + f_2 g_2 - f_3 g_3)\, d\Omega, \tag{8.7}$$

Minkowski's metric is *indefinite*. Indeed, by assuming that $f_3 = 0$, the product (f, f) is made non-negative [like the Dirichlet product (8.6)], while if $f_1 = f_2 = 0$, the product (f, f) becomes negative.

Minkowski's metric is ordinarily employed in the Minkowski space-time world of the theory of relativity and must be left to other, more specialized, studies. On the other hand, in various engineering problems, it is often convenient to operate in real vector spaces with a *positive semi-definite* inner product.

While it exceeds the scope and character of this book to go into a detailed analysis of such spaces, it is found that in many practical applications the distinction between a space with a positive-definite and that with a positive semi-definite metric may not be so vital.

With regard to the *positive semi-definiteness* of the inner product, the latter is required to satisfy the standard axioms (a)–(c) of Group C in Chapter 7, but axiom (d) is to be replaced by the more liberal

$$(x, x) \geq 0. \tag{8.7a}$$

In more graphic language, this implies that, if the length of a vector is identified with its norm (7.2), a vector with zero length need not be the zero vector.

In this book, we discuss a number of problems in which the inner product is assumed to be positive semi-definite. It turns out, however (not necessarily by accident), at least in the class of problems considered here, that it is not required to enter into the finer aspects of the theory of spaces with a positive semi-definite metric, inasmuch as only the simplest of the available tools—the Cauchy–Schwarz and Bessel inequalities—will be found adequate for our purposes.

Defining, as before, the norm of x by $\|x\| = (x, x)^{1/2}$, we have

$$\|(y, y)x - (y, x)y\|^2 = ((y, y)x - (y, x)y, (y, y)x - (y, x)y)$$
$$= (y, y)[(y, y)(x, x) - (y, x)^2]. \tag{8.7b}$$

Thus, if $(y, y) > 0$, the equality (8.7b) yields the Cauchy–Schwarz inequality

$$(x, y)^2 \leq (x, x)(y, y), \tag{8.7c}$$

the equality sign holding for $y = \alpha x$, with α a scalar. By reasoning similar to that leading to equation (7.21), it can also be shown that the Bessel inequality

$$\|x\|^2 \geq \sum_{k=1}^{\infty} (x, x^k)^2, \tag{8.7d}$$

where $(x^i, x^j) = \delta_{ij}$, $i, j = 1, 2, \ldots$, retains its validity in the positive semi-definite case.

There is a certain connection between the notion of a positive semi-definite inner product and a so-called *semi-norm*, reminiscent of the connection between the norm and the positive definite inner product, and which we wish now to examine.

We first note that the definition of a *semi-norm* differs from that of a norm simply in that requirement (a) in the definition (7.13) of the norm is discarded. Thus, denoting a semi-norm by the symbol $\|\cdot\|_{\text{sem}}$ [frequently, there is also used the notation $p(\cdot)$], we have (see Ref. 15, p. 71 and Ref. 30, p. 23)

$$\begin{aligned}
&\text{(b)} \quad \|\alpha x\|_{\text{sem}} = |\alpha|\,\|x\|_{\text{sem}}, \\
&\text{(c)} \quad \|x + y\|_{\text{sem}} \leq \|x\|_{\text{sem}} + \|y\|_{\text{sem}}.
\end{aligned} \tag{8.7e}$$

To this may be added the condition

$$(a^*) \quad \|x\|_{\text{sem}} \geq 0, \qquad\qquad (8.7\text{f})$$

stating that, in contrast to the norm, vanishing of the semi-norm of some x does not necessarily imply that $x = \theta$. Actually, the latter inequality follows from the requirements (b) and (c) above.[†] In fact, we have, by (c),

$$\|x - y\|_{\text{sem}} + \|y\|_{\text{sem}} \geq \|x\|_{\text{sem}}$$

or

$$\|x - y\|_{\text{sem}} \geq \|x\|_{\text{sem}} - \|y\|_{\text{sem}}.$$

Consequently, by (b),

$$\|x - y\|_{\text{sem}} = |-1| \, \|y - x\| \geq \|y\|_{\text{sem}} - \|x\|_{\text{sem}}.$$

As a result,

$$\|x - y\|_{\text{sem}} \geq |\, \|x\|_{\text{sem}} - \|y\|_{\text{sem}} |,$$

and, in particular,

$$\|x\| \geq 0,$$

which results if we set $y = \theta$ and observe that $\|\theta\|_{\text{sem}} = \|0x\|_{\text{sem}} = 0\|x\|_{\text{sem}} = 0$.

An example of a semi-normed space is found in the space $C'_{a \leq t \leq b}$ of those functions $x(t)$, of a single real variable t, which have a continuous derivative on $[a, b]$, if we set $\|x\|_{\text{sem}} = \int_a^b |dx/dt|^2 \, dt$. A slightly more general example of a semi-norm is exhibited by equation (11.15) infra (there is, of course, no difficulty in defining the associated space).

In boundary- or initial-value problems of a complicated type, such as those encountered in the theory of elasticity, the structure of the inner product may become complicated. As an illustration, consider the theory of thin elastic isotropic *plates* undergoing small deflections, in which the inner product of the deflections $u(x, y)$ and $v(x, y)$ is often taken in the form

$$(u, v) = \frac{D}{2} \int_{\Omega} [(\nabla^2 u)(\nabla^2 v) - (1 - v)(u_{,xx} v_{,yy} + u_{,yy} v_{,xx} - 2u_{,xy} v_{,xy})] \, dx \, dy.$$

$$(8.8)$$

Here, $\nabla^2 = \partial^2/\partial x^2 + \partial^2/\partial y^2$, Ω is the region occupied by the plate, D is its

---

[†] A conclusion identical with (a^*), but for the norm as such, can be drawn from conditions (b) and (c) in (7.13). That is, (a) in (7.13) need only be stated: "$\|x\| = 0$ implies $x = \theta$," the inequality "$\|x\| \geq 0$" being implied by (b) and (c).

bending rigidity, and v denotes Poisson's ratio. An alternative form of (8.8), following from *Clapeyron's theorem*, is (Ref. 31, p. 86)

$$(u, v) = \frac{1}{2} \int_{\Omega} q(u)v \ dx \ dy - \int_{C} \left[M_n(u)\frac{\partial v}{\partial n} - V_n(u)v \right] ds, \tag{8.9}$$

where $q(u)$ is the external load distributed over the surface of the plate, $M_n(u)$ and $V_n(u)$ are the bending moment and the transverse force, respectively, both acting at the contour C of the plate, and n is the external normal on C (Ref. 32, Sec. 22):

$$M_n(u) = -D\left[v\nabla^2 u + (1 - v)\left(\cos^2 \alpha \frac{\partial^2 u}{\partial x^2} + \sin^2 \alpha \frac{\partial^2 u}{\partial y^2} + \sin 2\alpha \frac{\partial^2 u}{\partial x \ \partial y}\right)\right], \tag{8.10a}$$

$$V_n(u) = -D\left\{\cos \alpha \frac{\partial(\nabla^2 u)}{\partial x} + \sin \alpha \frac{\partial(\nabla^2 u)}{\partial y} + (1 - v)\frac{\partial}{\partial s} \right.$$
$$\left. \times \left[\cos 2\alpha \frac{\partial^2 u}{\partial x \ \partial y} + \tfrac{1}{2} \sin 2\alpha\left(\frac{\partial^2 u}{\partial y^2} - \frac{\partial^2 u}{\partial x^2}\right)\right]\right\}. \tag{8.10b}$$

Here, α is the angle between the normal n and the x-axis and s is the coordinate measured along C.

The symmetry of the inner product in the form (8.9) is guaranteed by the *Rayleigh–Betti theorem*, according to which the right-hand side of (8.9) remains invariant under the interchange of the functions u and v (Ref. 31, p. 390). Concerning (8.8), it is of interest to note that this type of inner product represents, for $u \equiv v$, the potential energy of a plate subject to bending.

Expressions (8.8) and (8.9) are not the most involved examples of inner products; others of much more intricate forms have been used. For instance, the following one has been applied in linear viscoelastic problems[33]:

$$(f, g) = \int_{\Omega} \int_{0}^{\infty} \left\{ \frac{e^{-s_0 t}}{(t + 1)^3} [\mu * (f_{i,j} * g_{i,j} + f_{i,j} * g_{j,i}) + \lambda * f_{i,i} * g_{k,k}] \right.$$
$$\left. + \sum_{r=0}^{2} \frac{(r + 1)(r + 2)}{(2 - r)!} s_0^{2-r} \frac{e^{-s_0 t}}{(t + 1)^{r+3}} f_i * g_i \right\} dt \ d\Omega, \tag{8.10c}$$

where λ and μ are material parameters defined by the relaxation functions, s_0 is the abscissa of convergence of a Laplace transform, and an asterisk denotes the convolution operator,

$$a(t) * b(t) = \int_{0}^{t} a(\tau)b(t - \tau) \ d\tau. \tag{8.10d}$$

In studies of the linear theory of elasticity in three dimensions, the displacement vector $u(x, y, z)$ is represented by a triple of components $u_i(x, y, z)$, $i = 1, 2, 3$. In this manner, it is seen to have the character of a function vector of the third kind. In general anisotropic elasticity, we usually define the inner product by[†]

$$(u, v) = \frac{1}{2} \int_V c_{ijkl} u_{i,j} v_{k,l} \, dV, \tag{8.11}$$

where V is the region occupied by the body and the c_{ijkl}'s are the so-called elastic moduli. By virtue of the symmetry of the latter with respect to the pairs of indices ij and kl, this product obeys the required rule of symmetry (8.3).

An alternative form of the inner product applied in elastic problems displays its relationship to the states of deformation and stress, represented by the tensors of strain, e_{ij}, and stress, τ_{ij}, respectively. The latter two being connected by the generalized Hooke's law,

$$\tau_{ij} = c_{ijkl} e_{kl}, \tag{8.12}$$

we may consider a function vector denoted, say, by S (to emphasize our concern with elastic states), standing for either of the sets $\{e_{ij}\}$ or $\{\tau_{ij}\}$. If both are present, it is to be understood that one of them is expressed in terms of the other. According to our classification, the function vector S describing the *elastic state* of the body is a vector of the *sixth kind*, there being six independent components of each of the tensors e_{ij} and τ_{ij}.

Let us denote the function vectors associated with two different elastic states by S and S', and introduce an inner product in the form

$$(S, S') = \frac{1}{2} \int_V \tau_{ij} e'_{ij} \, dV$$

$$= \frac{1}{2} \int_V \tau'_{ij} e_{ij} \, dV, \tag{8.13}$$

the latter equality following from the reciprocal theorem. It is worth emphasizing that the self-inner product

$$(S, S) = \frac{1}{2} \int_V \tau_{ij} e_{ij} \, dV, \tag{8.14}$$

that is, the square of the norm of the vector S, represents the potential energy stored in the body subject to deformation. Inasmuch as it is shown in the

---

[†] The reader is reminded of the summation convention, observed here and later, and of the symbolism $u_{i,j}$ for the derivative $\partial u_i / \partial x_j$, for example.

linear theory of elastic isotropic bodies (to which we confine our attention) that the energy is a positive-definite quantity, the metric induced by (8.14) is positive definite as well.

Bearing in mind the linearized strain-displacement relations,

$$e_{ij} = \tfrac{1}{2}(u_{i,\,j} + u_{j,\,i}), \tag{8.15}$$

as well as equation (8.12), it is straightforward to cast equations (8.13) into a form similar to (8.11). The resulting metric, however, need not be positive definite.

As already noted, the specific form selected for the inner product is closely connected with the nature of the problem at hand. To illustrate this point, consider a *linear differential operator*,† L, which involves ordinary or partial derivatives. We write the associated differential equation in the symbolic form

$$Lu(P) = f(P), \tag{8.16}$$

where $u(P)$ is a function to be determined and $f(P)$ is a function preassigned. Both $u(P)$ and $f(P)$ are visualized as function vectors in a vector space. Furthermore, since $f(P)$ is treated as a vector, so is Lu.

Assume that the operator L is *symmetric*, i.e.,

$$(Lu,\ v)_H = (u,\ Lv)_H, \tag{8.17}$$

where the subscript denotes an inner product of the Hilbert type (8.2),

$$(Lu,\ v)_H = \int_\Omega Lu \cdot v \; d\Omega. \tag{8.17a}$$

Here, $Lu \cdot v \equiv \sum_i Lu_i\, v_i$, where the subscripts denote the components of the vectors Lu and v. In this sense, equation (8.17a) is a generalization of equation (8.2) (and of the familiar dot product of vectors). In what follows, we shall omit the dot symbol in those cases in which no confusion is likely to arise.

Assume that L is a *positive-definite operator*, that is,

$$(Lu,\ u)_H > 0 \qquad \text{for } u \neq \theta. \tag{8.18}$$

It is demonstrated in the *variational calculus* that the quadratic functional (Ref. 8, p. 75)

$$\Phi[u] \equiv (Lu,\ u)_H - 2(u,\ f)_H, \tag{8.19}$$

---

† An operator T is a mapping from a space $\mathscr{U}$ to a space $\mathscr{V}$ such that a vector $u \in \mathscr{U}$ is transformed into a vector $v \in \mathscr{V}$: $v = Tu$. If $T = L$, a linear operator, then $L[u_1 + u_2] = L[u_1] + L[u_2]$ and $L[\alpha u] = \alpha L[u]$.

where

$$(Lu, u)_H - 2(u, f)_H = \int_\Omega u \cdot (Lu - 2f) \, d\Omega, \tag{8.20}$$

reaches its minimum for $u = u_0$, the latter being a solution[†] of the equation (8.16). We recall that a quadratic functional involves products of linear terms, and a *functional* ("function of a function") is a mapping having as its domain a set of functions and as its range a set of scalars.[‡]

The minimum of $\Phi(u)$, realized by the solution u_0, is

$$\min \Phi[u] = \Phi[u_0] = -\||u_0\||_H^2, \tag{8.21}$$

where

$$\||u_0\||_H \equiv (Lu_0, u_0)_H^{1/2} \tag{8.21a}$$

is the norm of the vector u_0 in the Hilbert sense.

In certain stationary problems of mathematical physics, the inner product $(Lu, u)_H$ represents (often to within some scale factor) the potential energy stored in the body. In elasticity, this energy is the elastic energy produced by the action of external load. To demonstrate this claim, consider the governing equation of linear isotropic elasticity,

$$-[\mu\nabla^2 u + (\lambda + \mu) \, \text{grad div } u] = F, \tag{8.22}$$

where λ and μ are the Lamé constants, F is the vector of the body force per unit volume of the body, and u is the displacement vector. The preceding equation is given the more concise form

$$Lu = F \tag{8.23}$$

by introducing the operator

$$L \equiv -[\mu\nabla^2 + (\lambda + \mu) \, \text{grad div}]. \tag{8.23a}$$

Assume, for definiteness, that the work of the surface forces is zero because either: (a) the boundary Ω is fixed,

$$u = 0 \quad \text{on } \Omega, \tag{8.23b}$$

[†] It is emphasized that "solving (8.16)" means "finding a function which satisfies (8.16) and the *boundary conditions* of the given problem."

[‡] Examples of functions involving functions $u(x)$ of certain classes are: $\min_{a \leq x \leq b} |u(x)|$, $\int_a^b u(x) \, dx$, $u(x) + du(x)/dx|_{x=0}$, inner product (u, v), where v is a fixed vector. A functional $F[u]$ is said to be linear if $F[u + v] = F[u] + F[v]$ ("additivity") and $F[\alpha u] = \alpha F[u]$ for any scalar α ("homogeneity").

or (b) the boundary consists of two portions, Ω_1, on which the displacement is fixed, and Ω_2, which is free from external tractions, $t_{(n)}$,

$$u = 0 \quad \text{on } \Omega_1, \qquad t_{(n)} = 0 \quad \text{on } \Omega_2, \tag{8.23c}$$

with $\Omega = \Omega_1 + \Omega_2$.

Either of these assumptions leads to a corresponding inner product which is symmetric.

It is worth emphasizing that the boundary value problems consisting of the governing equation (8.22) and the boundary condition (8.23b) or (8.23c) are known, respectively, as the first and the third (or mixed) boundary value problems of linear elasticity (Ref. 31, p. 80).

Bearing the assumed boundary conditions in mind and applying *Clapeyron's theorem*, we find easily that (Ref. 31, p. 86)

$$\int_V uLu \, dV = 2 \int_V W(u) \, dV, \tag{8.24}$$

where

$$W(u) = \tfrac{1}{2}\tau_{ij}e_{ij} \tag{8.25}$$

is the strain energy per unit volume of the body, expressed in terms of the displacements.

Next, we recall the principle of minimum potential energy of an elastic system, known as *Lagrange's variational principle*, stating that of all elastic displacements u which satisfy the boundary conditions, those which also satisfy the equilibrium equations (8.22), i.e., the actual displacements u_0, make the potential energy of the system minimum. Symbolically,

$$\int_V [W(u) - uF] \, dV = \min \tag{8.26}$$

or, by virtue of (8.24),

$$\int_V u(Lu - 2F) \, dV = \min. \tag{8.27}$$

A comparison of the pair of equations (8.23) and (8.27) with the pair (8.16) and (8.20) reveals a similarity between their structures. Hence, by virtue of equation (8.21a), the norm of the actual solution u_0 is

$$\|\|u_0\|\|_H = (Lu_0, u_0)_H^{1/2}, \tag{8.28}$$

or, alternatively,

$$\|\|u_0\|\|_H = \left[2 \int_V W(u_0) \, dV \right]^{1/2}, \tag{8.28a}$$

by appeal to (8.24). This implies that the Hilbert product, in the form assumed above, equals twice the elastic strain energy stored in the body. This confirms our remark concerning the close relationship between the problem as such and the associated inner product.

As one more example, consider the previously discussed theory of bending of elastic isotropic plates.

We place the coordinate axes x and y on the middle plane of the plate and direct the z axis perpendicular to the latter. As derived in the theory of plates,

$$u_x = -w_{,x}z, \qquad u_y = -w_{,y}z, \tag{8.29}$$

where u_x and u_y are the displacements in the x and y directions, respectively, and $w = w(x, y)$ is the deflection. According to Kirchhoff's hypotheses, the normal stress τ_{zz} is disregarded and the shear stresses τ_{zx} and τ_{zy} are considered to be of secondary influence as compared with the stresses τ_{xx}, τ_{yy}, and τ_{xy}.

By virtue of Hooke's law,

$$\tau_{xx} = \frac{E}{1-v^2}\left(e_{xx} + ve_{yy}\right), \qquad \tau_{yy} = \frac{E}{1-v^2}\left(e_{yy} + ve_{xx}\right),$$

$$\tau_{xy} = \frac{E}{1+v}\,e_{xy}, \tag{8.30}$$

where, if desired, the strains can be expressed in terms of the displacements (8.29) via the relations (8.15).

The equations of equilibrium reduce, in the present case, to the single equation

$$D\nabla^4 w = q(x, y), \tag{8.31}$$

∇^4 denoting the two-dimensional biharmonic operator and $q(x, y)$ the intensity of the surface load.

The elastic strain energy per unit area of the plate is

$$W(u) = \tfrac{1}{2}\left(\tau_{xx}e_{xx} + \tau_{yy}e_{yy} + 2\tau_{xy}e_{xy}\right) \tag{8.32}$$

or, explicitly,

$$W(w) = \frac{D}{2}\left[(\nabla^2 w)^2 + 2(1-v)(w_{,xy}^2 - w_{,xx}w_{,yy})\right]. \tag{8.33}$$

The foregoing expression coincides with the integrand in equation (8.8), provided u and v are replaced by w. It is instructive to examine the particular case of a plate built-in on the contour C, so that

$$w = 0, \qquad \frac{\partial w}{\partial n} = 0 \qquad \text{on } C. \tag{8.34}$$

Integration of the last two terms in equation (8.33) by parts over the area of the plate yields the contour integral $-\int_C w_{,x}(d(w_{,y})/ds)\,ds$, vanishing on account of the first of conditions (8.34).

We next consider the identity

$$(\nabla^2 w)^2 = (w_{,x}\nabla^2 w)_{,x} + (w_{,y}\nabla^2 w)_{,y} - [w(\nabla^2 w)_{,x}]_{,x}$$
$$- [w(\nabla^2 w)_{,y}]_{,y} + w\nabla^4 w, \tag{8.34a}$$

which, in combination with the Gauss–Green formula, gives

$$\int_V (\nabla^2 w)^2\,dV = \int_C \nabla^2 w\,\frac{\partial w}{\partial n}\,ds - \int_C w\,\frac{\partial(\nabla^2 w)}{\partial n}\,ds + \int_V w\nabla^4 w\,dV. \tag{8.34b}$$

Since the contour integrals in the latter equation vanish, equation (8.33) becomes

$$\int_V W(w)\,dV = \frac{D}{2}\int_V w\nabla^4 w\,dV. \tag{8.35}$$

In view of the analogy between the conditions (8.34) and the zero-work conditions leading to equation (8.27) on the one hand, and the equations (8.24) and (8.35) on the other, we have

$$\int_V w(D\nabla^4 w - 2q)\,dV = \min, \tag{8.36}$$

this being a particular case of the more general equation (8.27), provided w, ∇^4, and q/D are put in place of u, L, and F, respectively. A similar conclusion is reached with regard to the pair of equations (8.23) and (8.31).

With equations (8.28) and (8.28a) in mind, the norm of the actual solution w_0 becomes

$$\||\,w_0\,\||_H = (D\nabla^4 w_0,\, w_0)_H^{1/2} \tag{8.37}$$

or, alternatively,

$$\||\,w_0\,\||_H = \left[2\int_V W(w_0)\,dV\right]^{1/2}. \tag{8.37a}$$

The last form of the norm demonstrates once more our remark about the close connections between the structure of an "appropriate" norm and the physical nature of the associated problem. This connection plays an important role in the approximate solution of a problem. To illustrate this point, suppose that we are interested in *torsion* of elastic bars. In the general case, our task consists in finding the so-called torsion function $\phi(x,\,y)$, related to the warping of cross sections and satisfying the plane Laplace equation within the cross section Ω,

$$\nabla^2\phi = 0, \tag{8.38}$$

while being subject to the condition

$$\frac{\partial \phi}{\partial n} = y \cos(n, x) - x \cos(n, y) \tag{8.38a}$$

on the boundary $\partial \Omega$ of Ω. The problem thus formulated is the well-known *Neumann* problem. We note that the right-hand side of (8.38a) is a known function of position on $\partial \Omega$.

It is demonstrated in the theory of torsion of bars that the only identically nonvanishing components of stress are the shears

$$\tau_{zx} = \mu \alpha(\phi_{,x} - y), \qquad \tau_{zy} = \mu \alpha(\phi_{,y} + x), \tag{8.39}$$

where μ is a Lamé constant and α is the angle of twist per unit length of the bar. Instead of prescribing the twisting moment M_z acting on the bar, it is often convenient to prescribe the angle α produced by M_z, and set $\alpha = 1$ for simplicity.

The elastic strain energy being

$$W(\phi) = \frac{1}{2\mu} \left(\tau_{zx}^2 + \tau_{zy}^2 \right), \tag{8.40}$$

we have the potential energy of torsion, per unit length of the bar,

$$U(\phi) = \frac{\mu}{2} \int_{\Omega} \left[(\phi_{,x} - y)^2 + (\phi_{,y} + x)^2 \right] d\Omega. \tag{8.41}$$

By the first Green formula for a harmonic function,

$$\int_{\Omega} \left(\phi_{,x}^2 + \phi_{,y}^2 \right) d\Omega = \int_{\partial\Omega} \phi \phi_{,n} \, ds, \tag{8.42}$$

and, consequently, applying the Gauss–Green theorem,

$$\int_{\Omega} \left(-y\phi_{,x} + x\phi_{,y} \right) d\Omega = - \int_{\partial\Omega} \phi \phi_{,n} \, ds. \tag{8.43}$$

Finally,

$$U(\phi) = \frac{\mu}{2} \left(P - \|\phi\|_D^2 \right), \tag{8.44}$$

where P is the polar moment of inertia of the cross section of the bar and $\|\phi\|_D$ is the metric in the Dirichlet sense,

$$\|\phi\|_D^2 = \int_{\Omega} \left(\phi_{,x}^2 + \phi_{,y}^2 \right) d\Omega. \tag{8.45}$$

The last integral, known as the *Dirichlet integral*, is associated with the inner product in the form (8.6),

$$(\phi, \psi)_D = \int_\Omega (\phi_{,x}\psi_{,x} + \phi_{,y}\psi_{,y}) \, d\Omega. \tag{8.46}$$

The contour of the bar being known, so also is the polar moment of inertia P. Thus, the norm of the function sought is, in fact, determined by the value of the potential energy alone. This should come as no surprise if one recalls the repeatedly mentioned influence of the physics of the problem on the structure of the norm.

With regard to the problem at hand, we note that, since $U(\phi) > 0$, equation (8.44) yields the inequality

$$\|\phi\|_D < P^{1/2}, \tag{8.47}$$

giving an upper bound for the norm of ϕ. For a circular cross section, the torsion function is constant and the norm zero [but not the strain energy (8.41)]. For a narrow rectangular or an elongated elliptic cross section, one can pose approximately $\phi = xy$; in this case, the strain energy is almost negligible and the norm is close to reaching its upper bound $P^{1/2}$.

On this, we end our discussion of the torsion problem. A more detailed analysis of the estimation of solutions of problems by means of bounds is postponed to Chapter 11.

Comment 8.1. It is clear that inner products such as (8.2), (8.6), and (8.6a) are all particular cases of the general form

$$(f, g) = \int_\Omega (w_1 f_1 g_1 + w_2 f_2 g_2 + \cdots + w_n f_n g_n) \, d\Omega, \tag{8.48}$$

where the scaling functions $w_i(\Omega)$ are known as *weighting* functions. Inner products of this form appear on certain occasions, for example, that in which the base vectors are selected as the so-called Legendre polynomials.

8.1. Theory of Quantum Mechanics

The ideas of function space have penetrated deeply into contemporary physics; there is probably no more dramatic example of their fertility and influence on practical applications than the theory of quanta, in which the entire structure of the microscopic is assumed to be patterned after the structure of Hilbert space.† The latter, as investigated in this theory, is taken to be a *complex* vector space; such a space differs from the real vector spaces discussed earlier (and tacitly assumed elsewhere throughout this book) in that the scalars involved are complex numbers (i.e., of the form $\alpha + i\beta$, where i is the imaginary unit and α and β are real numbers). In view of this fact, the following changes are made in Group C of the axioms, defining a real Hilbert space:‡

$$
\text{(a)} \qquad (u, v) = \overline{(v, u)},
$$

$$
\text{(b)} \qquad (\alpha u_1 + \beta u_2, v) = \alpha(u_1, v) + \beta(u_2, v), \tag{8.49}
$$

$$
\text{(c)} \qquad (u, \alpha v_1 + \beta v_2) = \bar{\alpha}(u, v_1) + \bar{\beta}(u, v_2).
$$

Here, an overbar denotes the complex conjugate [e.g., $\bar{c} = \alpha - i\beta$ if $c = \alpha + i\beta$, and $\bar{u}(x) = a(x) - ib(x)$ if $u(x) = a(x) + ib(x)$]. We assume that the functions of interest are *complex-valued functions of real variables*. Clearly, both the self-inner product, $(u, u) = \overline{(u, u)}$, and the norm, $\|u\| = \sqrt{(u, u)}$, are real numbers, as is the modulus, $|c| = \sqrt{c\bar{c}} = +\sqrt{\alpha^2 + \beta^2}$, of the complex number $c = \alpha + i\beta$. The Hilbert-type inner product (8.2) is now altered, becoming

$$
(u, v) = \int_{-\infty}^{\infty} u(t)\bar{v}(t) \, dt, \tag{8.50}
$$

while the inner product (5.13) of two complex n-tuples $u = (u_1, \ldots, u_n)$ and $v = (v_1, \ldots, v_n)$ takes the form

$$
(u, v) = \sum_{k=1}^{n} u_k \bar{v}_k, \tag{8.51}
$$

thus making the associated space into the so-called "unitary" space.

Now, if L is a linear operator acting on a Hilbert space, then the vector v resulting from the operation of L on the vector u will, in general, be linearly

---

† In this comment, we restrict ourselves to a cursory examination of the salient points of the mathematical fabric of quantum mechanics. To this end, we accept certain simplifying assumptions, e.g., the system has one degree of freedom, the operators involved possess discrete spectra, and the system is treated as a nonrelativistic one. The reader interested in becoming more thoroughly acquainted with the subject should consult any of the standard treatises, such as: Mandl,[34] Messiah,[35] Kemble,[36] and Powell and Crasemann.[37] An excellent introduction to quantum mechanics is to be found in the very readable book of Gillespie.[38] Certain points of view adopted in this comment were inspired by the last-cited work.

‡ Compare Chapter 7, the text following equation (7.1).

independent of u. If it is not, for some $u \neq \theta$, that is, if there is a complex number λ and a nonzero vector u for which

$$Lu = \lambda u, \tag{8.52}$$

then the number λ is called an *eigenvalue* of L; any vector u as in (8.52) is called an *eigenvector* of L, associated with λ (see our remark at the end of Chapter 10).

The operators appearing in quantum mechanics are of the so-called *Hermitian* type. The "hermiticity" of the operators is expressed in the condition

$$(u, Lv) = (Lu, v) \qquad \text{for all } u, v. \tag{8.53}$$

On account of the first of the axioms (8.49), it should be clear that hermiticity is a direct generalization of the property of symmetry of operators, defined earlier for real vector spaces [see equation (8.17)]. To illustrate this property, let us determine when the simple operator of multiplication by a complex number c is Hermitian. We have, referring to equation (8.49),

$$\begin{aligned} (u, cv) &= \bar{c}(u, v), \\ (cu, v) &= c(u, v). \end{aligned} \tag{8.54}$$

Thus, the multiplication operator determined by c is Hermitian if $c = \bar{c}$, that is, if c is real. Two properties of Hermitian operators are of special importance. First, we observe that the *eigenvalues* of an Hermitian operator *are real*. In fact, by virtue of (8.52), we have

$$\begin{aligned} (u, Lu) &= (u, \lambda u) \\ &= \bar{\lambda}(u, u), \end{aligned} \tag{8.55a}$$

and similarly,

$$\begin{aligned} (Lu, u) &= (\lambda u, u) \\ &= \lambda(u, u). \end{aligned} \tag{8.55b}$$

By hermiticity, we infer that

$$\lambda(u, u) = \bar{\lambda}(u, u), \tag{8.56}$$

and, if $u \neq \theta$, so $\lambda = \bar{\lambda}$: λ is real, as asserted.

The second property of Hermitian operators consists of the statement that the *eigenvectors* associated with two different eigenvalues *are orthogonal*. Indeed, let u_1 and u_2 be eigenvectors, and let λ_1 and λ_2, respectively, be the corresponding distinct eigenvalues. Then,

$$\begin{aligned} (u_1, Lu_2) &= (Lu_1, u_2) \\ (u_1, \lambda_2 u_2) &= (\lambda_1 u_1, u_2). \end{aligned} \tag{8.57}$$

But, since the eigenvalues are real, then

$$(\lambda_1 - \lambda_2)(u_1, u_2) = 0, \tag{8.57a}$$

whence, because $\lambda_1 \neq \lambda_2$ by hypothesis, we arrive at

$$(u_1, u_2) = 0, \tag{8.58}$$

as claimed.

In quantum mechanics, it is assumed that the set of eigenvectors associated with an Hermitian operator is *complete*. This means that any vector in the underlying Hilbert space can be represented as a (perhaps infinite) sum of the eigenvectors.† Inasmuch as a set of eigenvectors corresponding to distinct eigenvalue and each of unit norm is orthonormal, it constitutes a basis for the space.

Proceeding now to a review of the essential features of quantum mechanics, we must mention three fundamental concepts appearing in the formal theory of this discipline. The first of these is the notion of the physical *state* of a system. It is assumed here that in knowing this state, we know everything that can possibly be known about the physical aspect of the system. The second fundamental concept is that of the *observables* of the system, the latter being simply dynamical variables associated with the system, such as position or momentum, or a function of these, e.g., the energy of the system.

The last idea involves the operation of *measurement* performed on the system. In classical understanding, an observable *has* always a definite value, and a measurement of an observable amounts simply to a *registering* of its current value. The viewpoint of quantum mechanics on this matter is radically different and surprising. According to this position, an observable has, generally speaking, *no* objective value independent of observer; its value is, in a sense, "developed" by the very act of measuring. In a similar vein, the *state* of a system, which in classical mechanics is *identified* with the current values of observables of the system (position and momentum, for example) becomes now clearly distinguished from the observables at hand. And so, while everything that is of interest about the physical aspects of a system is assumed to be obtainable from the associated state, other postulates are needed in order to clarify what things can be learned and how these facts can be deduced from the known state.

The just-characterized tenets of the formal aspects of quantum mechanics find their apparent reflection in the postulational basis of the theory. The latter begins with a central postulate asserting the existence of a one-to-one correspondence between the properties of a physical system and

---

† Compare the remarks on completeness in Chapter 7, preceding equation (7.22).

the properties of a Hilbert space. This means, in particular, the following: (1) Each *state* of a given system is represented by a vector, say, ψ, in a vector space. Any possible state ψ of the system belongs to the unit sphere, that is, $\|\psi\| = \sqrt{(\psi, \psi)} = 1$. This is often called the "normalization" condition. It is also assumed that two normalized vectors differing only by a scalar factor of unit modulus represent the same physical state: $\psi_1 \sim \psi_2$ if $\psi_2 = \alpha\psi_1$, where $|\alpha|^2 = 1$.[†] (2) To each observable, O_A, say, of a given system, there corresponds a linear Hermitian operator L_A in a Hilbert space, endowed with a complete orthonormal set of eigenvectors A_i, $i = 1, 2, \ldots$, and a corresponding set of eigenvalues $\lambda_i{}^A$, $i = 1, 2, \ldots$, so that

$$L_A A_i = \lambda_i{}^A A_i, \qquad i = 1, 2, \ldots . \tag{8.59}$$

Conversely, to each operator L_A with the foregoing properties in a Hilbert space, there corresponds an observable O_A of a physical system. Perhaps most surprising here is the additional postulate that no measurement of O_A can give values differing from the eigenvalues $\lambda_i{}^A$, $i = 1, 2, \ldots$. Of course, since the eigenvectors form an orthonormal basis for the space, any state vector can be represented by the linear combination

$$\psi_A = \sum_i (\psi_A, A_i)A_i, \tag{8.60}$$

where

$$(\psi_A, A_i) = \int_{-\infty}^{\infty} \psi_A(x)\overline{A_i(x)}\, dx, \tag{8.60a}$$

and x represents some argument of the functions involved. (3) The third fundamental axiom of quantum mechanics concerns the probability that a measurement of O_A will yield the eigenvalue $\lambda_i{}^A$. This probability[‡] is assumed to be given by $|(\psi_A, A_i)|^2$.

The foregoing discussion clearly illustrates how deeply the language and formulations of function spaces pervade the realm of quantum mechanics.[§] A more detailed treatment lying beyond the scope of this comment, we close with the following application.

Let a system with a single degree of freedom, in the form of a particle of mass m, move along the x-axis. It is postulated in quantum mechanics that

---

[†] It is easy to show that, in this case, also the norm of ψ_2 equals unity.

[‡] As already noted, we assume, for simplicity, that the spectrum of the operator is discrete. If the distribution of the eigenvalues is continuous, the matter becomes more involved. Compare, e.g., Gillespie (Ref. 38, Sec. 4.6b).

[§] Much is owed here to J. von Neumann's work.[39]

the observables, position x and momentum $p = mv(v = dx/dt)$, are associated with the operators (Ref. 38, p. 86)

$$L_x = x, \qquad (8.61a)$$

$$L_p = -i\hbar \frac{d}{dx}, \qquad (8.61b)$$

respectively, where $\hbar = h/2\pi$ and h is Planck's constant. It is postulated, moreover, that with every physical system, there is associated a *Hamiltonian* (or energy) operator L_H, corresponding to the observable total *energy* of the system. As such, it is a Hermitian operator possessing a complete orthonormal set of eigenvectors H_i and a corresponding set of real eigenvalues λ_i^H. Hence,

$$L_H H_i = \lambda_i^H H_i, \qquad i = 1, 2, \dots. \qquad (8.62)$$

For the one-degree-of-freedom system under investigation, the Hamiltonian operator is

$$L_H = \frac{1}{2m}(L_p)^2 + V(L_x), \qquad (8.63)$$

where V is a potential function. Combining equations (8.61) and (8.63) yields

$$L_H = -\frac{\hbar^2}{2m}\frac{d^2}{dx^2} + V(x). \qquad (8.64)$$

The foregoing operator is now inserted into equation (8.62) to give the celebrated so-called time-independent version of *Schrödinger's* equation,

$$-\frac{\hbar^2}{2m}\frac{d^2}{dx^2}H_i(x) + V(x)H_i(x) = \lambda_i^H H_i(x), \qquad i = 1, 2, \dots. \qquad (8.65)$$

This is, of course, the eigenvalue equation for the Hamiltonian operator L_H.

Problems

1. Show that if $f[x]$ is a linear functional (i.e., $f[x + y] = f[x] + f[y], f[\alpha x] = \alpha f[x]$, for any two vectors x and y and any scalar α), then $f[0] = 0$.

2. (Riesz representation theorem). Show that a linear functional $f[x]$ in $\mathscr{E}_n$ can be represented as a scalar product in the sense of equation (6.22).

3. Let the linear operator L in $\mathscr{E}_2$ transform the vector $x^1 = (12, 5)$ into $y^1 = (3, -4)$ and the vector $x^2 = (2, 2)$ into $y^2 = (8, 3)$. Find the vector $y^3 = Lx^3 = (y_1, y_2)$, where $x^3 = (14, 8)$.

4. The operation of one operator upon another, or the *composition* of two operators, is defined by the operation of the first operator on the output of the second. In $\mathscr{E}_2$, let the operator L_1 transform any vector $x = (x_1, x_2)$ into $y = (0, x_1)$, and let the operator L_2 transform any vector $\bar{x} = (\bar{x}_1, \bar{x}_2)$ into $\bar{y} = (\bar{x}_1, 0)$. Find the compositions of L_1 acting on L_2, $L_1 L_2$, and L_2 acting on L_1, $L_2 L_1$.

5. Decide whether the form $(x, y) = \int_0^t x(\tau)y(t - \tau)\, d\tau$ is, or is not, an inner product in the sense of Group C of the properties in Chapter 7.

6. Define the inner product

$$(u, v) = \frac{\mu}{2} \int_V \left[u_{i,j}(v_{i,j} + v_{j,i}) + \frac{2v}{1 - 2v} u_{i,i} v_{j,j} \right] dV.$$

Show: (a) that this inner product is symmetric; (b) that (u, u) is positive definite (cf. equations (8.15), (8.25), and Hooke's law $\tau_{ij} = 2\mu e_{ij} + \lambda e_{kk} \delta_{ij}$); (c) by appeal to (8.22) and the divergence theorem, derive the Rayleigh–Betti reciprocal theorem.

7. Let an operator L in $\mathscr{E}_n$ carry the orthonormal vectors $\{e_i\}$ into the orthonormal vectors $\{f_i\}$, according to the formula $f_i = Le_i$. Show that L (called a *unitary operator*) does not change the inner product of any two elements of $\mathscr{E}_n$, that is, $(Lx, Ly) = (x, y)$. In particular, $\|Lx\| = \|x\|$ (an operator possessing the latter property is called an *isometry*).

8. Suppose that L is a linear operator on an inner product space, for which the sequence $\{Lx^m\}$ converges to Lx whenever the sequence $\{x^m\}$ converges to x (this is just a criterion for the *continuity* of L). Show that if $\lim_{m \to \infty} x^m = x$ and $\lim_{m \to \infty} y^m = y$, then $\lim_{m, n \to \infty} (Lx^m, y^n) = (Lx, y)$.

9

Some Geometry of Function Space

To avoid punctiliousness, which would obscure rather than clarify the essentials, we agreed in Chapter 6† to make no distinction between a linear manifold and a subspace, giving both the common name "subspace." Practically, this amounts to assuming that every manifold of interest is closed.

It is not difficult to convince oneself that a subspace of a vector space is itself a vector space, inasmuch as it satisfies all the requirements imposed on the latter. We are thus in a position to determine the dimension of a subspace by the number of independent vectors, spanning the subspace. Of the countless examples of subspaces of function spaces, we cite only the following series, the members of which get successively smaller: the set of functions of x continuous at a single point of the interval $[x_a, x_b]$; the set of functions of x continuous in the entire interval $[x_a, x_b]$; the set of all polynomials in x with real coefficients; and, finally, the set of multiples αx of x with α being a real factor.

By its very definition, each subspace includes the zero vector. Yet, it is sometimes convenient to generalize the term and study "subspaces" devoid of the origin (i.e., of the zero vector). Such subspaces are said to be *translated*, or *shifted*, and the operation of converting a "regular" subspace space into a translated one is called a translation, or shift, of the subspace.‡

From a geometric point of view, a translation of a subspace $\mathscr{E}_1$ or $\mathscr{E}_2$ in the space $\mathscr{E}_3$, for instance, is equivalent to a parallel translation of a line or a plane passing through the origin to a new location. This is shown in Figure 9.1a, in which the line $\mathscr{S}_1$, passing through the origin O in the plane $\mathscr{S}_2$, is shifted parallel to itself to the location $\bar{\mathscr{S}}_1$.

---

† Compare the text preceding equation (6.18).

‡ Some authors employ the term "linear variety" instead of "translated subspace," e.g., Taylor.[11]

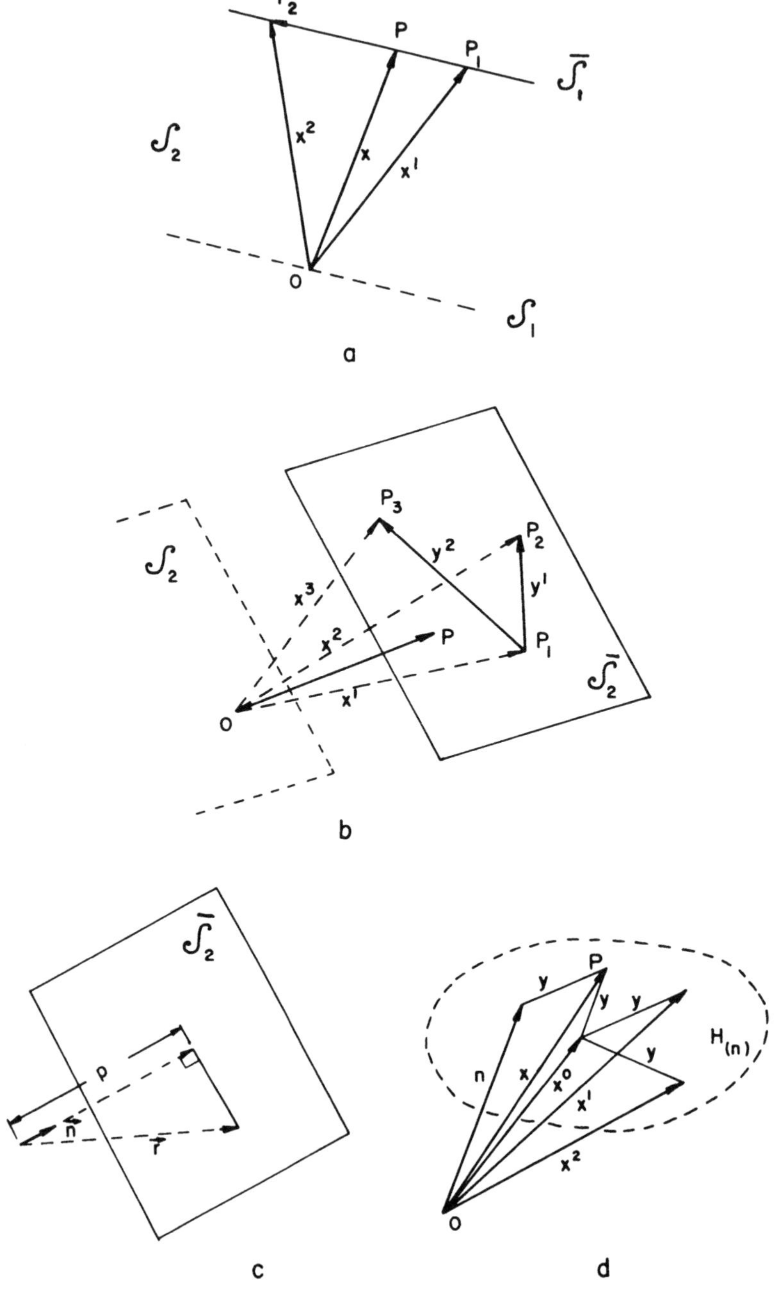

Figure 9.1. Illustration for subspaces.

At this stage, it is convenient to distinguish between two categories of function vectors: those having their seat *within* a given subspace, and those located *outside* the subspace, possibly serving as position vectors of points *in* the subspace.† To illustrate this idea, consider the following example.

A translated subspace $\mathscr{S}_1$ (Figure 9.1a) passes through the points P_1 and P_2 with linearly independent position vectors x^1 and x^2, respectively. Let x be the position vector of a generic point P on $\mathscr{S}_1$. Directly from Figure 9.1a,

$$\overrightarrow{P_1 P} = x - x^1, \qquad \overrightarrow{PP_2} = x^2 - x. \tag{9.1}$$

Since $\overrightarrow{P_1 P}$ and $\overrightarrow{PP_2}$ are collinear vectors, then $\overrightarrow{PP_2} = \alpha \overrightarrow{P_1 P}$, where $-\infty < \alpha < \infty$. Thus, from the preceding equations,

$$x = \frac{1}{\alpha + 1}(\alpha x^1 + x^2). \tag{9.2}$$

This can also be written as

$$x = \alpha_1 x^1 + \alpha_2 x^2, \qquad \text{where } \alpha_1 + \alpha_2 = 1. \tag{9.3}$$

As α varies from $-\infty$ to ∞, the extremity of the vector x sweeps the entire line $\mathscr{S}_1$. Hence, either of the equivalent equations (9.2) or (9.3) is the equation of $\mathscr{S}_1$ in vector form. With regard to the vectors x^1, x^2, and x, none of them lies on the line $\mathscr{S}_1$, so that all three might intuitively be considered as "outsiders," that is, as vectors *not in* $\mathscr{S}_1$, but in a space $\mathscr{S}_2$, of higher dimension than the given subspace $\mathscr{S}_1$. On the other hand, vectors such as $\overrightarrow{P_1 P_2}$ "reside" *in* the subspace $\mathscr{S}_1$ and are members of $\mathscr{S}_1$.

Some authors take a different point of view[41] in assuming that all vectors in a function space emanate from a common origin (i.e., are bound). In that case, the statement that a vector *belongs* to a translated subspace means that merely the *tip* of the vector is in this subspace. Consequently, the vectors x^1, x^2, and x^3 in Figure 9.1b are considered to lie *in* the subspace $\mathscr{S}_2$. While no doubt admissible, such an interpretation turns out to be inappropriate for our future studies. Of course, the analytical aspect of the matter remains unaffected by this, or other, geometric illustration.

Returning to the question of "*intrinsic*" and "*extrinsic*" vectors with respect to the subspace $\mathscr{S}_1$, we note that the "extrinsic" representation (9.3) of $\mathscr{S}_1$ can be converted into a kind of "semi-intrinsic" representation by means of the "intrinsic" vector

$$y^1 = x^2 - x^1. \tag{9.4}$$

---

† We follow here the classification of Synge (Ref. 40, Sec. 2.4).

This gives

$$x = x^1 + \alpha_2 y^1, \tag{9.5}$$

where now y^1 is an intrinsic vector and the coefficient α_2 is arbitrary. If the space origin O is shifted, say, to the point P_1, then $x^1 = 0$, and the equation of the line $\mathscr{S}_1$ acquires a totally "intrinsic" character, $x = \alpha_2 y^1$.

Just as equations (9.3) and (9.5) are two alternative representations of a translated one-dimensional subspace $\mathscr{S}_1$, so also are the equations

$$x = \alpha_1 x^1 + \alpha_2 x^2 + \alpha_3 x^3, \qquad \alpha_1 + \alpha_2 + \alpha_3 = 1, \tag{9.6}$$

where $\{x^1, x^2, x^3\}$ is a set of linearly independent vectors, and

$$x = x^1 + \alpha_2 y^1 + \alpha_3 y^2 \tag{9.7}$$

are two alternative representations of a translated two-dimensional subspace $\mathscr{S}_2$ (Figure 9.1b). In equation (9.7), α_2 and α_3 are arbitrary coefficients, and y^1 and y^2 are two "intrinsic" linearly independent vectors.

Carrying our generalization a step further, we assign to a *translated subspace* $\mathscr{S}_n$ of dimension n the pair of equations

$$x = \alpha_1 x^1 + \alpha_2 x^2 + \cdots + \alpha_{n+1} x^{n+1}, \tag{9.8}$$

$$\alpha_1 + \alpha_2 + \cdots + \alpha_{n+1} = 1, \tag{9.8a}$$

where $\{x^1, \ldots, x^{n+1}\}$ is a set of linearly independent vectors or, alternatively,

$$x = x^1 + \alpha_1 y^1 + \alpha_2 y^2 + \cdots + \alpha_n y^n, \tag{9.9}$$

where, in the last equation, the coefficients α_i are unrestricted and the y^i's are linearly independent "intrinsic" vectors. The translated subspace $\mathscr{S}_n$ itself is imagined in the first case as passing through the tips of the vectors x^k, $k = 1$, $2, \ldots, n + 1$, and in the second case to include the vectors y^k, $k = 1, 2, \ldots, n$.

Up to this point, our analysis has involved translated subspaces of *finite* dimensions which, by analogy to ordinary planes, might be visualized as *many-dimensional "planes"* and said to be *translated n-planes*. Examples of *translated* subspaces of *infinite* dimension are provided by *hyperplanes, $H_{(n)}$*, where the subscript $n = 1, 2, \ldots$ does not stand for the dimension, but for the so-called *class* of the hyperplane. As in n-spaces, hyperplanes are reminiscent of ordinary planes. Their definition, however, cannot be found in equations such as (9.8) or (9.9), but in the so-called normal equation of a plane (Figure 9.1c),

$$\vec{r} \cdot \vec{n} = p, \tag{9.10}$$

where $\vec{r}$ is a position vector, $\vec{n}$ is a unit vector normal to the plane, and p is the distance from the origin to the plane. Consider now any Hilbert space. The defining equation of a hyperplane of the *first class*, $H_{(1)}$, is

$$(x, x^1) = \alpha_1, \tag{9.11}$$

where x^1 is a fixed vector $x^1 \neq \theta$ and α_1 is a fixed scalar (Figure 9.1d). Although modeled on equation (9.10), the last equation has a much broader sense. First, the selected vector x^1 is not necessarily perpendicular to the hyperplane; second, the scalar α_1 is not necessarily the distance from the origin to $H_{(1)}$.

The *infinite* dimensionality of $H_{(1)}$, as compared to the finite dimensionality of a translated n-plane, $\mathscr{S}_n$, follows from the defining equation. Indeed, for n preassigned function vectors y^k lying in $\mathscr{S}_n$, an equation such as (9.9) determines completely the functions x populating this subspace. On the other hand, the class of functions x defined by equation (9.11) is so extensive that its complete representation requires an infinite number of linearly independent vectors. This is easily shown by assuming that, for instance, the function vector x^1 is constant and its value is equal to α_1. Suppose, for definiteness, that we are interested in the space of square integrable functions,† $\mathscr{L}_2$. Then, with x denoting a function $f(x)$, say, equations (8.2) and (9.11) yield

$$\int_a^b f(x)\, dx = 1 \tag{9.12}$$

as the single condition imposed on $f(x)$. It is obvious that this condition is satisfied by an *infinite* number of linearly independent functions of a single variable, that is a representation of the function vectors in $H_{(1)}$ requires an infinite number of linearly independent functions. Accordingly, the dimension of $H_{(1)}$ is infinite.

An object more general than $H_{(1)}$ is a hyperplane $H_{(n)}$, of order $n > 1$, defined by the system of equations

$$(x, x^\nu) = \alpha_\nu, \qquad \nu = 1, 2, \ldots, n, \tag{9.13}$$

where the x^ν's are vectors considered to be "extrinsic" with respect to $H_{(n)}$, their tips marking points in $H_{(n)}$. The vectors of the set $\{x^\nu\}$ are assumed to be linearly independent.

Hyperplanes, because they contain a whole straight line if they contain two points of the line, are reminiscent of plane-like structures. Inasmuch as containing the whole straight line through any two of its points is a stronger requirement than *convexity* of a system (mentioned before in Chapter 6), hyperplanes are *a fortiori* convex structures.

---

† Compare the footnote referring to the text preceding equation (10.19), infra.

We note in passing that any linear manifold is convex (Ref. 11, p. 130). The converse, however, is not always true. This follows, for example, from the second of equations (9.3), preventing free choice of the coefficients.

As we have seen in Chapter 5, it is always possible to replace a set of linearly independent vectors, say $\{x^v\}$, by an orthonormal set, $\{i^v\}$, derived from the former. After such an operation, equations (9.13) become (with possibly different α_v's)

$$(x, i^v) = \alpha_v, \qquad v = 1, 2, \ldots, n. \tag{9.14}$$

For a selected vector x^0 and a generic vector x, each with its tip in $H_{(n)}$, the vector (cf. Figure 9.1d)

$$y = x - x^0, \tag{9.15}$$

joining the extremities of x and x^0, lies in $H_{(n)}$. However, both x and x^0 obey conditions (9.14), so that

$$(y, i^v) = 0, \qquad v = 1, 2, \ldots, n. \tag{9.15a}$$

This leads immediately to the equations

$$x = x^0 + y \quad \text{and} \quad (y, i^v) = 0, \qquad v = 1, 2, \ldots, n, \tag{9.16}$$

as an alternative description of a hyperplane $H_{(n)}$. Figure 9.1d depicts the situation. It is seen that the positions of space points, such as the point P, are identified by the tips of the intrinsic vectors y radiating from the tip of an extrinsic vector x^0. This representation is reminiscent of the representation (9.9) of space points in $\mathscr{S}_n$.

In carrying out our analysis, we have repeatedly used the concepts of *perpendicularity* of a vector to, and the *distance* of a point from, a subspace—without appropriate definition of either. It is now necessary to make up for these omissions, assuming at the start that the spaces considered are inner product spaces.

We first recall that, in the ordinary three-space, a vector can be resolved into two components, one perpendicular and the other parallel to a given plane; similarly, a vector in a linear space can be decomposed into two vectors: one perpendicular to a given subspace, and another lying in the subspace. Let us first consider plane-like structures, namely, an n-subspace in the strict sense, $\mathscr{S}_n$, a translated subspace, $\bar{\mathscr{S}}_n$, and a hyperplane, $H_{(n)}$. To illustrate, we concentrate our attention on a subspace $\mathscr{S}_n$ represented by equation (9.9). In this case, it is convenient to replace the set $\{y^k\}$ by a set of intrinsic orthonormal vectors $\{i^k\}$,

$$x = x^0 + \sum_{k=1}^{n} \beta_k i^k, \tag{9.17}$$

with x^1 replaced by x^0.

As already noted, the sense of the preceding equation is that a vector x, extrinsic with respect to the given n-subspace $\bar{\mathscr{S}}_n$, can be represented by a sum of two vectors: one, x^0, not in $\mathscr{S}_n$, and another represented by a linear combination of the intrinsic vectors i^k. It should be clear that the choice of the vector x^0 is left to our discretion, and nothing prevents us from making it *perpendicular* to the vectors i^k. We then set

$$x^0 \equiv n \tag{9.18}$$

and require that

$$(x^0, i^k) \equiv (n, i^k) = 0, \qquad \text{for } k = 1, 2, \ldots, n. \tag{9.19}$$

We next form the inner product of equation (9.17) with a generic vector i^k, obtaining

$$\beta_k = (x, i^k), \qquad k = 1, 2, \ldots, n, \tag{9.20}$$

so that, finally,

$$x = n + \sum_{k=1}^{n} (x, i^k) i^k. \tag{9.21}$$

In this equation, n is a vector orthogonal to $\mathscr{S}_n$ and the sum represents a vector in $\mathscr{S}_n$. From this point of view, it is indeed natural to interpret the vector sum

$$\sum_{k=1}^{n} (x, i^k) i^k \tag{9.22}$$

geometrically as the *orthogonal projection* of x on $\mathscr{S}_n$. The particular terms in this sum, $(x, i^k) i^k$, play the part of the orthogonal projections of x on the i^k-axes (in a vectorial sense), while the products (x, i^k) are the scalar values of these projections.

It is now tempting to extend the concept of orthogonality to subspaces of infinite dimension. Examples of these are hyperplanes represented by the system of equations (9.16), where, again, n denotes the order of the hyperplane (and not its dimension!). The set $\{i^\nu\}$ appearing in the system (9.16) consists of vectors *extrinsic* with respect to the hyperplane $H_{(n)}$, and it would seem natural to represent the vector x^0 in (9.16) by a linear combination of the orthonormal vectors i^ν,

$$x^0 = \sum_{\nu=1}^{n} \gamma_\nu i^\nu \tag{9.23}$$

with γ_ν as coefficients.

The question now arises concerning whether the vector x^0 defined in this manner is actually the position vector of a point in $H_{(n)}$. If this is so, then x^0 must satisfy equations (9.14) defining the hyperplane. Substitution of expression (9.23) into equations (9.14) convinces us that this actually is the case provided that $\gamma_v = \alpha_v$ for every $v = 1, 2, \ldots, n$. Thus, the equations (9.16) defining $H_{(n)}$ are

$$x = x^0 + y, \qquad (y, i^v) = 0, \tag{9.24}$$

where $v = 1, 2, \ldots, n$, and

$$x^0 = \sum_{v=1}^{n} \alpha_v i^v. \tag{9.24a}$$

The last two of the preceding equations give

$$(x^0, y) = \sum_{v=1}^{n} (y, i^v) = 0, \tag{9.25}$$

so we can assert that

$$x^0 \equiv n, \tag{9.25a}$$

where n is a vector orthogonal to a vector y in $H_{(n)}$. It is now possible to assume that the vector $x^0 \equiv n$ in equation (9.24) is fixed, while the vectors x and y are arbitrary, the first extrinsic and the second intrinsic. It follows that the vector n is orthogonal to *every* vector y in $H_{(n)}$ and, consequently, *orthogonal* to $H_{(n)}$. By analogy with equation (9.22), one can now think of the vector y, defined by equations (9.24) and (9.25), as the orthogonal projection of x on $H_{(n)}$. In this way, we have twice resolved a vector into two components: one perpendicular and one "parallel" to an n-linear space [equations (9.21) and (9.22)], and one perpendicular and one "parallel" to a hyperplane [equations (9.24) and (9.25)].

We can now generalize our results in considering the *orthogonality* of a *vector* to an *arbitrary subspace*, independent of its structure and dimensionality. Denoting the subspace by $\mathscr{S}$, it is said that a *vector* x is *orthogonal* to the *subspace* $\mathscr{S}$, written $x \perp \mathscr{S}$, if the vector is orthogonal to every vector in $\mathscr{S}$.

With this definition in mind, we turn our attention to the important question of the orthogonality of two subspaces. We say that *two subspaces*, $\mathscr{S}'$ and $\mathscr{S}''$, are *orthogonal* to each other, denoted by $\mathscr{S}' \perp \mathscr{S}''$, if $x' \perp x''$ for every x' in $\mathscr{S}'$ and every x'' in $\mathscr{S}''$. Examples of mutually orthogonal subspaces are provided by two perpendicular lines, by a plane and a line perpendicular to it, but not by two perpendicular planes, as it is easy to conclude.

As still another example, consider two finite-dimensional subspaces, $\mathscr{S}_m$ and $\mathscr{S}_n$, in the form (9.17), say,

$$x = x_0 + \sum_{k=1}^{m} \beta_k i^k$$

$$\bar{x} = \bar{x}_0 + \sum_{k=1}^{n} \bar{\beta}_k j^k,$$

(9.26)

where $\{i^k\}$ and $\{j^k\}$ are two respective orthonormal bases. If $\mathscr{S}_m \perp \mathscr{S}_n$, then each vector $y = \sum_{k=1}^{m} \beta_k i^k$ lying in $\mathscr{S}_m$ must be perpendicular to each vector $\bar{y} = \sum_{k=1}^{n} \bar{\beta}_k j^k$ lying in $\mathscr{S}_n$,

$$(y, \bar{y}) = 0 = \left(\sum_{k=1}^{m} \beta_k i^k, \sum_{k=1}^{n} \bar{\beta}_k j^k \right).$$

(9.27)

The preceding equation must hold for any choice of $m + n$ coefficients β_k and $\bar{\beta}_k$. It follows that

$$(i^\rho, j^\sigma) = 0$$

(9.27a)

for all $\rho = 1, 2, \ldots, m$ and all $\sigma = 1, 2, \ldots, n$, producing the conditions for orthogonality of $\mathscr{S}_m$ and $\mathscr{S}_n$.

Returning to the general case, let $\mathscr{S}$ be a subspace of a Hilbert space, $\mathscr{H}$. It can be shown that every element x in $\mathscr{H}$ can be represented in a *unique* way in the form (Ref. 23, p. 158)

$$x = n + y,$$

(9.28)

where n is a vector orthogonal to $\mathscr{S}$, $n \perp \mathscr{S}$, and y is called the *orthogonal projection* of x on $\mathscr{S}$. It is demonstrated that all vectors n with $n \perp \mathscr{S}$ form a subspace, $\mathscr{S}^\perp$, called the *orthogonal complement* of $\mathscr{S}$, such that

$$\mathscr{H} = \mathscr{S} \oplus \mathscr{S}^\perp,$$

(9.29)

where $\oplus$ denotes the so-called *direct sum*† of $\mathscr{S}$ and $\mathscr{S}^\perp$; equation (9.29) is known as a *decomposition* formula for the Hilbert space. In words, this formula expresses the important fact that a Hilbert space can be decomposed into two orthogonal subspaces (without any remainder!). We shall make use of this conclusion in Chapter 12 during the course of our discussion of the hypercircle method (Ref. 42, p. 450).

---

† A vector space $\mathscr{V}$ is said to be the *direct sum* of two of its subspaces $\mathscr{S}'$ and $\mathscr{S}''$ if every vector z in $\mathscr{V}$ can be written uniquely as $z = x + y$ where $x \in \mathscr{S}'$ and $y \in \mathscr{S}''$.

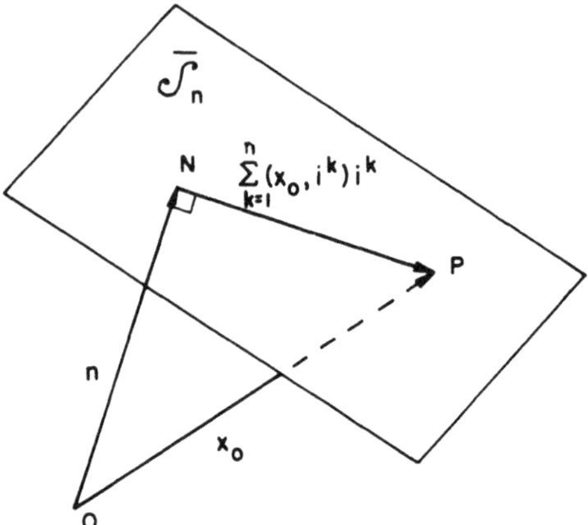

Figure 9.2. Illustration for orthogonal projection of x_0.

We now wish to show that a normal to a subspace of *finite dimension*, $\mathscr{S}_n$, drawn from a point outside $\mathscr{S}_n$ has the property of being the *shortest distance* from the point to the subspace.

Represent the subspace in the form (9.17) and write

$$\|x\|^2 = (x, x)$$

$$= (x^0)^2 + 2 \sum_{k=1}^{n} \beta_k (x^0, i^k) + \sum_{k=1}^{n} \beta_k^2. \tag{9.30}$$

Finding the shortest distance of $\mathscr{S}_n$ from the origin is equivalent to minimizing the foregoing expression with respect to the coefficients β_k, upon keeping the vector x^0 fixed. This yields

$$\beta_k = -(x^0, i^k), \qquad k = 1, 2, \ldots, n. \tag{9.31}$$

Consequently, the vector n minimizing (9.30) is

$$n = x^0 - \sum_{k=1}^{n} (x^0, i^k) i^k. \tag{9.32}$$

This coincides with the result (9.21) and implies that the line segment joining the origin with the closest point of $\mathscr{S}_n$ is normal to $\mathscr{S}_n$, while the vectorial sum in (9.32) represents the *orthogonal* projection of x^0 on $\mathscr{S}_n$ (Figure 9.2).

What was just proved for finite-dimensional n-subspaces is also true for hyperplanes. To show this, take the inner product of equation (9.24) with itself (for $x^0 = n$) to obtain

$$\|x\|^2 = \|n\|^2 + \|y\|^2, \tag{9.33}$$

after making use of condition (9.25). Clearly, $\|x\|$ reaches its minimum for $\|y\| = 0$, when it becomes equal to $\|n\|$. Thus, the normal n to $H_{(n)}$ [equation (9.25a)] is the shortest "line segment" joining the origin with the hyperplane.

In addition to planelike structures (n-subspaces, hyperplanes), it is often useful to consider subsets having geometric structure similar to that of a *sphere*. These are the *finite-dimensional spheres*, denoted here generically by $\mathscr{B}_n$, and their generalization to infinite dimensions, the *hyperspheres*, $\mathscr{B}_\infty$.

An *n-sphere*, $\mathscr{B}_n$, is a set in $\mathscr{E}_n$ with the property that each of its points is at the same distance from a fixed point, selected as the center of the sphere. By the very definition, an n-sphere is a direct idealization of an ordinary sphere in Euclidean three-space. Its two- and one-dimensional counterparts are a circle and a pair of points, respectively. The equation of an n-sphere with center at the space origin is

$$\|x\| = R^2, \tag{9.34}$$

where R is the radius of the sphere. By virtue of equation (9.17), and bearing in mind the location of the origin, we conclude that equation (9.34) is equivalent to the pair of equations,

$$x = \sum_{k=1}^{n} \beta_k i^k \tag{9.35}$$

and

$$\beta_1{}^2 + \beta_2{}^2 + \cdots + \beta_n{}^2 = R^2. \tag{9.36}$$

While an n-sphere is reminiscent of an ordinary sphere, it would be hazardous to interpret this similarity in a literal sense. To illustrate, consider a two-space, $\mathscr{E}_2$, including a two-sphere, that is, a circle $\mathscr{B}_2$ (Figure 9.3a). Evidently, the circle divides the space $\mathscr{E}_2$ into two separate regions: the inside and the outside of $\mathscr{B}_2$. A three-dimensional being, however, would conclude that this separation is not as definite as it seems to a two-dimensional being residing in $\mathscr{E}_2$. Clearly, it is possible to pass continuously from the inside to the outside of the circle by marching into the space surrounding $\mathscr{B}_2$. With reference to a three-space, therefore, the difference between the interior and the exterior of $\mathscr{B}_2$ becomes irrelevant. In a similar sense, a pair of points representing a one-sphere fails to enclose a line segment in $\mathscr{E}_1$. In general, we can state that any n-sphere, observed from a surrounding space of dimension higher than n, fails to provide an impervious enclosure, although it provides such an enclosure in the n-space.

If an *infinite-dimensional space*, the matter is different, for if we wish to isolate a portion of such a space, we must utilize a *hypersphere*, $\mathscr{B}_\infty$, as one that, being an infinite-dimensional structure, is able to provide an actual enclosure. A hypersphere as such contains the space origin; if it does not, it is called a *translated* hypersphere.

A question may now arise as to whether an abstract construct, such as a hypersphere, may serve any practical purpose. The discussion in Chapter 12, devoted to the method of the hypercircle, should convince the reader that the answer is in the affirmative. Indeed, it is shown that the concept of a hypersphere is of real help in arriving at approximate solutions to involve boundary-value problems. At this point, it should suffice to give the following example.

Let it be required to find an approximate solution to a boundary-value problem to within a prescribed accuracy, δ, say. Inasmuch as a function in a function space is represented by a vector or, specifically, by the tip of the vector, the problem to be examined may be formulated as follows: it is required to find a point P_{approx} in the neighborhood of the point P_{exact} representing the exact solution, the radius of the neighborhood, δ, being prescribed. Clearly, the location of P_{exact} in unknown, but it may be assumed

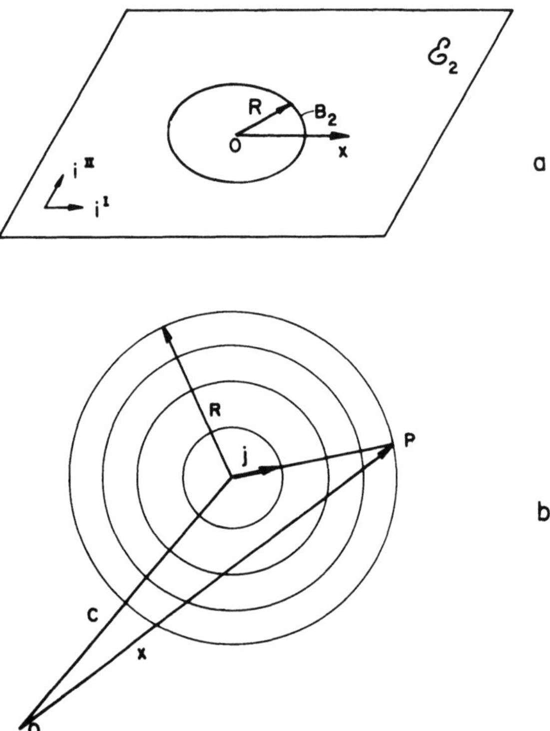

Figure 9.3. Spheres and hyperspheres.

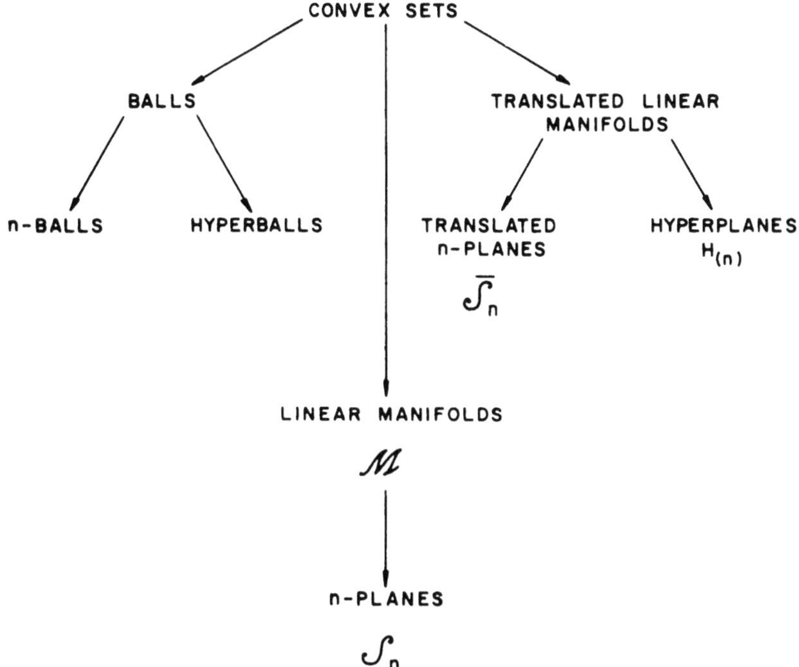

Figure 9.4. Certain classes of subspaces.

that one is seeking a function belonging to a certain class, such as the class of continuous functions. Since the space of such functions is infinite-dimensional, the δ-neighborhood in question must be represented by a hypersphere of radius δ in order to provide an uninterrupted enclosure.

Now let c be the position of the center of a hypersphere and let R be its radius. The equation of the hypersphere is then

$$\|x - c\|^2 = R^2, \tag{9.37}$$

and it divides an infinite-dimensional space into an *inside*, $\|x - c\| < R^2$, and an *outside*, $\|x - c\| > R^2$.

In the present case, there is no simple expression, similar to that given by equation (9.35) for a point on an n-sphere, for the position vector of a point on the hypersphere. However, if j is an arbitrarily directed unit vector emanating from the center of the hypersphere, the equation alternative to equation (9.37) is

$$x = c + Rj, \qquad \|j\| = 1. \tag{9.38}$$

Clearly, if the exact solution to a problem is represented by the tip of the vector c and $R = \delta$ is the admissible error, then every point such that $\|x - c\| \leq \delta$ corresponds to an acceptable approximate solution.

The hypercircle method, mentioned above, narrows the search for an approximate solution by the requirement that the representation point of the latter lie on a hypercircle, i.e., on the intersection of a hypersphere with a hyperplane.

Instead of pursuing this matter further at present, we conclude this chapter with a diagram (Figure 9.4) displaying the relations among certain families of convex subsets, arranged according to the increasing complexity of their geometric structure. By a *ball* is meant here a "solid" sphere, that is, a

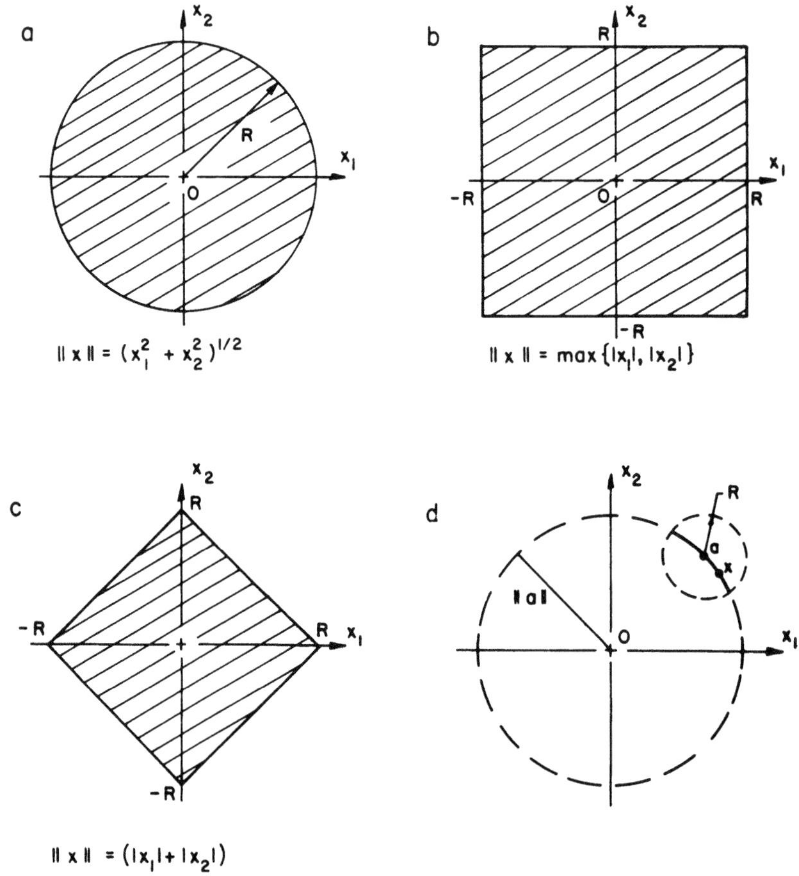

Figure 9.5. Closed balls (disks) $\|x\| \leq R$.

spherical "boundary," plus the interior. For example, $\|x - c\| \leq R$, instead of (9.37), is the equation of a hyperball. In contrast to such a closed ball, an open ball ($\|x - c\| < R$) includes only interior points. It is important to note that balls assume various geometrical aspects, depending on the choice of norm (or metric). For example, in a Euclidean plane x_1, x_2, a closed ball (disk) $\|x\| \leq R$ becomes the interior plus the circumference of a circle or one of two squares, depending on whether one defines the norm by $\|x\| = (x_1{}^2 + x_2{}^2)^{1/2}$, $\|x\| = \max\{|x_1|, |x_2|\}$, or $\|x\| = (|x_1| + |x_2|)$, respectively; here $x \equiv x_1$, x_2 (Figure 9.5). Note that in an abstract sense, the distance of the point A, for example, from the center of the ball in Figure 9.5b is R, not $R/(2)^{1/2}$, as would appear from our geometrical intuition.

A somewhat surprising result is obtained if one considers an open ball (open disk) on a Euclidean plane, the center of the ball being at a point $a \neq \theta$ and its radius $R < \|a\|$. Let the distance function in the plane be defined by (Ref. 43, p. 133)

$$\delta(a, x) = \begin{cases} \|a\| + \|x\| & \text{if } \|x\| \neq \|a\|, \\ \|a - x\| & \text{if } \|x\| = \|a\|, \end{cases}$$

where $\|x\|$ is the usual Euclidean norm. Now if $\|x\| \neq \|a\|$, then, by definition, $\delta(a, x) = \|a\| + \|x\| < R$ or $\|x\| < R - \|a\| < 0$, which is impossible. On the other hand, if $\|x\| = \|a\|$, then, the point x lies on the circle with radius $\|a\|$ and, since $\|a - x\| < R < \|a\|$, the ball reduces to an arc on the circle (Figure 9.5, heavy line).

Problems

1. Using the infinite sequence $\{f^n(t) = t^n\}$, show that the dimension of the space $C_{-1 \leq t \leq 1}$ is infinite.

2. Show that a finite-dimensional Hilbert space is the direct sum of any subspace $\mathcal{S}$ and its orthogonal complement $\mathcal{S}^\perp$ [cf. (9.29)].

3. Verify that if the norms of two vectors in an inner product space are equal, then their sum and difference are mutually orthogonal. Give a geometric interpretation.

4. Show that the dimension of the direct sum of two finite-dimensional subspaces equals the sum of the dimensions of the subspaces.

5. Show that $\mathscr{E}_3$ is the direct sum of any coordinate plane and the remaining coordinate axis.

6. Show that the subspaces of the preceding exercise are mutually orthogonal.

7. Let $\mathscr{S}$ be the (two-dimensional) subspace of $\mathscr{E}_4$ with base vectors $f = (f_1, f_2, f_3, f_4)$ and $g = (g_1, g_2, g_3, g_4)$. Find a basis for the orthogonal complement $\mathscr{S}^\perp$ of $\mathscr{S}$. Take $f = (2, 4, 6, 4)$ and $g = (4, 8, 14, -2)$.

8. Let $Pr_{\mathscr{S}} x$ denote the orthogonal projection of x on a subspace $\mathscr{S}$ in $\mathscr{H}$ (see equation (13.3c) infra). Verify the reciprocal relation $(Pr_{\mathscr{S}} x^1, x^2) = (x^1, Pr_{\mathscr{S}} x^2)$.

10

Closeness of Functions. Approximation in the Mean. Fourier Expansions

It is well known that most problems of applied mechanics either cannot be solved rigorously, or the effort to obtain such solutions is too great to justify the work expended. Under these circumstances, there is no choice but to employ methods leading to approximate solutions. The question then arises concerning the evaluation of the accuracy of an approximate solution, that is, the *closeness* of the approximating function-vector to the exact solution-vector.

If the quantities under comparison are numbers, a natural measure of their closeness is the absolute value of their difference. If, on the other hand, the quantities to be compared are functions, the values of which vary from point to point, a decision on how to assess their nearness is not as straightforward. To define the closeness of two functions of position, say $f(P)$ and $F(P)$, by the absolute value of their difference, $|f(P) - F(P)|$, without indicating the point of comparison is meaningless. A more appropriate criterion is the *maximum* deviation of the functions from each other,

$$d = \max_{P \in \Omega} |f(P) - F(P)|, \tag{10.1}$$

where P varies over the common domain of definition of these functions, Ω. It is clear that, in general, the smaller the value of d, the better one function approximates the other. Specifically, if $d < \varepsilon$, where ε is a positive number, usually small, then we say that the function $f(P)$, for instance, *approximates* the function $F(P)$ *uniformly* with accuracy ε:

$$|f(P) - F(P)| < \varepsilon \qquad \text{for all } P \text{ in } \Omega. \tag{10.2}$$

In a geometric interpretation, the criterion of uniform approximation expresses the fact that the differences of the corresponding (infinitely many) values $f(P_i)$ and $F(P_i)$ of the components of the vectors $f(P)$ and $F(P)$ remain less than ε; here, P_i is a generic point in the domain Ω.

To achieve a uniform approximation in practical applications is often tedious. Moreover, there always arises the question as to whether such an approximation is the most desirable. Figure 10.1 illustrates the situation for a function of a single variable. The function $F(x)$ is to be approximated in the interval $[a, b]$; functions $f_1(x)$ and $f_2(x)$ are the approximating functions and ε denotes the permissible maximum deviation. It is apparent that, in a global sense, the function $f_2(x)$—violating the condition of the uniform approximation—represents the function $F(x)$ "better" than the uniformly approximating function $f_1(x)$. This happens despite the fact that, in a small vicinity of the point $x = c$, the function $f_2(x)$ displays a relatively large deviation from $F(x)$.

Instead of a uniform approximation, it is frequently more practical to resort to the so-called *approximation in the mean*. This type of approximation of a function has some features reminiscent of the mean of a set of numbers. We recall that the *mean of order p*, $\bar{\alpha}(p)$, of a set of numbers α_i, $i = 1, 2, \ldots, n$, is defined by

$$\bar{\alpha}(p) = \left[\frac{1}{n} \sum_{k=1}^{n} (\alpha_k)^p\right]^{1/p}. \tag{10.3}$$

For $p = 1$ we have the *arithmetic mean*, while $p = 2$ yields the *mean square*. The arithmetic mean has the obvious defect of being misleading if the numbers involved have different signs so that they cancel each other. A similar

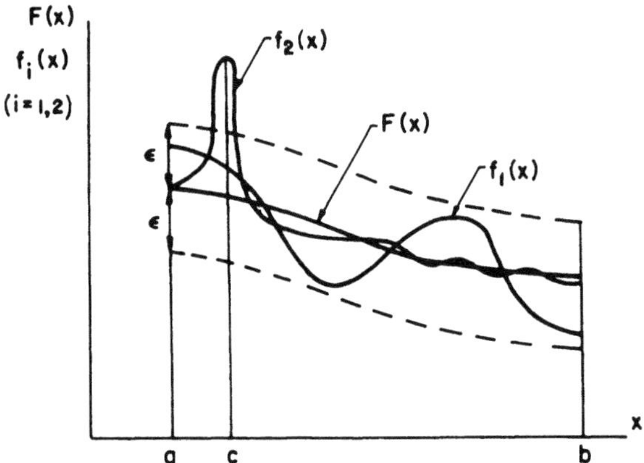

Figure 10.1. Approximation of the function $F(x)$.

remark applies to all odd p. Of the even values of p, the order $p = 2$ is the simplest and most convenient, and is used almost exclusively.

The idea of the mean-square deviation of two collections of n numbers $\{\alpha_i\}$ and $\{\beta_i\}$,

$$\left[\frac{1}{n}\sum_{i=1}^{n}(\alpha_i - \beta_i)^2\right]^{1/2}, \tag{10.4}$$

is easily extended to the *mean-square deviation* of functions. In the latter case, the summation is replaced by an integration, the formula (10.4) becoming

$$\bar{d} = \left[\frac{1}{\Omega}\int_{\Omega}[f(P) - F(P)]^2\, d\Omega\right]^{1/2}. \tag{10.5}$$

In order to clarify the relation between the mean-square deviation of functions and the distance between functions, we write, referring to equation (7.13a),

$$d(f, F) = \|f - g\|, \tag{10.6}$$

or, alternatively,

$$d(f, F) = (f - F, f - F)^{1/2}. \tag{10.7}$$

For definiteness, we use the Hilbert inner product (8.2), so that the preceding equation becomes

$$d(f, F) = \left[\int_{\Omega}[f(P) - F(P)]^2\, d\Omega\right]^{1/2}. \tag{10.8}$$

The right-hand sides of equations (10.5) and (10.8) are identical except for the multiplicative factor $(1/\Omega)^{1/2}$. By convention, we suppress this factor and call the right-hand side of equation (10.8) the *mean-square* deviation of the functions $f(P)$ and $F(P)$. In a geometric interpretation, therefore, the *mean-square* deviation acquires the same meaning as the *distance* between functions (here with respect to the Hilbert metric). It is an adequate measure of *closeness*, often called briefly the *mean distance* between functions:

$$\|f - F\|_M = \left[\int_{\Omega}[f(P) - F(P)]^2\, d\Omega\right]^{1/2}. \tag{10.9}$$

As repeatedly stressed in Chapter 8, there exist intimate connections between an inner product and the physical aspect of the corresponding problem. Specifically, the norm

$$\||u_0\||_H = (Lu_0, u_0)_H^{1/2}$$

$$= \left[2\int_{V} W(u_0)\, dV\right]^{1/2} \tag{10.10}$$

of the exact solution, u_0, to an elastic problem, with the positive-definite[†] operator, L, acting on u_0, was in Chapter 8 found to represent, in some scale, the stored potential energy, $W(u_0)$.[‡] The preceding equation leads to a special definition of closeness, which we denote by

$$\||f - F\||_H = (L(f - F), (f - F))_H^{1/2}, \tag{10.11}$$

or, explicitly, by

$$\||f - F\||_E = \left[\int_V ((f - F), L(f - F))\, dV \right]^{1/2}, \tag{10.12}$$

after replacing the subscript H by E. We then say that $\||f - F\||_E$ represents the *energy distance* between the functions. Accordingly,[§]

$$\||u\||_E = \left[\int_\Omega (u, L(u))\, d\Omega \right]^{1/2} \tag{10.13}$$

is said to represent the *energy norm* of a function $u(P)$, $P\varepsilon\Omega$.

There is an obvious connection between the notion of *distance* (or *closeness*) and the idea of convergence understood as a diminution of distance between a sequence of functions, say f_n, and a given function f.[¶] With the three types of distance examined above [equations (10.2), (10.9), and (10.12)], we associate three types of convergence: *uniform convergence, convergence in the mean*, and *energy convergence*.

It is said that the sequence $f_n(P)$, $n = 1, 2, \ldots$, converges to $f(P)$ *uniformly* over a domain Ω if for each $\varepsilon > 0$, no matter how small,

$$| f_n(P) - f(P) | < \varepsilon \qquad \text{for all } n > n(\varepsilon), \tag{10.14}$$

provided the number $n(\varepsilon)$, independent of P, is large enough. Moreover, this condition has to be satisfied for all P in Ω. We can express this fact differently by saying that if with increasing n,

$$\max | f_n(P) - f(P) | \to 0, \tag{10.15}$$

P in Ω, then

$$f_n(P) \overset{U}{\to} f(P), \tag{10.15a}$$

where the appended U denotes uniform convergence. Uniform convergence is a stringent requirement imposed on a sequence of functions. It is often

---

† See the definition (8.18). Clearly, the present notation $f(P)$ should not be confused with the right-hand member of equation (8.16).

‡ Compare equation (8.28a) as well as Mikhlin (Ref. 8, Sec. 8).

§ Compare also Stakgold (Ref. 24, p. 524), where the energy norm is denoted by $\|u\|_a$.

¶ Compare Figure 6.1 and the pertinent remarks in Chapter 6.

hard to achieve, and since it will find no application in this exposition, there is no need to discuss it further.

The two other types of convergence, that is, convergence in the mean and energy convergence, are closely related to function space concepts. To illustrate, let us replace the function $f(P)$ in equation (10.9) by a sequence $f_n(P)$, $n = 1, 2, \ldots$, and replace $F(P)$ by $f(P)$. The definition of convergence then becomes

$$\| f_n(P) - f(P) \|_M \to 0 \tag{10.16}$$

or

$$f_n(P) \overset{M}{\to} f(P), \tag{10.16a}$$

provided $n \to \infty$. Although it is said that either of the equations above expresses *convergence in the mean*, strictly speaking, the order of convergence in the mean is $p = 2$.

A similar argument implies that either of the conditions

$$\vertiii{ f_n(P) - f(P) }_E \to 0 \tag{10.17}$$

or

$$f_n(P) \overset{E}{\to} f(P) \tag{10.17a}$$

defines *convergence in energy* as $n \to \infty$.

Convergence in the mean constitutes a particular case of convergence in energy. In fact, if the operator L is replaced by the identity operator (leaving a vector unchanged), equation (10.17) reduces to equation (10.16).

If the operator L stands for multiplication by a scalar function $\alpha(P)$, the criterion of convergence in energy becomes [cf. equation (10.11)]

$$\int_\Omega \alpha(P)[f_n(P) - f(P)]^2 \, d\Omega \to 0, \tag{10.18}$$

for $n \to \infty$, and defines convergence in the mean with a weighting function $\alpha(P)$.

It is evident that equations such as (10.16), (10.17), or (10.18) may also serve as definitions of corresponding convergences of a *series* of function vectors, provided the symbol $f_n(P)$ is interpreted as a partial sum of the series.

The form of the mean-square distance (10.8) [like that of the Hilbert inner product (8.2)] indicates that the finiteness of the norm depends on the *square integrability* of the functions involved. The corresponding space† $\mathscr{L}_2$

---

† Mentioned earlier parenthetically, e.g., in connection with equations (7.6) and (9.12); it is, occasionally, called a *Lebesgue* space and denoted in the single-variable case by $\mathscr{L}_2(a, b)$, where (a, b) indicates the interval of square integrability: $\int_a^b [f(t)]^2 \, dt < \infty$.

of square integrable functions includes, among others, the important class of continuous and piecewise continuous functions, but also many discontinuous functions. Relevant examples are contained in the improper Riemann integrals

$$\int_0^1 \frac{dt}{t^{1/4}} = \tfrac{4}{3} \quad \text{and} \quad \int_0^1 \frac{dt}{t^{1/2}} = 2, \tag{10.19}$$

the integrands of which are infinite at the lower limit. The space $\mathscr{L}_2$ is of major interest in studies of the so-called *generalized Fourier series*. These are expansions in terms of sets of orthogonal functions,† from which most bases for vector spaces are constructed. Although with no intention to analyze Fourier representations in detail, let us focus our attention on the simple example of a function of a single variable, $f(t)$, $a \le t \le b$, supposed to be of class $\mathscr{L}_2$. We select a complete set of orthonormal functions $\phi_k(t)$, all of class $\mathscr{L}_2$, and consider the expansion

$$f(t) = c_1 \phi_1(t) + c_2 \phi_2(t) + \cdots, \tag{10.20}$$

where

$$\begin{aligned} c_k &= (f, \phi_k) \\ &= \int_a^b f(t)\phi_k(t)\, dt, \qquad k = 1, 2, \ldots, \end{aligned} \tag{10.21}$$

are the Fourier coefficients. The series (10.20) is called the *generalized Fourier series* of $f(t)$ with respect to the set $\{\phi_k(t)\}$. Classical examples of sets of orthonormal functions are the following popular systems.

(a) The classical trigonometric system,

$$\frac{1}{(2\pi)^{1/2}}, \frac{\cos t}{\pi^{1/2}}, \frac{\sin t}{\pi^{1/2}}, \ldots, \frac{\cos nt}{\pi^{1/2}}, \frac{\sin nt}{\pi^{1/2}}, \ldots, \tag{10.22a}$$

orthogonal in the interval $[0, 2\pi]$ or, in view of the periodicity of trigonometric functions, in any interval of length 2π. A more general form of this system,

$$\frac{1}{(2l)^{1/2}}, \frac{1}{l^{1/2}}\cos\frac{\pi t}{l}, \frac{1}{l^{1/2}}\sin\frac{\pi t}{l}, \ldots, \frac{1}{l^{1/2}}\cos\frac{n\pi t}{l}, \frac{1}{l^{1/2}}\sin\frac{n\pi t}{l}, \ldots, \tag{10.22b}$$

represents a system orthonormal in an arbitrary interval of length $2l$.

---

† The general form of the Fourier coefficients is given by equations (7.19).

From the preceding systems, other orthonormal systems are derived, for example, a pure sine system,

$$\left(\frac{2}{\pi}\right)^{1/2}\sin t,\ \left(\frac{2}{\pi}\right)^{1/2}\sin 2t,\ \ldots,\qquad(10.22c)$$

and a pure cosine system,

$$\left(\frac{1}{\pi}\right)^{1/2},\ \left(\frac{2}{\pi}\right)^{1/2}\cos t,\ \left(\frac{2}{\pi}\right)^{1/2}\cos 2t,\ \ldots,\qquad(10.22d)$$

both orthogonal in the interval $[0, \pi]$. Either of the systems above is *complete*[†] in the space $\mathscr{L}_2$ and can serve as a basis for this space.[‡] It does not seem superfluous to recall that removal of even a single term from either of the systems above would deprive it of the property of being a basis (at least from a theoretical viewpoint), *even though* such a curtailed system would still include an infinite number of elements.

(b) Bessel functions of the first kind of an arbitrary positive order p,

$$\frac{2^{1/2}J_p(\lambda_1 t)}{lJ_{p+1}(\lambda_1 l)},\ \frac{2^{1/2}J_p(\lambda_2 t)}{lJ_{p+1}(\lambda_2 l)},\ \ldots,\qquad(10.22e)$$

also form an orthonormal system in the interval $[0, l > 0]$ with the weighting function t. This means that

$$\frac{2}{l^2 J_{p+1}(\lambda_i l)J_{p+1}(\lambda_k l)}\int_0^l tJ_p(\lambda_i t)J_p(\lambda_k t)\,dt = \delta_{ik},\qquad(10.22f)$$

where $\lambda_1 < \lambda_2 < \lambda_3 < \cdots$ are the positive roots of the equation

$$J_p(\lambda l) = 0.\qquad(10.22g)$$

A series formed of the preceding functions is known as *Fourier–Bessel* series.

(c) Legendre polynomials,

$$\tfrac{1}{2}P_0(t),\ \tfrac{3}{2}P_1(t),\ \ldots,\ (n + \tfrac{1}{2})^{1/2}P_n(t),\ \ldots,\qquad(10.22h)$$

form a system orthonormal in the interval $[-1, 1]$ and complete in $\mathscr{L}_2(-1, 1)$. The first few of the Legendre polynomials are

$$P_0(t) = 1,\ P_1(t) = t,\ P_2(t) = \tfrac{1}{2}(3t^2 - 1),$$
$$P_3(t) = \tfrac{1}{2}(5t^3 - 3t),\ P_4(t) = \tfrac{1}{8}(35t^4 - 30t^2 + 3).\qquad(10.22i)$$

A corresponding series is called a *Legendre–Fourier* series.

[†] A *maximal* orthonormal set in an inner-product space is called *complete*.
[‡] More exactly, for the spaces $\mathscr{L}_2(0, 2\pi)$, $\mathscr{L}_2(0, 2l)$, and $\mathscr{L}_2(0, \pi)$, respectively.

(d) Still another system† of orthonormal functions is provided by the set

$$f_n(x) = \frac{e^{inx}}{(2\pi)^{1/2}},\tag{10.22j}$$

where n is an integer and the interval of orthogonality is $[-\pi, \pi]$.

We conclude our list of certain of the common orthonormal systems by recalling that the latter arise rather naturally from the so-called *eigenvalue* problems. For instance, trigonometric, Bessel, and Legendre functions spring from differential equations constituting particular cases of the Sturm–Liouville equation (Ref. 44, Chap. 3),

$$\frac{d}{dt}\left[p(t)\frac{dy}{dt}\right] + [q(t) + \lambda r(t)] = 0.\tag{10.23}$$

Here, the coefficient functions $p(t)$, $q(t)$, and $r(t)$ are continuous functions of the variable t, and λ is a parameter with values in the discrete set of so-called *eigenvalues* [cf., e.g., equation (10.22g)]. Solutions of the eigenvalue problems associated with the corresponding eigenvalues are known as *eigenfunctions*. In the cases mentioned above, the eigenfunctions are just the four orthonormal systems (a)–(d). As such, they are reminiscent in many ways of the coordinate vectors of rectangular reference frames, while the corresponding Fourier coefficients play the part of components of vectors under decomposition along the axes of the frame.

Problems

1. Determine the function to which the sequence $\{f_n(t) = t^n\}$ converges pointwise in the interval $[0, 1]$. Draw a picture. Is the convergence uniform?

2. Let the sequence $\{f_n\}$ be given by $f_n(x) = n$ for $0 < x < 1/n$, $f_n(x) = 0$ at all remaining points of the interval $0 \leq x \leq 1$. Show that the fact that $f_n(x) \rightarrow f(x) = 0$ at every point of the interval does not imply convergence to $f(x)$ in the mean.

3. A sequence $\{f_n\}$ in a Hilbert space is said to converge *weakly* to f if $\lim_{n \to \infty} (f_n, g) = (f, g)$ for all elements g in the space. The sequence converges *strongly* to f if $\lim_{n \to \infty} \|f_n \rightarrow f\| = 0$. Show that a strongly convergent sequence also converges weakly to the same limit.

---

† This system is listed principally for the purpose of acknowledging the existence of complex-vector spaces, for which Groups A and B of the axioms of Chapter 7 need be altered only by allowing the scalars to be complex numbers. The axioms of Group C, defining an inner product, must be partially modified as well.

4. Find the arithmetic mean and the mean-square distances between the functions $f(x) = x^2$ and $g(x) = x^3$ in the interval $-1 \le x \le 1$ [cf. formula (10.9)].

5. Show that a necessary and sufficient condition for an orthonormal set to be complete in a Hilbert space [cf. the footnote after equation (10.22b)] is that for any vector x in the space the Parseval equality be satisfied.

6. Let the set $\{1/2^{1/2}, \cos t, \sin t, \ldots, \cos nt, \sin nt\}$ be a basis for the subspace $\mathscr{S}$ of $\mathscr{L}_2(-\pi, \pi)$. What is the dimension of $\mathscr{S}$? Is the set orthonormal in the interval $-\pi \le t \le \pi$? By (7.17), the vector f_n in $\mathscr{S}$ nearest to a given element x, i.e., the one for which $\|x - f_n\|$ is a minimum, is represented by $f_n = \alpha_0(1/2^{1/2}) + \alpha_1 \cos t + \beta_1 \sin t + \cdots + \alpha_n \cos nt + \beta_n \sin nt$ where $\{\alpha_i\}, \{\beta_i\}$ are the Fourier coefficients of x. Find the latter, taking the inner product in the form $(x, y) = (1/\pi) \int_{-\pi}^{\pi} xy \, dt$. Write the corresponding Bessel inequality.

7. Show that the set of eigenvectors of a linear operator L, corresponding to the eigenvalue λ, i.e., the set of solutions of $Lx = \lambda x$ in the linear space $\mathscr{V}$ in which L acts [cf. equation (10.23)], constitutes a subspace $\mathscr{S}$ of $\mathscr{V}$.

8. The adjoint of an operator L in a Hilbert space is the operator L^* such that $(Lx, y) = (x, L^*y)$ for all x, y (see equation (11.82) infra). Show that any eigenvector of L belonging to the eigenvalue λ is orthogonal to any eigenvector of L^* belonging to λ^* if $\lambda \ne \lambda^*$.

11

Bounds and Inequalities

For all the pessimism of Lord Russell, there seems little doubt that a search
for an approximate solution, where an exact solution cannot be obtained, is
both theoretically defensible and practically unavoidable. This fact,
however, does not absolve us from the obligation of estimating the error of
the approximation solution, either by a direct computation or, more often,
by finding the so-called *bounds*. In this chapter, we devote our attention to
the second question, by examining some methods of constructing bounds for
solutions of boundary value problems.†

Two relations are essential in our discussion: the Cauchy–Schwarz
inequality (7.14),

$$(f, g)^2 \leq (f, f)(g, g), \tag{11.1}$$

and the Bessel inequality (7.21),

$$\sum_{k=1}^{n} (f, i^k)^2 \leq (f, f); \tag{11.2}$$

here, f and g are arbitrary vectors and $\{i^k\}$ is a set of orthonormal vectors. As
the space under consideration, we choose a Hilbert space.

As already noted, the geometric content of the Cauchy–Schwarz
inequality is that the length of the orthogonal projection f' of a vector f on a
vector g (Figure 11.1) is not greater than the length of f. Bessel's inequality,
on the other hand, expresses the fact that the square of the length of a vector
is not less than a partial sum of squares of its components along the axes of a
Cartesian rectangular frame; equality occurs when the sum includes all
components, in which case we arrive at the Pythagorean equation (5.19).

---

† The exposition is mainly based on the memoir of Diaz.[29] The reader interested specifically in
the question of bounds will find much important material in this notable work.

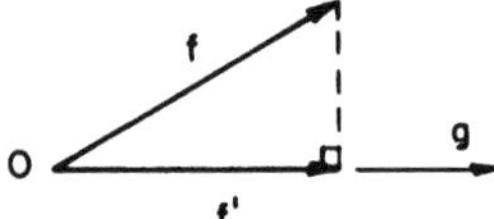

Figure 11.1. Illustration for the Cauchy–Schwarz inequality.

Figure 11.2 illustrates the situation for a vector f with vectorial components $f^k = (f, i^k)i^k$, $k = 1, 2, 3$. Clearly, $\|f\|^2 > \|f^1\|^2 + \|f^2\|^2$, but $\|f\|^2 = \sum_{k=1}^{3} \|f^k\|^2$. Generally speaking, Bessel's inequality furnishes a *lower bound* for the norm of a vector. If the coordinate vectors, say g^k, are not normalized, we can write the inequality (11.2) in the form

$$\sum_{k=1}^{n} \frac{(f, g^k)^2}{(g^k, g^k)} \le (f, f). \tag{11.3}$$

For a single vector, i.e., if $n = 1$, (11.3) reduces to the Cauchy–Schwarz inequality (11.1).

Having the lower bound provided by Bessel's inequality, it is natural to look for an *upper bound* for the norm of a vector. This can be accomplished as follows (Figure 11.3).

Let $\{i^k\}$ denote a set of n orthonormal function vectors selected so that the function of interest, f, is orthogonal to the subspace $\mathcal{S}$ spanned by these vectors (recall that orthogonal vectors are linearly independent),

$$(f, i^k) = 0 \qquad \text{for } k = 1, 2, \ldots, n. \tag{11.4}$$

In Figure 11.3, the subspace $\mathcal{S}$ is visualized as the plane $\mathcal{S}_2$ of the vectors i^1 and i^2. A right triangle, OFG, is then constructed with f as a leg and a vector g as its hypotenuse. We have

$$(g - f, f) = 0, \tag{11.5}$$

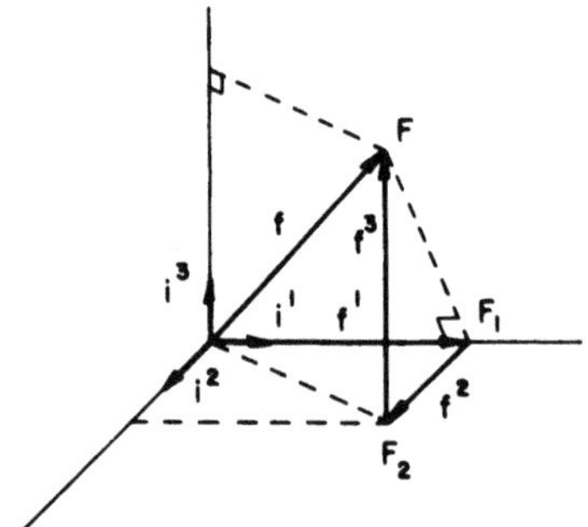

Figure 11.2. Illustration for the Bessel inequality.

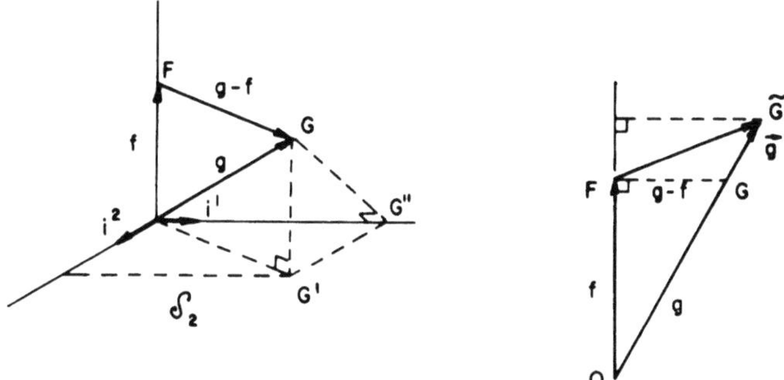

Figure 11.3. Illustration for the upper bound for f.

so that

$$\|g - f\|^2 = \|g\|^2 - \|f\|^2. \tag{11.6}$$

But

$$(\overline{OG'})^2 = \|g - f\|^2 \geq (\overline{OG''})^2 = (g, i^1)^2 \tag{11.7}$$

or, generally, in a many-dimensional space,

$$\|g - f\|^2 \geq \sum_{k=1}^{n} (g, i^k)^2. \tag{11.8}$$

Figure 11.3 displays the situation in three-space, in which $\|g - f\|^2 > (g, i^1)^2$, $\|g - f\|^2 = (g, i^1)^2 + (g, i^2)^2$. Combining (11.6) and (11.8) gives immediately

$$\|g\|^2 - \|f\|^2 \geq \sum_{k=1}^{n} (g, i^k)^2$$

or, changing the notation,

$$(f, f) \leq (g, g) - \sum_{k=1}^{n} (g, i^k)^2. \tag{11.9}$$

This inequality provides an *upper bound* for the norm of the vector f. Evidently, all the more there is

$$(f, f) \leq (g, g), \tag{11.9a}$$

showing that an upper bound for the norm of an arbitrary function f—however crude it might be—is furnished by *any* function g, provided the difference $g - f$ between the two functions is orthogonal to the function of interest f. From the graph displayed in Figure 11.3, it is apparent that the condition (11.5) is actually too restrictive, our argument remaining valid if this condition is replaced by a weaker one, namely,

$$(g - f, f) \geq 0. \tag{11.10}$$

Indeed, in this case, the angle OFG may turn out to be obtuse ($\measuredangle\, OF\tilde{G}$ in Figure 11.3), and the norm of $\tilde{g}$, now replacing the norm of g, may be larger than the latter $(\overrightarrow{OG} > \overline{OG})$. Consequently, the inequalities (11.9) and (11.9a) hold also in the more general case after replacing g by $\tilde{g}$.

Upon collecting results, we conclude that, for any two numbers m and n, there is

$$\left\| \sum_{k=1}^{m} (f, g^k)g^k \right\|^2 \leq \|f\|^2 \leq \left\| g - \sum_{k=1}^{n} (g, h^k)h^k \right\|^2, \tag{11.11}$$

where $\{g^k\}$ is an arbitrary set of *orthonormal* vectors, g is a function such that $(g - f, f) \geq 0$, and $\{h^k\}$ is a set of *orthonormal* vectors orthogonal to the vector of interest, f. Increasing the numbers m and n sharpens the bounds. This improvement process can be carried as far as we wish, inasmuch as, in general, the lower bounds increase and the upper bounds decrease monotonically with increasing m and n, respectively. A hidden catch, however, is that the bounds include the (unknown) function f, the norm of which they are meant to approximate, and one is caught in a vicious circle. To escape from the difficulty and render the inequality (11.11) useful, it is necessary to appeal to the information provided by the differential equation and the boundary conditions of the problem at hand.

As a simple illustration, consider the Neumann problem

$$\nabla^2 f = 0 \quad \text{in } \Omega, \tag{11.12}$$

and

$$\frac{\partial f}{\partial n} = h \quad \text{on } \partial\Omega, \qquad \int_{\partial\Omega} h\, ds = 0,\dagger \tag{11.13}$$

where ∇^2 is the plane Laplacian and h is a function prescribed on the boundary $\partial\Omega$ of Ω. As a first step, we introduce an appropriate inner product, appearing as a Dirichlet-type product (8.6). Indeed, by Green's first identity, we have

$$\int_{\Omega} g\nabla^2 f\, dx\, dy + \int_{\Omega} (f_{,x}g_{,x} + f_{,y}g_{,y})\, dx\, dy = \int_{\partial\Omega} g\frac{\partial f}{\partial n}\, ds, \tag{11.14}$$

$\dagger$ This condition is imposed to ensure the existence of a solution of the Neumann problem. (Solution is unique to within an additional constant only.)

and it is natural to take the second area integral, i.e., the *bilinear Dirichlet integral*, as the inner product of f and g. It follows from this that the norm of the function f is represented by the Dirichlet integral,†

$$\|f\|^2 = \int_\Omega (f_{,x}^2 + f_{,x}^2)\, dx\, dy \tag{11.15}$$

and, in view of equations (11.12) and (11.13), equation (11.14) takes the form

$$(f, g) = \int_{\partial\Omega} gh\, ds. \tag{11.16}$$

We thus conclude that, for a given function g, the inner product (f, g) is known, and the inequality (11.3) yields a lower bound for the norm of the solution vector as

$$\frac{(\int_{\partial\Omega} gh\, ds)^2}{\|g\|^2} \le \|f\|^2, \tag{11.17}$$

where the norm of g is given by a Dirichlet integral like (11.15).

A derivation of an upper bound for the norm of f is slightly more complicated, as shown by the following procedure (Ref. 45, p. 107).

We start with the Gauss–Green theorem

$$\int_\Omega (f_{,x}g_1 + f_{,y}g_2)\, dx\, dy + \int_\Omega f(g_{1,x} + g_{2,y})\, dx\, dy = \int_{\partial\Omega} f(g_1 n_x + g_2 n_y)\, ds, \tag{11.17a}$$

where g_1 and g_2 are two functions, so far unspecified, and n_x and n_y are the components of n. For $g_1 \equiv f_{,x}$ and $g_2 \equiv f_{,y}$ the foregoing equation becomes

$$\|f\|^2 = \int_{\partial\Omega} f\, \frac{\partial f}{\partial n}\, ds, \tag{11.17b}$$

after referring to equation (11.12). From the Cauchy–Schwarz inequality, there follows

$$\left[\int_\Omega (f_{,x}g_1 + f_{,y}g_2)\, dx\, dy \right]^2 \le \|f\|^2 \int_\Omega (g_1{}^2 + g_2{}^2)\, dx\, dy. \tag{11.17c}$$

We now assume that

$$g_{1,x} + g_{2,y} = 0 \quad \text{in } \Omega \tag{11.17d}$$

---

† We note that the inner product (11.14) is positive semi-definite, and the norm (11.15) is, in fact, a semi-norm (cf. their respective definitions in Chapter 8). In the present case, it is alternatively conceivable to introduce a norm on a family of equivalent classes of functions, declaring two functions equivalent if, and only if, they differ by a constant.

and [note that $\int_{c\Omega} (g_1 n_x + g_2 n_y) \, ds = 0$ by the second condition (11.13)]

$$g_1 n_x + g_2 n_y = \frac{\partial f}{\partial n} \quad \text{on } \partial\Omega. \tag{11.17e}$$

In view of these assumptions, equation (11.17a) reduces to

$$\int_\Omega (f_{,x} g_1 + f_{,y} g_2) \, dx \, dy = \int_{c\Omega} f \frac{\partial f}{\partial n} \, ds, \tag{11.17f}$$

so that finally, by (11.17b) and (11.17f), the inequality (11.17c) provides the desired upper bound,

$$\|f\|^2 \leq \int_\Omega (g_1{}^2 + g_2{}^2) \, dx \, dy. \tag{11.17g}$$

Conditions (11.17d) and (11.17e), imposed on the functions g_1 and g_2, are somewhat restrictive, and it is convenient to replace the first of them by the requirement that $g_1 = g_{,y}$ and $g_2 = -g_{,x}$, where g is some function suitable for our purposes. This function obeys the following boundary condition:

$$\frac{\partial g}{\partial s} = g_{,x} \frac{\partial x}{\partial s} + g_{,y} \frac{\partial y}{\partial s} = -g_{,x} n_y + g_{,y} n_x = \frac{\partial f}{\partial n} \quad \text{on } C. \tag{11.17h}$$

With this in mind, the inequality (11.17g) becomes

$$\|f\|^2 \leq \int_\Omega (g_{,x}^2 + g_{,y}^2) \, dx \, dy, \tag{11.18}$$

and in combination with the inequality (11.17), furnishes two-sided bounds for the norm of the function of interest. However, there remains the question of just how close the bounds actually approach the exact value of $\|f\|$.

At this point, it may be in order to verify the fact that our choice of the norm for the Neumann problem, in the form of a bilinear Dirichlet integral, was not accidental, but resulted from certain natural connections existing between the Laplace equation and the Dirichlet integral. In fact, it is demonstrated in the variational calculus that the Laplace equation is a differential description of a class of phenomena possessing an alternate characterization using an integral approach, which consists of making the value of the Dirichlet integral stationary. As a simple illustration, consider the one-dimensional problem of minimizing the functional

$$I[y] = \int_0^1 \left(\frac{dy}{dx}\right)^2 dx \tag{11.19}$$

under the conditions that

$$y(0) = 0, \quad y(1) = 1. \tag{11.20}$$

Suppose that $y_0 = y_0(x)$ is the function which actually minimizes $I[y]$, and that

$$\bar{y}(x) = y_0(x) + \varepsilon\eta(x) \tag{11.21}$$

is any admissible function that satisfies the conditions (11.20) and serves as a comparison function. Here, $\eta(x)$ is an arbitrary, sufficiently regular, function with $\eta(0) = \eta(1) = 0$, and ε a parameter assumed to be small as compared with unity. By replacing y by $\bar{y}$, we make the functional (11.19) a function of ε, a stationary value of which occurs for $\varepsilon = 0$. This yields

$$\int_0^1 \frac{dy_0}{dx}\frac{d\eta}{dx}\,dx = 0, \tag{11.22}$$

and, since η vanishes at $x = 0$ and $x = 1$ and (11.22) holds for all functions η described above, then[†] after integrating by parts,

$$\frac{d^2 y_0}{dx^2} = 0. \tag{11.23}$$

The foregoing equation is the so-called Euler–Lagrange equation associated with the variational problem $\delta I[y] = 0$. It is evident that the boundary value problem (11.23) and (11.20) represents a *Dirichlet problem* in its simplest form. On the other hand, equation (11.19) turns out to yield the Dirichlet metric (11.15) reduced to a one-dimensional case.

As a general remark, it is important to remember that the connection between the norm and the actual values of the function of interest is somewhat loose, inasmuch as the norm here is an integral, rather than a pointwise, description of a function.[‡]

Returning to the matter of the accuracy of the bounds derived heretofore, we now have to examine certain procedures enabling one to improve the bounds.

To begin, let us attempt to determine bounds for a function f in terms of the following: a given function c, a set of n orthonormal functions $\{h^i\}$, and a set of n scalars $\{\alpha_i\}$ representing the projections of f on the vectors h^i:

$$(f, h^i) = \alpha_i, \qquad i = 1, 2, \ldots, n. \tag{11.24}$$

It is required that

$$\|f - c\|^2 = r^2, \tag{11.25}$$

where r is a given constant.

---

† This follows from the *basic lemma* of the calculus of variations, e.g., Weinstock (Ref. 46, p. 16).

‡ It should also be noted that, since the functionals minimized (such as here $I[y]$) are "level" at the exact solution (here y_0), it is understandable that $I[y]$ provides a better estimate for $I[y_0]$ than any approximate (admissible though it be) solution $\bar{y}$ does for y_0.

Under these assumptions, it can be shown that (Ref. 29, p. 11)

$$\sum_{i=1}^{n} \alpha_i^2 + \left\{ \left[\|c\|^2 - \sum_{i=1}^{n} (c, h^i)^2 \right]^{1/2} - \left[r^2 - \sum_{i=1}^{n} (\alpha_i - (c, h^i))^2 \right]^{1/2} \right\}^2 \leq \|f\|^2$$

$$\leq \sum_{i=1}^{n} \alpha_i^2 + \left\{ \left[\|c\|^2 - \sum_{i=1}^{n} (c, h^i)^2 \right]^{1/2} + \left[r^2 - \sum_{i=1}^{n} (\alpha_i - (c, h^i))^2 \right]^{1/2} \right\}^2.$$

$$(11.26)$$

We omit the proof of the foregoing inequality, inasmuch as for our purposes it is rather more appropriate to exhibit some of its geometric aspects. We first identify equation (11.25) as the equation of a *hypersphere* (9.37), $\mathscr{B}_\infty$, of radius r and center at the tip of the vector c. We observe that equations (11.24) describe a *hyperplane* of class n, $H_{(n)}$. If these two intersect, the "curve" of intersection is a *hypercircle*, $\Gamma_{(n)}$, of class n because one of the meeting sets is of this class (Figure 11.4 illustrates the situation). Note that equations (11.24) and (11.25) clearly demonstrate that f is the position vector of a point common to $\mathscr{B}_\infty$ and $H_{(n)}$, so that the tips of all such f's trace the hypercircle $\Gamma_{(n)}$.

In order to form a mental image of the resulting configuration, it is convenient to draw a corresponding sketch in two dimensions and confine the number of h^i vectors to a single one, say h^k (Figure 11.5). A hypersphere is then represented by a circle ($\mathscr{B}_\infty'$ or $\mathscr{B}_\infty''$ in Figure 11.5) and a hyperplane by a line ($H_{(n)}$). The hypercircle of intersection reduces to a pair of points (A_1' and A_2' for $\mathscr{B}_\infty'$) symmetrically located with respect to the subspace $\mathscr{S}_n = \{\sum_{k=1}^{n} c_k h^k\}$ (a line in Figure 11.5), or to a pair A_1'' and A_2'', if the location of $\mathscr{B}_\infty''$ is asymmetric. The line $\mathscr{S}_n$, in the case considered, becomes the subspace $\mathscr{S}_1$ of all vectors $c_k h^k$, where h^k is a unit vector perpendicular to

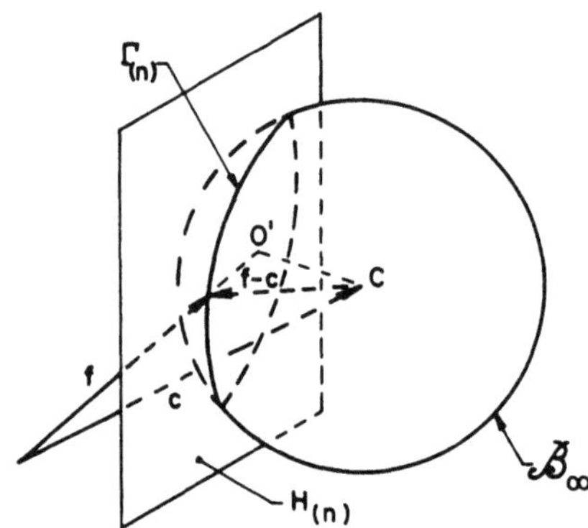

Figure 11.4. An intersection of a hyperplane with a hypersphere.

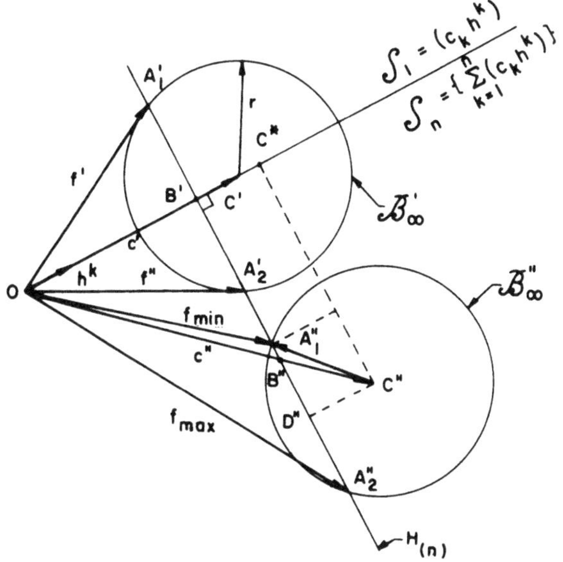

Figure 11.5. Bounds for a function f.

the line $H_{(n)}$ and c_k is a scalar. The perpendicularity of the lines $\mathcal{S}_1$ and $H_{(n)}$ follows from the fact that the tip of every f-vector lies on $H_{(n)}$, while the orthogonal projection $\alpha_k = (f, h^k)$ of any f-vector on h^k is the same for all vectors f. A similar conclusion is reached if a vector in $H_{(n)}$ is represented as a difference of two position vectors; for example,

$$y \equiv \overrightarrow{A_1'A_2'} = f'' - f'. \tag{11.27}$$

Applying (11.24) to f'' and f' and subtracting gives

$$(y, h^k) = 0, \tag{11.27a}$$

as asserted. In the location $\mathcal{B}_\infty'$, the center C' of the sphere lies on $\mathcal{S}_1$ and is identified by the position vector

$$\overrightarrow{OC'} \equiv c' = (c', h^k)h^k. \tag{11.28}$$

Now, the distance $\overline{OB'}$ is, by (11.24),

$$\overline{OB'} = (f', h^k) = \alpha_k \tag{11.29}$$

and

$$(\overline{B'A'})^2 = r^2 - (\overline{B'C'})^2$$
$$= r^2 - [(c', h^k) - \alpha_k]^2. \tag{11.30}$$

Hence,

$$\| f' \|^2 = (\overline{OB'})^2 + (\overline{B'A'})^2$$
$$= \alpha_k{}^2 + r^2 - [(c', h^k) - \alpha_k]^2. \tag{11.30a}$$

In the situation considered, the vector c in the inequality (11.26) coincides with the present vector c' and, by (11.28) and for a single vector h^k, the inequality reduces to an equality coinciding with equation (11.30a). It should be clear that, for a number of h^k-vectors greater than one, the right-hand side of equation (11.30a) must be summed over the index k. It is interesting to note that, for all the superficial simplicity of Figure 11.5, the pair of points $A_1{}'$ and $A_2{}'$ stands for a hypercircle, the dimensionality of which is finite. It is also apparent that every vector f (including, of course, vectors f' and f'') whose tip lies on the hypercircle is determined by the formula

$$\| f' \|^2 = \| f'' \|^2 = \sum_{k=1}^{n} [\alpha_k{}^2 + r^2 - (\alpha_k - (c, h^k))^2]^{1/2}. \tag{11.31}$$

In the general case, the center of the hypersphere does not lie on the line $\mathscr{S}_n$ (sphere $\mathscr{B}_\infty{}''$ in Figure 11.5), and there exist two different limit positions (producing two bounds), $f_{\min}$ and $f_{\max}$, of the vector f. It seems sufficient to derive one of them, say, the lower bound for the norm $\| f \|$, represented by the segment $\| f_{\min} \| = \overline{OA_1{}''}$.

By inspecting the simplified model displayed in Figure 11.5, we find

$$\overline{OB'} = (f_{\min}, h^k) = \alpha_k,$$
$$\overline{OC^*} = (c, h_k), \qquad \overline{C''D''} = (c'', h^k) - \alpha_k,$$
$$(\overline{OA''})^2 = (\overline{OB'})^2 + (\overline{B'A''})^2, \tag{11.32}$$
$$= \alpha_k{}^2 + (\overline{C''C^*} - \overline{A''D''})^2,$$
$$= \alpha_k{}^2 + \{[(\overline{OC''})^2 - (\overline{OC^*})^2]^{1/2} - [r^2 - (\overline{C''D''})^2]^{1/2}\}^2.$$

The last of the foregoing equations takes the form

$$\| f_{\min} \|^2 = \alpha_k{}^2 + \{[\|c''\|^2 - (c'', h^k)^2]^{1/2} - [r^2 - ((c'', h^k) - \alpha_k)^2]^{1/2}\}^2, \tag{11.33}$$

which, after summation over the index k and suppressing of the double primes, coincides with the left-hand side of inequality (11.26). If we disregard the sums in the latter inequality, that is, make all α_i's and h^i's vanish, we obtain

$$[\|c\| - r]^2 \le \| f \|^2 \le [\|c\| + r]^2. \tag{11.34}$$

The two-dimensional sketch displayed in Figure 11.6, representing the foregoing inequality, illustrates a well-known theorem of plane geometry,

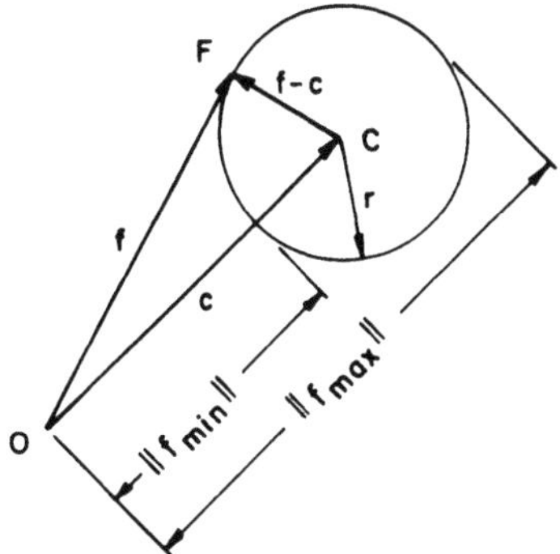

Figure 11.6. Illustration for inequality (11.34).

namely, that the length of a tangent to a circle from an exterior point is no less than the distance from the point to the circle, and no greater than the length of the secant from the point through the center of the circle. The function space aspect of this observation is that, by selecting an arbitrary function, say c, one can construct bounds for the norm of the unknown function f, provided the norm r of the difference of the two is given [cf. (11.25)]. In our simplified model, the upper bound in the inequality (11.34) is most naturally deduced from the triangle OFC in Figure 11.6 (in which $\|f - c\| = r$), after an appeal to the triangle inequality. The lower bound, on the other hand, is a direct consequence of the fact that a side of a triangle is not less than the difference of the two remaining sides.

It is worth noticing that the distance C^*C'' in Figure 11.5, equal to $\|c'' - (c'', h^k)h^k\|$ for $n = 1$, becomes $\|c'' - \sum_{k=1}^{n} (c'', h^k)h^k\|$ for $n > 1$; consequently, the equality $\|c'' - \sum_{k=1}^{n} (c'', h^k)h^k\| = 0$ would imply that the center of the hypersphere lies on the line (or rather in the subspace) $\mathscr{S}_n$.

In applications, it is practical to modify the form of the inequality (11.26). This is necessitated by the unwelcome circumstance that the bounds involve the vector c and the scalars $\{\alpha_i\}$, all of which depend on the unknown function f, via equations (11.24) and (11.25) (provided r is preassigned). The dilemma may be circumvented by the following procedure. We select two vectors, g and h, and two orthonormal sets, $\{\bar{g}^k\}$ and $\{\bar{h}^k\}$, obeying the following conditions:

$$(g - f, h - f) = 0, \tag{11.35a}$$

$$(g - f, \bar{h}^k) = 0, \tag{11.35b}$$

$$(h - f, \bar{g}^i) = 0, \tag{11.35c}$$

$$(\bar{g}^i, \bar{h}^k) = 0, \tag{11.35d}$$

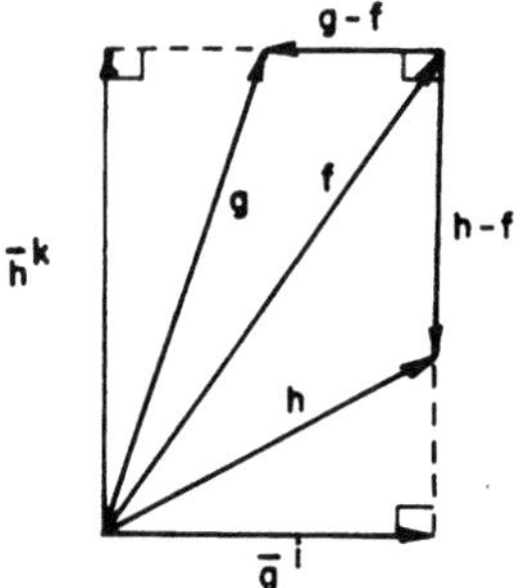

Figure 11.7. Illustration for relations (11.36).

where $i = 1, 2, \ldots, s$ and $k = 1, 2, \ldots, t$. The conditions above are represented graphically by the sketch shown in Figure 11.7.

Upon casting equation (11.35a) into the form

$$\left(f - \frac{g+h}{2} - \frac{g-h}{2}, f - \frac{g+h}{2} + \frac{g-h}{2} \right) = 0$$

and expanding, we arrive at the equation

$$\left\| f - \frac{g+h}{2} \right\|^2 = \left\| \frac{g-h}{2} \right\|^2. \tag{11.36}$$

With this in mind, we set

$$c = \frac{g+h}{2}, \qquad r = \left\| \frac{g-h}{2} \right\|, \tag{11.37}$$

and $n = s + t$, $\bar{g}^i = h^i$ for $i = 1, 2, \ldots, s$ and $\bar{h}^k = h^{s+k}$ for $k = 1, 2, \ldots, t$. From (11.35b) and (11.35c), there follow $(f, \bar{g}^i) = (h, \bar{g}^i)$ for $i = 1, 2, \ldots, s$ and $(f, \bar{h}^k) = (g, \bar{h}^k)$ for $k = 1, 2, \ldots, t$. In this way, the unknown vector f is replaced by the given vectors g and h. Inequality (11.26) takes now the somewhat clumsy, but some serviceable, form

$$F + (G^{1/2} - H^{1/2})^2 \leq \| f \| \leq F + (G^{1/2} + H^{1/2})^2, \tag{11.38}$$

where

$$F = \sum_{i=1}^{s} (h, \bar{g}^i)^2 + \sum_{k=1}^{t} (h, \bar{h}^k),$$

$$G = \left\| \frac{g+h}{2} \right\|^2 - \sum_{i=1}^{s} \left(\frac{g+h}{2}, \bar{g}^i \right)^2 - \sum_{k=1}^{t} \left(\frac{g+h}{2}, \bar{h}^k \right)^2, \tag{11.39}$$

$$H = \left\| \frac{g-h}{2} \right\|^2 - \sum_{i=1}^{s} \left(\frac{g-h}{2}, \bar{g}^i \right)^2 - \sum_{k=1}^{t} \left(\frac{g-h}{2}, \bar{h}^k \right)^2.$$

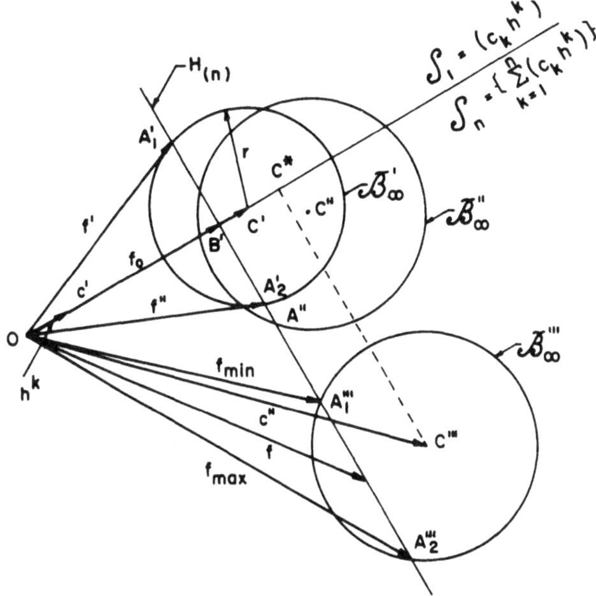

Figure 11.8. Bounds for a function f.

If $s = t = 0$, i.e., the sets of orthonormal vectors are disregarded, the just-written inequality simplifies to

$$\left(\left\|\frac{g+h}{2}\right\| - \left\|\frac{g-h}{2}\right\|\right)^2 \leq \|f\|^2 \leq \left(\left\|\frac{g+h}{2}\right\| + \left\|\frac{g-h}{2}\right\|\right)^2; \quad (11.40)$$

this is an inequality companion to (11.34).

Let us now examine a *second* important inequality, a geometric interpretation of which is given by the sketch in Figure 11.8, representing a more general version of inequality (11.26), inasmuch as it involves at the same time the surface and the interior of the hypersphere. The intersection of the latter with a hyperplane constitutes now a solid circle (a disk). On the symbolic diagram shown in Figure 11.8, a hypersphere $\mathscr{B}_\infty'$ is represented by a circle, a hyperplane $H_{(n)}$ by a line, and their intersection by the *chord* $A_1'A_2'$ (rather than by a pair of points A_1' and A_2' as in Figure 11.5). The set of functions admitted for competition is now considerably richer than before, due to the fact that it includes vectors whose tips lie on the entire chord $A_1'A_2'$. As a result of this, the equality (11.25) is replaced by the inequality

$$\|f - c\|^2 \leq r^2. \quad (11.41)$$

If the center C' of the hypersphere lies on $\mathcal{S}_1$, then the vectors f associated with the intersection of $\mathscr{B}_\infty'$ and its interior and $H_{(n)}$, and at greatest distance from the zero vector (at the origin O), are those vectors whose tips lie on the circumference of the hypercircle of intersection. These vectors obey the earlier condition (11.25). A similar conclusion is reached with regard to the locations $\mathscr{B}_\infty''$ and $\mathscr{B}_\infty''''$ of the hypersphere: the "longest" vector f associated with the intersection is the one whose tip lies on the circumference of the hypercircle (f_{max} in Figure 11.8). In both situations, therefore, the norm of the unknown vector f is bounded from above, and the bound coincides with the upper bound in (11.26) or (11.38). Accordingly,

$$\|f\|^2 \leq \sum_{i=1}^n \alpha_i{}^2 + \left\{ \left[\|c\|^2 - \sum_{i=1}^n (c, h^i)^2 \right]^{1/2} + \left[r^2 - \sum_{i=1}^n (\alpha_i - (c, h^i))^2 \right]^{1/2} \right\}^2,$$

$$(11.42)$$

or

$$\|f\|^2 \leq F + (G^{1/2} + H^{1/2})^2, \qquad (11.43)$$

where F, G, and H are defined by (11.39).

In order to find a lower bound, it is necessary to examine two cases: (a) the line $\mathcal{S}_1$ meets the disk of intersection (point B', hyperspheres $\mathscr{B}_\infty'$ and $\mathscr{B}_\infty''$), and (b) the line $\mathcal{S}_1$ misses this disk (hypersphere $\mathscr{B}_\infty''''$). Since the vector $\overrightarrow{OB'} \equiv f_0$ is the orthogonal projection of every f vector on the subspace $\mathcal{S}_1$, we have $f_0 = (f, h^k)h^k$ for $n = 1$ and $f_0 = \sum_{i=1}^n (f, h^i)h^i$ in the general case when $n > 1$. In the latter event, there is

$$\|f_0\|^2 = \sum_{i=1}^n \alpha_i{}^2, \qquad (11.44)$$

where the coefficients α_i are defined by (11.24). Let us now examine more closely each of the cases (a) and (b) in turn.

(a) In this case, f_0 is the vector closest to the zero vector at the origin O. Thus, there is always $\|f\| \geq \|f_0\|$, and (11.43) becomes

$$\sum_{i=1}^n \alpha_i{}^2 \leq \|f\|^2 \leq F + (G^{1/2} + H^{1/2})^2. \qquad (11.45)$$

The lower bound is here immediately identified with that given by the Bessel inequality (7.21).

(b) In this case, the hypersphere is at the location $\mathscr{B}_\infty''''$ and the situation does not differ from that displayed in Figure 11.5 (hypersphere $\mathscr{B}_\infty''$ of that figure). Consequently, the inequality (11.26) and its modification (11.38) remain true. This completes the discussion.

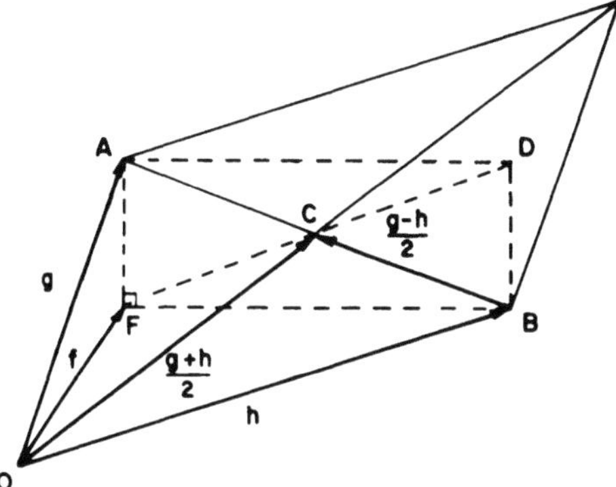

Figure 11.9. Upper bounds for a function f.

For $s = t = 0$, whence the sets h^i and g^i are disregarded, the pertinent inequality coincides with the inequality (11.40), provided

$$\left\| \frac{g+h}{2} \right\| - \left\| \frac{g-h}{2} \right\| > 0. \tag{11.46}$$

If this expression is zero or negative, however, the lower bound in (11.40) must be replaced by

$$0 \le \|f\|^2, \tag{11.47}$$

and the result is trivial. The inequality (11.46) is represented graphically in Figure 11.9. From the triangle OFC, it is not hard to infer that

$$\|f\| \le \left\| \frac{g+h}{2} \right\| + \overline{FC}. \tag{11.48}$$

Similarly, from the rectangle $ADBF$, we find that $\overline{FC} = \|(g-h)/2\|$. This furnishes the upper bound already appearing in the inequality (11.40), but missing in (11.47).

It is interesting to give some thought to the role of *bounds* versus that of *approximation in the mean*, examined in Chapter 10. To illustrate, let f be a vector whose bounds, respectively, approximation, are to be determined. We select a set of n orthonormal vectors $\{g^i\}$ and determine the coefficients c_i such that the combination $\sum_{i=1}^{n} c_i g^i$ gives the best mean approximation for f within the subspace spanned by the g^i's. Evidently,

$$\left\| f - \sum_{i=1}^{n} c_i g^i \right\|^2 = \|f\|^2 - 2 \sum_{i=1}^{n} (f, g^i) c_i + \sum_{i=1}^{n} c_i^2. \tag{11.49}$$

By adding and subtracting $\sum_{i=1}^{n} (f, g^i)^2$, we obtain

$$\left\| f - \sum_{i=1}^{n} c_i g^i \right\|^2 = \| f \|^2 - \sum_{i=1}^{n} (f, g^i)^2 + \sum_{i=1}^{n} [c_i - (f, g^i)]^2, \quad (11.50)$$

showing that the best approximation in the mean is achieved if the coefficients c_i are selected as

$$c_i = (f, g^i). \tag{11.51}$$

Consequently,

$$f \approx \sum_{i=1}^{n} (f, g^i) g^i \tag{11.52}$$

is the best approximation in the mean of f in terms of the orthonormal set $\{g^i\}$.

We now form the inner product of the foregoing approximate equality with itself to obtain

$$(f, f) \approx \sum_{i=1}^{n} \sum_{k=1}^{n} (f, g^i)(f, g^k)(g^i, g^k) \tag{11.53}$$

or, since $(g^i, g^k) = \delta^{ik}$,

$$(f, f) \approx \sum_{i=1}^{n} (f, g^i)^2. \tag{11.54}$$

The right-hand member of this formula coincides with the lower bound in the inequality (11.45) [compare equations (11.24)]; we thus conclude that the lower bound is actually furnished by the best mean approximation $\sum_{i=1}^{n} c_i g^i$.

The procedures for constructing bounds discussed above are, of course, but a few examples of the many general lines of approach possible. Apart from these, it is often practicable to adopt an ad hoc procedure especially adjusted to the problem in hand. The following examples illustrate this point.

Example 11.1. A cylindrical bar of arbitrary simply connected cross section Ω is subjected to torsion by terminal couples M_z. It is required to find the torsional rigidity $D = M_z/\alpha$ of the bar, where α is the twist per unit length.

Let the axis, i.e., the locus of centroids of the cross sections of the bar, coincide with the z axis of a Cartesian rectangular system x, y, z, and let the material of the bar display a *general rectilinear anisotropy* such that the planes parallel to the xy plane are planes of elastic symmetry. As in the isotropic case [cf. the text preceding equation (8.38)], the problem may be formulated in terms of either of two functions: Prandtl's stress function $\psi = \psi(x, y)$ or the torsion function $\phi = \phi(x, y)$. In terms of the former, the only identically nonvanishing stress components have the form

$$\tau_{zx} = \alpha\psi_{,y}, \qquad \tau_{zy} = -\alpha\psi_{,x}. \tag{11.55}$$

The stress–strain relations being now

$$2e_{zx} = a_{55}\tau_{zx} + a_{45}\tau_{zy},$$
$$2e_{zy} = a_{44}\tau_{zy} + a_{45}\tau_{zy}, \tag{11.56}$$

where the a_{ij}'s ($i, j = 4, 5$) are elastic compliances, the single compatibility equation becomes

$$a_{44}\psi_{,xx} - 2a_{45}\psi_{,xy} + a_{55}\psi_{,yy} = -2 \quad \text{in } \Omega, \tag{11.57}$$

and is supplemented by the boundary condition

$$\psi = 0 \quad \text{on } \partial\Omega. \tag{11.58}$$

This leaves us with a Dirichlet problem for the generalized Poisson equation (11.57).

In terms of the torsion function, the stress–strain relations are (Ref. 47, Sect. 29)

$$\tau_{zx} = \frac{\alpha}{\Delta}[a_{44}(\phi_{,x} - y) - a_{45}(\phi_{,y} + x)],$$
$$\tau_{zy} = \frac{\alpha}{\Delta}[a_{55}(\phi_{,y} + x) - a_{45}(\phi_{,x} - y)], \tag{11.59}$$

where $\Delta = a_{44}a_{55} - a_{45}^2$. The stress here must satisfy a single equation of equilibrium. This leads to the following equation for the torsion function:[†]

$$a_{44}\phi_{,xx} - 2a_{45}\phi_{,xy} + a_{55}\phi_{,yy} = 0 \quad \text{in } \Omega. \tag{11.60}$$

The boundary condition (8.38a) now becomes

$$(a_{44}\phi_{,x} - a_{45}\phi_{,y})\cos(n, x) - (a_{45}\phi_{,x} - a_{55}\phi_{,y})\cos(n, y)$$
$$= (a_{45}x + a_{44}y)\cos(n, x) - (a_{55}x + a_{45}y)\cos(n, y) \quad \text{on } \partial\Omega, \tag{11.61}$$

producing, in combination with the preceding equation, a Neumann problem for a generalized Laplace equation.

[†] For isotropy, $a_{45} = 0$ and $a_{44} = a_{55} = 1/\mu$, so that equations (11.59) and (11.60) reduce to the former equations (8.39) and (8.38), respectively.

It is not difficult to verify that the torsional rigidity of the bar is[48]

$$D = \frac{1}{\Delta}(P - D_1), \tag{11.62}$$

where

$$P = \int_\Omega (a_{55}x^2 + 2a_{45}xy + a_{44}y^2)\, dx\, dy \tag{11.62a}$$

and

$$D_1 = \int_\Omega [(a_{45}x + a_{44}y)\phi_{,x} - (a_{55}x + a_{45}y)\phi_{,y}]\, dx\, dy. \tag{11.62b}$$

The first of the preceding expressions is a known quantity depending on the form of the cross section; the second involves the unknown torsion function.

In order to apply a vector space approach based on a positive definite metric,† we introduce an inner product in the form

$$(u,\, v) = \int_\Omega [a_{44}u_{,x}v_{,x} - a_{45}(u_{,x}v_{,y} + u_{,y}v_{,x}) + a_{55}u_{,y}v_{,y}]\, dx\, dy, \tag{11.63}$$

where u and v are sufficiently regular functions of x and y. The preceding bilinear form is symmetric in u and v and positive definite when: (a) u and v are identified with Prandtl's function ψ, or (b) when they are identified with the torsion function ϕ, both of these functions satisfying the conditions (11.58) and (11.61) on the boundary $\partial\Omega$, respectively.

In fact, the strain energy per unit length of the twisted bar,

$$V = \frac{1}{2}\int_\Omega (a_{44}\tau_{zy}^2 + 2a_{45}\tau_{zx}\tau_{zy} + a_{55}\tau_{zy}^2)\, dx\, dy \tag{11.64}$$

in case (a) becomes

$$V = \frac{\alpha^2}{2}\int_\Omega [a_{44}(\psi_{,x})^2 - 2a_{45}\psi_{,x}\psi_{,y} + a_{55}(\psi_{,y})^2]\, dx\, dy, \tag{11.65}$$

which, on account of its physical meaning, is a positive-definite function (Ref. 48, p. 327). Thus, by virtue of equation (11.65), (ψ, ψ) is positive definite, as claimed.

---

† We note that it would suffice to employ a positive semi-definite metric, since, as demonstrated in Chapter 8 [cf. equation (8.7c)], the Cauchy–Schwarz inequality—the only vector space tool used in the problem under discussion—holds for spaces with a positive semi-definite metric.

In case (b) we have

$$(\phi, \phi) = \int_\Omega [a_{44}(\phi_{,x})^2 - 2a_{45}\phi_{,x}\phi_{,y} + a_{55}(\phi_{,y})^2]\, dx\, dy. \qquad (11.66)$$

Since (11.66) has the same form as (11.65), we can assert that the form (11.66) is also positive definite. A direct verification of the latter fact can be given as follows.

There are reasons for concluding[48] that a_{44} and a_{55} are both positive and that $|a_{45}| < a_{44}$ if, for definiteness, we assume that $a_{44} < a_{55}$. Then, if $a_{45} > 0$, (ϕ, ϕ) is certainly greater than the non-negative quantity

$$\int_\Omega a_{44}(\phi_{,x} - \phi_{,y})^2\, dx\, dy \geq 0. \qquad (11.67)$$

On the other hand, if $-a_{45} > 0$, then (ϕ, ϕ) is greater than

$$\int_\Omega |a_{45}|(\phi_{,x} + \phi_{,y})^2\, dx\, dy > 0. \qquad (11.68)$$

It follows that $(\phi, \phi) \geq 0$, and hence that the inner product (11.66) satisfies the requirements imposed on a metric.

Modeling our discussion on the isotropic case[49], we find bounds for D from the Cauchy–Schwarz inequality (11.1),

$$(u, v)^2 \leq (u, u)(v, v), \qquad (11.69)$$

where u and v are vectors.

A lower bound is obtained by selecting a sufficiently regular function $f = f(x, y)$ satisfying Prandtl's condition (11.58). Upon setting $u = \psi$ and $v = f$ in equation (11.63) and using equations (11.57) and (11.58), we arrive at the relation

$$(\psi, f) = 2 \int_\Omega f\, dx\, dy. \qquad (11.70)$$

Likewise, putting $u = \psi$ and $v = f$ in equation (11.69) and recalling that[31]

$$D = 2 \int_\Omega \psi\, dx\, dy, \qquad (11.71)$$

we have

$$D \geq \frac{4(\int_\Omega f\, dx\, dy)^2}{\int_\Omega [a_{44}(f_{,x})^2 - 2a_{45} f_{,x} f_{,y} + a_{55}(f_{,y})^2]\, dx\, dy}, \qquad (11.72)$$

where the right-hand member is a known quantity. The upper bound is obtained by taking recourse to the torsion function. We first select a sufficiently regular, but otherwise arbitrary, function $g = g(x, y)$. We then pose $u = \phi$ and $v = g$ in the equation

$$\int_{\Omega} v[a_{44}u_{,xx} - 2a_{45}u_{,xy} + a_{55}u_{,yy}] \, dx \, dy + (u, v)$$

$$= \int_{\partial\Omega} v[a_{44}u_{,x} - a_{45}u_{,y}) \cos(n, x) + (a_{55}u_{,y} - a_{45}u_{,x}) \cos(n, y)] \, ds,$$

$$(11.73)$$

obtained easily from (11.63). Next, applying equations (11.60) and (11.61), we arrive at

$$(g, \phi) = \int_{\partial\Omega} g[(a_{45}x + a_{44}y) \cos(n, x) - (a_{55}x + a_{45}y) \cos(n, y)] \, ds,$$

$$(11.74)$$

where, for a given form of the cross section, the value of the integral is known. It is found convenient to cast the preceding expression into the form

$$(g, \phi) = \int_{\Omega} [(a_{45}x + a_{44}y)g_{,x} - (a_{55}x + a_{45}y)g_{,y}] \, dx \, dy, \quad (11.75)$$

and set $u = \phi$ and $v = g$ in the Cauchy–Schwarz inequality to obtain

$$(\phi, \phi) \geq \frac{(g, \phi)^2}{(g, g)}, \tag{11.76}$$

the numerator in this inequality being positive definite for the same reason as stated earlier for expression (11.66).

The right-hand member is here a known quantity; the left-hand member coincides with D_1 by equation (11.62b) if the latter is transformed by appeal to the Gauss–Green theorem and equation (11.60). Consequently, by virtue of equation (11.62), we have explicitly

$$D \leq \frac{1}{a_{44}a_{55} - a_{45}^2}$$

$$\times \left[P - \frac{\{\int [(a_{45}x + a_{44}y)g_{,x} - (a_{55}x + a_{45}y)g_{,y}] \, dx \, dy\}^2}{\int [a_{44}(g_{,x})^2 - 2a_{45}g_{,x}g_{,y} + a_{55}(g_{,y})^2] \, dx \, dy} \right]. \quad (11.77)$$

The preceding formula furnishes an upper bound for the torsional rigidity. If the bounds fail to be sufficiently close, the estimates may be improved by using any of the known methods guaranteeing successive sharpening of bounds[50]. Instead of discussing this matter from a general point of view, we content ourselves with the following numerical example.

The faces of an orthotropic bar of rectangular cross section, axb, are parallel to the planes of elastic symmetry.

We assume the function f in the form

$$f = \left(x^2 - \frac{a^2}{4}\right)\left(y^2 - \frac{b^2}{4}\right),$$

satisfying condition (11.58). The g function, associated with the warping of the cross section via the torsion function, should, for better results, be taken in the form

$$g = xy, \tag{11.77a}$$

exhibiting the expected symmetry. With these assumptions in mind and using the notation

$$a_{44} = \frac{1}{G_{23}}, \qquad a_{55} = \frac{1}{G_{13}}, \qquad c^2 = \frac{a^2}{b^2}\frac{G_{23}}{G_{13}},$$

we have

$$G_{13}\,ab^3\beta(c) \le D \le G_{13}\,ab^3\beta^*(c),$$

where $\beta(c)$ and $\beta^*(c)$ are known functions of the ratio c (clearly $a_{45} = 0$ for an orthotropic material). Table 11.1 lists the bounds as functions of the parameter c. These are compared with the exact values calculated by Saint-Venant.[51]

It is of interest to note that for $c = 1$ and the torsion function in the form (11.77a), the Rayleigh–Ritz method in combination with Lagrange's variational principle yields the same value 0.167 for $\beta^*(1)$ as that listed in the table, that is, an upper bound for D. This result seems to confirm the conclusion of Diaz and Weinstein[45, 50] that the Rayleigh–Ritz method, used for

Table 11.1. Values of the functions $\beta(c)$ and $\beta^*(c)$

c	$\beta(c)$ lower bound	Saint-Venant solution	$\beta^*(c)$ upper bound	Mean value	Error of the mean value
1	0.139	0.141	0.167	0.153	8.5°₀
1.25	0.169	0.172	0.207	0.186	8.2
1.50	0.192	0.196	0.231	0.212	8.2
1.75	0.209	0.214	0.251	0.230	7.5
2.00	0.222	0.229	0.267	0.245	7.0
2.50	0.239	0.249	0.287	0.263	5.6

solving Dirichlet and Neumann problems for the Laplace equation, always furnishes an upper bound for the solution.

As a second demonstration of the usefulness of the machinery afforded by inner product spaces, we apply their theory to the solution of a problem involving bending of an anisotropic plate. We are interested here in finding bilateral bounds for the exact solution, and make use of a (practically) positive semi-definite metric[52]. This problem, slightly more difficult than the preceding ones, assumes knowledge of singular solutions of partial differential equations [see, e.g., Ref. 53].

Example 11.2. We first note that while the familiar nonhomogeneous biharmonic equation[24, 54-57]—and more applied equation[48, 59, 60]—

$$\frac{\partial^4 w}{\partial x^4} + 2\frac{\partial^4 w}{\partial x^2\,\partial y^2} + \frac{\partial^4 w}{\partial y^4} = f(x,\,y)$$

has been investigated rather extensively, much less attention has been devoted to the generalizations of this equation appearing, for example, in the theory of thin elastic anisotropic plates.

These are, in particular, the equations of the forms

$$D_{11}\frac{\partial^4 w}{\partial x^4} + 4D_{16}\frac{\partial^4 w}{\partial x^3\,\partial y} + 2(D_{12} + 2D_{66})\frac{\partial^4 w}{\partial x^2\,\partial y^2}$$

$$+ 4D_{26}\frac{\partial^4 w}{\partial x\,\partial y^3} + D_{22}\frac{\partial^4 w}{\partial y^4} = q$$

and

$$D_1\frac{\partial^4 w}{\partial x^4} + 2D_3\frac{\partial^4 w}{\partial x^2\,\partial y^2} + D_2\frac{\partial^4 w}{\partial y^4} = q, \tag{11.78}$$

where the D_{ij}'s and D_i's denote constants interpreted as bending rigidities of a plate experiencing a deflection $w = w(x,\,y)$ under the action of a transverse load $q(x,\,y)$. The last equation describes an "orthotropic" plate—the term coined by M. T. Huber.

The governing equations above can be supplemented by a variety of *boundary conditions*. The most common are as follows:

(a) Clamped, or built-in, contour $\partial\Omega$:

$$w = 0, \quad \frac{\partial w}{\partial n} = 0 \quad \text{on } \partial\Omega, \tag{11.78a}$$

where n denotes the outer normal on the contour $\partial\Omega$.

(b) Freely supported contour:

$$w = 0, \quad M_n = 0 \quad \text{on } \partial\Omega, \tag{11.78b}$$

where M_n denotes the bending moment acting in the plane perpendicular to the contour at each point.

(c) Free contour (compare with (8.10a) and (8.10b) for isotropy):

$$M_n = 0, \quad V_n = Q_n + \frac{\partial M_{nt}}{\partial s} = 0 \qquad \text{on } \partial\Omega. \tag{11.78c}$$

Here, V_n is the transverse force on the contour, computed as the sum of the shear force Q_n and the derivative of the twisting moment M_{nt} in the direction tangent to the contour.

(d) Contour acted upon by distributed bending moment m and vertical force p, both reckoned per unit length of the contour:

$$M_n = m, \quad V_n = p \qquad \text{on } \partial\Omega. \tag{11.78d}$$

(e) Contour deformed due to the distributed deflection $\bar{w}$ and inclination $\bar{\alpha}$ imposed on the contour:

$$w = \bar{w}, \quad \frac{\partial w}{\partial n} = \bar{\alpha} \qquad \text{on } \partial\Omega. \tag{11.78e}$$

The solution of *boundary value problems* involving such generalized harmonic operators[47, 61-65] is relatively difficult; often, exact solutions cannot be found via known techniques. This is especially true when the boundary of the domain occupied by the plate is irregular. In such cases, the derivation of bilateral bounds for the solution seems to be of help, inasmuch as it evidences the error involved in any approximate solution which can be obtained.

As a starting point, we assume the existence and the uniqueness of the solution of the boundary value problems associated with equation (11.78). With this in mind, we describe a method for obtaining bilateral bounds for the solution function $w(x, y)$ at a point (x_0, y_0) of a plane domain Ω with boundary $\partial\Omega$, the point (x_0, y_0) being given. We follow here, basically, the imaginative procedure proposed by Diaz and Greenberg for the conventional biharmonic boundary value problem, so that the conclusions of the present reasoning may be considered as a generalization of the results obtained by these authors in the paper referred to above.

Making use of the idea of Huber, we introduce the notation

$$\varepsilon^4 = D_1/D_2, \qquad \rho = D_3/(D_1 D_2)^{1/2}$$

and cast equation (11.78) into the form

$$\varepsilon^4 \frac{\partial^4 w}{\partial x^4} + 2\rho\varepsilon^2 \frac{\partial^4 w}{\partial x^2 \, \partial y^2} + \frac{\partial^4 w}{\partial y^4} = p, \tag{11.79}$$

where $p = q/D_2$. The assumption of a solution of the homogeneous version of the preceding equation in the form $w = w(x + \mu y)$, where μ is, in general, a complex parameter, leads to the characteristic equation

$$\mu^4 + 2\rho\varepsilon^2\mu^2 + \varepsilon^4 = 0,$$

whose solutions depend on the value of the parameter ρ. These are found to be, for $\rho > 1$,

$$\mu_{1,2,3,4} = \pm\varepsilon i[\rho^2 \pm (\rho^2 - 1)^{1/2}]^{1/2},$$

for $\rho = 1$,

$$\mu_{1,2} = \mu_{3,4} = \varepsilon i\rho,$$

and, for $\rho < 1$,

$$\mu_{1,2} = \pm\alpha i, \qquad \mu_{3,4} = \pm\beta i,$$

i denoting the imatinary unit and

$$\alpha, \beta = \left(\frac{i + \rho}{2}\right)^{1/2} \pm i\left(\frac{i - \rho}{2}\right)^{1/2}.$$

Clearly,

$$\alpha^2 + \beta^2 = \varepsilon^2, \qquad \alpha^2\beta^2 = \varepsilon^4,$$

so that equation (11.79) can be reduced to the form

$$\nabla_\alpha{}^2\nabla_\beta{}^2 w = p,$$

where

$$\nabla_\alpha{}^2 \equiv \alpha^2\frac{\partial^2}{\partial x^2} + \frac{\partial^2}{\partial y^2}, \qquad \nabla_\beta{}^2 \equiv \beta^2\frac{\partial^2}{\partial x^2} + \frac{\partial^2}{\partial y^2}.$$

In what follows, we confine our attention to the case $\rho \leq 1$, of more practical interest; this shall be subsequently justified. Using the Gauss–Green theorem, we arrive at the formula

$$\int_\Omega \nabla_\alpha{}^2\phi\nabla_\beta{}^2\psi \, d\Omega = \int_\Omega \phi\nabla_\alpha{}^2\nabla_\beta{}^2\psi \, d\Omega + \int_{\partial\Omega}\left[\nabla_\beta{}^2\psi\frac{\partial_\alpha\phi}{\partial n} - \phi\frac{\partial_\alpha}{\partial n}(\nabla_\beta{}^2\psi)\right] ds,$$

$$(11.79a)$$

generalizing the first classical Green identity. Here and later,

$$\frac{\partial_\alpha}{\partial n} \equiv \alpha^2 l\frac{\partial}{\partial x} + m\frac{\partial}{\partial y}, \qquad \frac{\partial_\beta}{\partial n} \equiv \beta^2 l\frac{\partial}{\partial x} + m\frac{\partial}{\partial y}$$

with l and m denoting the direction cosines of the unit outer normal n to the contour $\partial\Omega$.

Upon adding to (11.79a) its complementary identity, we arrive at the symmetric form

$$\frac{1}{2}\int_\Omega (\nabla_\alpha{}^2\phi\nabla_\beta{}^2\psi + \nabla_\beta{}^2\phi\nabla_\alpha{}^2\psi)\, d\Omega = \int_\Omega \phi\nabla_\alpha{}^2\nabla_\beta{}^2\psi\, d\Omega$$

$$+ \frac{1}{2}\int_{\partial\Omega}\left\{ \nabla_\alpha{}^2\psi\,\frac{\partial_\beta\phi}{\partial n} + \nabla_\beta{}^2\psi\cdot\frac{\partial_\alpha\phi}{\partial n} - \phi\left[\frac{\partial_\alpha}{\partial n}(\nabla_\beta{}^2\psi) + \frac{\partial_\beta}{\partial n}(\nabla_\alpha{}^2\psi)\right]\right\}\, ds. \quad (11.79b)$$

We note that, by Cauchy's inequality for integrals,

$$\left|\int_\Omega \phi(\Omega)\psi(\Omega)\, d\Omega\right|^2 \le \int_\Omega \phi^2(\Omega)\, d\Omega \int_\Omega \psi^2(\Omega)\, d\Omega,$$

we have

$$\left(\int_\Omega \nabla_\alpha{}^2\phi\nabla_\beta{}^2\psi\, d\Omega\right)^2 \le \int_\Omega (\nabla_\alpha{}^2\phi)^2\, d\Omega \int_\Omega (\nabla_\beta{}^2\psi)^2\, d\Omega.$$

Let us now introduce the symmetric inner product on a vector space of functions continuous in the closed domain Ω together with all their partial derivatives up to the fourth order inclusive,

$$(\phi,\psi) \equiv \frac{1}{2}\int_\Omega (\nabla_\alpha{}^2\phi\nabla_\beta{}^2\psi + \nabla_\alpha{}^2\psi\nabla_\beta{}^2\phi)\, d\Omega$$

producing the square of the norm of the function ψ in the form

$$(\psi,\psi) = \int_\Omega \nabla_\alpha{}^2\psi\nabla_\beta{}^2\psi\, d\Omega. \quad (11.79c)$$

Evidently the preceding self-product can in general be *positive, negative,* or *zero*; for our further derivations, however, it will be essential that it remain *non-negative*. If we limit ourselves to the most practical applications, we can easily convince ourselves that our requirement is fulfilled over a broad range of real materials, from plywood to boron–epoxy composites. For each of these strikingly different materials, the parameter $\rho < 1$ (Table 11.2), and this condition is sufficient for the positive semi-definiteness of the

Table 11.2. Values of ε^4 and ζ

Material	$D_1/h^3/12$	$D_2/h^3/12$	$D_3/h^3/12$	ε^4	ρ
Plywood [birch with bakelite glue[119], kgf cm^{-2}]	1.7×10^5	0.14×10^5	0.183×10^5	12.1	0.375
Boron–epoxy composite[66] (psi)	30.97×10^6	3.54×10^6	3.02×10^6	0.74	0.289

product (11.79c). In order to prove this, one only needs to set $(\partial^2\psi/\partial x^2)/(\partial^2\psi/\partial y^2) \equiv z$ and conclude that

$$\nabla_\alpha^2\psi\nabla_\beta^2\psi = \left(\frac{\partial^2\psi}{\partial y^2}\right)^2 (\varepsilon^4 z^2 + 2\rho\varepsilon^2 z + 1) \geq 0$$

provided $\rho \leq 1$.

We now arrive at a useful inequality by introducing two functions h_1 and h_2 such that, with no field equation for h_1 and no boundary condition for h_2,

$$h_1 = w \qquad \text{on } \partial\Omega, \tag{11.80a}$$

$$\frac{\partial_\alpha h_1}{\partial n} = \frac{\partial_\alpha w}{\partial n}, \qquad \frac{\partial_\beta h_1}{\partial n} = \frac{\partial_\beta w}{\partial n} \qquad \text{on } \partial\Omega \tag{11.80b}$$

and

$$\nabla_\alpha^2\nabla_\beta^2 h_2 = \nabla_\alpha^2\nabla_\beta^2 w \qquad \text{in } \Omega. \tag{11.80c}$$

For definiteness, we also assume that the function w obeys the following boundary conditions:

$$w = f, \quad \frac{\partial w}{\partial x} = g_1, \quad \frac{\partial w}{\partial y} = g_2 \qquad \text{on } \partial\Omega.$$

Evidently,

$$(h_1 - h_2, h_1 - h_2) = (h_1 - w, h_1 - w) + (h_2 - w, h_2 - w),$$

since the product $(h_1 - w, h_2 - w)$ vanishes on account of the identity (11.79b) and the conditions (11.80a)–(11.80c). This implies that

$$\left.\begin{array}{c}(h_1 - w, h_1 - w) \\ (h_2 - w, h_2 - w)\end{array}\right| \leq (h_1 - h_2, h_1 - h_2). \tag{11.80d}$$

Following Mossakowski, we select a singular solution, $\bar{s}$, of the equation

$$\nabla_\alpha^2\nabla_\beta^2 w = \frac{1}{D_2}\, \delta(x - 0, y - 0),$$

where δ is the Dirac function (see Chapter 16 for a definition), in the form

$$\bar{s} = C_1\beta s_1(x, y) + C_2\alpha s_2(x, y),$$

where

$$s_1 = (x^2 - \alpha^2 y) \ln(x^2 + \alpha^2 y^2) - 4xy \arctan \frac{\alpha y}{x},$$

$$s_2 = (x^2 - \beta^2 y) \ln(x^2 + \beta^2 y^2) - 4xy \arctan \frac{\beta y}{x}.$$

We interpret this solution as the deflection of a plate loaded by a concentrated unit force applied at the origin of the coordinates x, y. From the condition of static equilibrium, it follows that

$$C_1 = -C_2 = -\frac{1}{8\pi D_0(\alpha^2 - \beta^2)},$$

where $D_0 = \varepsilon^2 D_2$.

It is not difficult to verify that, for $\rho = 1$ and $\alpha = \beta = \varepsilon$,

$$\bar{s} = \frac{1}{16\pi\varepsilon D_0}(x^2 + \varepsilon^2 y^2) \ln(x^2 + \varepsilon^2 y^2),$$

so that for the isotropic case, we recover the classical singularity of the biharmonic equation,

$$\bar{s} = \frac{1}{8\pi D} r^2 \ln r.$$

A lengthy, but routine, calculation, using the identity (11.79b), leads to the following representation for the value of w at the origin:

$$w(0, 0) = \frac{\alpha\beta}{D_0} \bar{\Omega},$$

where

$$\bar{\Omega} = \int_\Omega \bar{s} \nabla_\alpha{}^2 \nabla_\beta{}^2 w \, d\Omega$$

$$+ \frac{1}{2} \int_{\partial\Omega} \left\{ \nabla_\alpha{}^2 w \frac{\partial_\beta \bar{s}}{\partial n} + \nabla_\beta{}^2 w \frac{\partial_\alpha \bar{s}}{\partial n} - \bar{s} \left[\frac{\partial_\alpha}{\partial n}(\nabla_\beta{}^2 w) + \frac{\partial_\beta}{\partial n}(\nabla_\beta{}^2 w) \right] \right\} ds$$

$$- \frac{1}{2} \int_{\partial\Omega} \left\{ \nabla_\alpha{}^2 \bar{s} \frac{\partial_\beta w}{\partial n} + \nabla_\beta{}^2 \bar{s} \frac{\partial_\alpha w}{\partial n} - w \left[\frac{\partial_\alpha}{\partial n}(\nabla_\beta{}^2 \bar{s}) + \frac{\partial_\beta}{\partial n}(\nabla_\alpha{}^2 \bar{s}) \right] \right\} ds. \quad (11.81a)$$

Following Diaz and Greenberg, we introduce three auxiliary functions h_0, h_3, and h_4, such that

$$\nabla_\alpha^2 \nabla_\beta^2 h_0 = 0 \qquad \text{in } \Omega,$$

$$h_0 = -\bar{s}, \quad \frac{\partial h_0}{\partial x} = -\frac{\partial \bar{s}}{\partial x}, \quad \frac{\partial h_0}{\partial y} = -\frac{\partial \bar{s}}{\partial y} \qquad \text{on } \partial\Omega,$$

$$h_3 = h_0, \quad \frac{\partial h_3}{\partial x} = \frac{\partial h_0}{\partial x}, \quad \frac{\partial h_3}{\partial y} = \frac{\partial h_0}{\partial y} \qquad \text{on } \partial\Omega, \qquad (11.81\text{b})$$

$$\nabla_\alpha^2 \nabla_\beta^2 h_4 = 0 \qquad \text{in } \Omega.$$

Now, by the Cauchy–Schwarz inequality,

$$(h_1 - w, h_3 - h_0)^2 \le (h_1 - w, h_1 - w)(h_3 - h_0, h_3 - h_0),$$

or, by appeal to (11.80d),

$$(h_1 - w, h_3 - h_4)^2 \le (h_1 - h_2, h_1 - h_2)(h_3 - h_4, h_3 - h_4). \quad (11.81\text{c})$$

Green's identity, after some manipulation, implies that

$$(h_1 - w, h_3 - h_0) = (h_1, h_3) - \int_\Omega h_3 \nabla_\alpha^2 \nabla_\beta^2 w \, d\Omega$$

$$- \frac{1}{2} \int_{\partial\Omega} \left\{ \nabla_\alpha^2 w \frac{\partial_\beta h_3}{\partial n} + \nabla_\beta^2 w \frac{\partial_\alpha h_3}{\partial n} - h_3 \left[\frac{\partial_\alpha}{\partial n}(\nabla_\beta^2 w) + \frac{\partial_\beta}{\partial n}(\nabla_\alpha^2 w) \right] \right\} ds.$$

The only expression in the preceding equation involving the unknown function w is the contour integral. It is easily eliminated, however, by observing that it coincides with the first contour integral in equation (11.81a). Making use of the inequality (11.81c), we thus arrive at the basic inequality,

$$\left[\frac{\alpha\beta}{D_0} w(0, 0) - b \right]^2 \le ac,$$

where

$$a = (h_1 - h_2, h_1 - h_2), \qquad c = (h_3 - h_4, h_3 - h_4),$$

$$b = \int_\Omega (\bar{s} + h_3)\nabla_\alpha^2 \nabla_\beta^2 w \, d\Omega - \frac{1}{2} \int_{\partial\Omega} \left\{ \nabla_\alpha^2 \bar{s} \frac{\partial_\beta w}{\partial n} + \nabla_\beta^2 \bar{s} \frac{\partial_\alpha w}{\partial n} \right.$$

$$\left. - w \left[\frac{\partial_\alpha}{\partial n}(\nabla_\beta^2 \bar{s}) + \frac{\partial_\beta}{\partial n}(\nabla_\alpha^2 \bar{s}) \right] \right\} ds - (h_1, h_3).$$

Analogous reasoning, using inequality (11.80d), leads to the second basic inequality,

$$\left[\frac{\alpha\beta}{D_6} w(0, 0) - b' \right]^2 \le ac, \qquad (11.81\text{d})$$

where

$$b' = \int_{\Omega} \bar{s} \nabla_x{}^2 \nabla_\beta{}^2 w \, d\Omega$$

$$- \frac{1}{2} \int_{\partial\Omega} \left\{ \nabla_x{}^2 \bar{s} \frac{\partial_\beta w}{\partial n} + \nabla_\beta{}^2 \bar{s} \frac{\partial_x w}{\partial n} - w \left[\frac{\partial_x}{\partial n} (\nabla_\beta{}^2 \bar{s}) + \frac{\partial_\beta}{\partial n} (\nabla_x{}^2 \bar{s}) \right] \right\} ds$$

$$- \frac{1}{2} \int_{\partial\Omega} \left\{ \nabla_x{}^2 h_4 \frac{\partial_\beta w}{\partial n} + \nabla_\beta{}^2 h_4 \frac{\partial_x w}{\partial n} - w \left[\frac{\partial_x}{\partial n} (\nabla_\beta{}^2 h_4) + \frac{\partial_\beta}{\partial n} (\nabla_x{}^2 h_4) \right] \right\} ds$$

$$+ \frac{1}{2} \int_{\partial\Omega} \left\{ \nabla_x{}^2 h_2 \frac{\partial_\beta \bar{s}}{\partial n} + \nabla_\beta{}^2 h_2 \frac{\partial_x \bar{s}}{\partial n} - \bar{s} \left[\frac{\partial_x}{\partial n} (\nabla_\beta{}^2 h_2) + \frac{\partial_\beta}{\partial n} (\nabla_x{}^2 h_2) \right] \right\} ds$$

$$+ (h_2, h_4).$$

We now turn our attention to a numerical example in which we set, for definiteness and simplicity of writing, $\rho = 1$; this implies that $\alpha = \beta = \varepsilon$. We also assume that the plate is circular, of radius 1, that it is acted upon by a uniform load q, and is built-in on the boundary in the sense that

$$w = 0, \quad \frac{\partial w}{\partial x} = \frac{\partial w}{\partial y} = 0 \qquad \text{on } \partial\Omega.$$

The origin of the Cartesian coordinates x, y (or of the polar coordinates r, ϕ) we locate at the center of the plate. If in the inequality (11.81d) one considers b' as an approximate value of $w(0, 0)$, then, in the case under consideration,

$$w(0, 0) \approx \frac{D_2}{\varepsilon^2} \frac{q}{D_2} \left\{ \int_{\Omega} \bar{s} \, d\Omega \right\} + \int_{\partial\Omega} \left\{ \nabla_\varepsilon{}^2 h_2 \frac{\partial_\varepsilon \bar{s}}{\partial n} - \bar{s} \frac{\partial_\varepsilon}{\partial n} (\nabla_\varepsilon{}^2 h_2) \right\} ds.$$

A lengthy, but straightforward, calculation† gives

$$w(0, 0) \approx \frac{q}{16\varepsilon^3 D_2} \left\{ -\frac{1 + \varepsilon^2}{8} + \tfrac{1}{4}[2\bar{\alpha} \ln w + (1 - \varepsilon)^2] \right\}$$

$$+ \frac{C}{2} \frac{1}{\varepsilon} \left[(\gamma + \delta - 4\bar{\alpha})(1 + \ln \omega) + (\gamma - \delta)\frac{1 - \varepsilon}{1 + \varepsilon} \right]$$

$$- \frac{1}{\varepsilon^3} \left[\varepsilon^2(\gamma + \delta)\ln \omega + \varepsilon^2(\gamma - \delta)\frac{1 - \varepsilon}{1 + \varepsilon} \right] + (\delta + \varepsilon^4\gamma - \varepsilon^2\gamma - \varepsilon^2 \delta)I \Big\},$$

---

† It may be of interest to note that, of the two seldom-seen integrals $\int_0^{2\pi} [\sin^2 x/(p + q \cos x)] \, dx$ and I [see (11.81e)] appearing in this calculation, the first was meticulously analyzed by Brodovitskii, while the second does not seem to lend itself to evaluation in a simple form. See Brodovitskii.[67]

where

$$\bar{\alpha} = \frac{1 + \varepsilon^2}{2}, \qquad \bar{\beta} = \frac{1 - \varepsilon^2}{2}, \qquad \gamma = 3\varepsilon^2 + 1, \qquad \delta = \varepsilon^2 + 3,$$

$$\omega = \left(\frac{1 + \varepsilon}{2}\right)^2, \qquad C = q/8D_2(3\varepsilon^4 + 2\varepsilon^2 + 3), \tag{11.81e}$$

$$I = \frac{1}{4} \int_0^{2\pi} \sin^2 2\psi \, \ln(\bar{\alpha} + \bar{\beta} \cos 2\psi) \, d\psi,$$

and we have selected

$$h_2 = C(1 - x^2 - y^2)^2,$$

with $C = q/8D_2(3\varepsilon^4 + 2\varepsilon^2 + 3)$, in order to satisfy the condition (11.80c).

The conditions imposed on the functions h_1 and h_4 are satisfied if we simply set $h_1 = h_4 = 0$ identically in $\Omega + \partial\Omega$.

For an *isotropic* material $\bar{\alpha} = \varepsilon = 1$, $\bar{\beta} = 0$, $\gamma = \delta = 4$, $D_0 = D$, and $C = q/64D$. Conditions (11.81b) are then obeyed by the function $h_3 = q(1 - r^2)/16\pi D$. With these in mind, the inequality (11.81d) reduces to

$$\left(w(0, 0) - \frac{q}{64D}\right)^2 = 0,$$

demonstrating that—by sheer accident—b' turns out to be the exact value of the deflection at the center of the plate (cf., e.g., Timoshenko,[32] p. 60).

Let us next consider an orthotropic plate with a pronounced difference in its principle bending rigidities characterized by the parameter $\varepsilon = 3$, that is, by $D_1/D_2 = 81$. With the same assumptions as before concerning the functions h_1, h_2, and h_4, we set

$$h_3 = -\frac{1}{16\pi\varepsilon^3 D_2} (\bar{\alpha} + \bar{\beta} \cos 2\phi) \ln(\bar{\alpha} + \bar{\beta} \cos 2\phi)$$

in $\Omega + \partial\Omega$, thus complying with the boundary conditions (11.81b). A lengthy calculation leads to the following inequality:

$$\left(w(0, 0) - 0.032 \frac{q}{64D_2}\right)^2 < 0.001\left(\frac{q}{64D_2}\right)^2.$$

The exact value of the deflection[47] is $w(0, 0) = 0.030q/64D_2$, so that by taking $w(0, 0) \approx b'$, we commit an error not exceeding 6.7%. The resulting bound ($|0.030q/64D_2 - 0.032q/64D_2| \leq 0.01q/64D_2$) can be improved by any of the known iteration procedures; see, e.g., Ref. 58, Sect. 3.

11.1. Bounds for a Solution at a Point

11.1.1. The L^*L Method of Kato–Fujita

So far in this chapter, procedures for obtaining bounds for integral ("mean-square") approximations of unknown vectors have been given. In this section we intend to present a method proposed by Kato and Fujita for deriving pointwise bounds for solutions of boundary value problems[68-71].

We first recall that a linear differential operator L^* is called the *adjoint* of a linear differential operator L if

$$(Lu, v) = (u, L^*v). \tag{11.82}$$

The preceding equation serves to define L^*. In case $L^* = L$, we say that the operator L is *self-adjoint*. If the boundary conditions associated with L^* differ from those associated with L, some authors say that L^* is the *formal* adjoint of L and that L is *formally* self-adjoint if $L = L^*$. We shall not distinguish, however, between formal and exact adjoits at this point (see, however, Chapter 15, Sec. 15.3, as well as Refs. 22 and 72).

It is shown that if L^* is the adjoint of L, then $H \equiv L^*L$ is a self-adjoint operator in a Hilbert space containing the domain of L[73].

To illustrate, set $L = d/dx$. To find the adjoint, we write for some interval $a \leq x \leq b$, with $u = v = 0$ at $x = a$ and $x = b$,

$$(Lu, v) = \int_a^b vu'\, dx$$

$$= vu\Big|_a^b - \int_a^b uv'\, dx = - \int_a^b uv'\, dx. \tag{11.83}$$

But $(u, Lv) = \int_a^b uv'\, dx$ and, comparing with (11.83), we find that $L^* = -d/dx$. Now $H = L^*L = d^2/dx^2$ and

$$(Hu, v) = - \int_a^b u''v\, dx$$

$$= \int_a^b u'v'\, dx. \tag{11.84}$$

This result, when compared with

$$(u, Hv) = - \int_a^b uv''\, dx$$

$$= \int_a^b u'v'\, dx, \tag{11.84a}$$

proves our assertion.

The central theorem of the Kato–Fujita method concerning linear differential equations of the class

$$L^*Lu = f, \tag{11.85}$$

where f is a known vector, is as follows:

Let u_0 be the *solution* of the equation (11.85) subject to certain boundary conditions. We set $\alpha = (u_0, g)$ where g is a given vector, and choose vectors h, h_0, h_1, and h_2 such that h and h_0 are any conveniently selected vectors and

$$L^*h_1 = f, \qquad L^*h_2 = g. \tag{11.86}$$

Then the following inequality obtains[69]:

$$|\alpha - \tfrac{1}{2}(\beta + \gamma)| \leq \tfrac{1}{2}\,\delta\varepsilon, \tag{11.87}$$

where

$$\begin{aligned}
\beta &= (h, g) + (f, h_0) - (Lh, Lh_0),\\
\gamma &= (h_1, h_2),\\
\delta &= \|Lh_0 - h_2\|, \qquad \varepsilon = \|Lh - h_1\|.
\end{aligned} \tag{11.87a}$$

It should be noted that many problems in mathematical physics can be reduced to the form (11.85), so that the Kato theorem has a wide range of application.

As an example, assume that $Lu \equiv \operatorname{grad} u$, where u is a function of the variables x and y ranging over a domain Ω with the boundary $\partial\Omega$. We have, for $v = [v_x, v_y]$ and $\vec{i}$ and $\vec{j}$ as coordinate vectors,

$$\begin{aligned}
(Lu, v) &= \int_\Omega \left(\vec{i}\frac{\partial u}{\partial x} + \vec{j}\frac{\partial u}{\partial y}\right) \cdot (\vec{i}v_x + \vec{j}v_y)\, dx\, dy\\[4pt]
&= \int_\Omega \left(\frac{\partial u}{\partial x} v_x + \frac{\partial u}{\partial y} v_y\right) dx\, dy\\[4pt]
&= \int_\Omega \left[(uv_x)_{,x} + (uv_y)_{,y}\right] dx\, dy - \int_\Omega u(v_{x,x} + v_{y,y})\, dx\, dy\\[4pt]
&= \int_{\partial\Omega} [uv_x n_x + uv_y n_y]\, ds - \int_\Omega u \operatorname{div} v\, dx\, dy.
\end{aligned} \tag{11.88}$$

Disregarding the line integral,† we infer that

$$L^*v = -\operatorname{div} v. \tag{11.89}$$

---

† For L^* to be an exact adjoint of L, it suffices that $u = 0$ on $\partial\Omega$.

Consequently, $L^*Lu = f$ becomes

$$-\text{div grad } u = f$$

or
$$-\nabla^2 u = f, \tag{11.90}$$

which is the familiar Poisson equation. By a similar argument, it is shown that $Lu \equiv \nabla^2 u$ implies that $L^*v = \nabla^2 u$, whence $L^*Lu = f$ becomes

$$\nabla^2\nabla^2 u = f \quad \text{in } \Omega,$$

$$u = \frac{\partial u}{\partial n} = 0 \quad \text{on } \partial\Omega. \tag{11.91}$$

Following Fujita, we apply Kato's theorem to the Poisson problem

$$-\nabla^2 u = f \quad \text{in } \Omega, \tag{11.92}$$

$$u = 0 \quad \text{on } \partial\Omega. \tag{11.92a}$$

As shown above, $L^* = -\text{div}$ in the present case. For simplicity, we also assume that $f = 1$. We now take $g = \delta(P)$, the Dirac function with singularity at a preassigned point $P = (x, y)$. Consequently,†

$$\alpha = u_0(P). \tag{11.93}$$

In view of equation (11.86), we select, with $\vec{r} = \overrightarrow{PQ}$, where Q is a point in Ω,

$$h_1 = -\vec{r}/2 \tag{11.94}$$

and $(r = |\overrightarrow{PQ}|)$

$$h_2 = -\frac{1}{2\pi} \text{ grad ln } r, \tag{11.95}$$

the latter being a singular solution of the equation $-\text{div } h = \delta(P)$ (Ref. 72, p. 71). Furthermore, we put

$$h = 0 \quad \text{and} \quad h_0 = \begin{cases} -\dfrac{1}{2\pi} \ln r & \text{for } r \le 1, \\[2mm] 0 & \text{for } r > 1. \end{cases} \tag{11.96}$$

---

† Compare a definition and the third property of Dirac function, equations (16.3), in Chapter 16.

With these in mind, we evaluate β and γ from equations (11.87a), finding

$$\beta = (f, h_0) = -\int_0^1 \ln r \cdot r \, dr$$

$$= \tfrac{1}{4}, \tag{11.97a}$$

$$\gamma = (h, h_2) = \frac{1}{4\pi} \int_\Omega d\Omega, \tag{11.97b}$$

where Ω is the preassigned domain of integration. Imagine that the latter is bounded by an ellipse $x^2/a^2 + y^2/b^2 = 1$ and set $b = 1$ for convenience. Then the "approximation" of $\alpha = u_0(0)$ for P at the origin is

$$u_0(0) \approx \frac{\beta + \gamma}{2}$$

$$= \frac{a + 1}{8}, \tag{11.98}$$

the "error" being measured by the value of $\varepsilon\delta/2$. This gives for $a = 1$, 2, and 3, $u_0(0) \approx 0.25$, 0.375, and 0.45, respectively. The pertinent exact values[74] are 0.25, 0.4, and 0.45, so that the relative errors of our crude approximation are 0%, 6.2%, and 8.9%, respectively, which are practically tolerable.

The approximation $\alpha \approx \tfrac{1}{2}(\beta + \gamma)$ can, if desired, be considerably improved by introducing additional trial vectors and representing the functions h, h_0, h_1, and h_2 as linear combinations of these vectors with coefficients determined by minimizing the values of δ^2 and ε^2 from the last two of equations (11.87a).

Fujita found that a close correspondence exists between the L^*L-method and the method of the hypercircle examined in the next chapter. In the language of the theory of elasticity, the vector h_1 corresponds to the vector S defining the elastic state of the body; the conditions of compatibility satisfied by S translate into the requirement that h_1 belongs to the range of the operator L; the equations of equilibrium have their counterpart in the first of equations (11.86); finally, the solution vector S_0 corresponds to the vector Lu_0.

11.1.2. The Diaz–Greenberg Method

The Diaz–Greenberg procedure for obtaining pointwise bounds for solutions of boundary value problems consists in making use of Green's function or tensor. Below, the method is illustrated with an example of a two-dimensional nonhomogeneous biharmonic problem.

Consider the boundary value problem

$$\nabla^2\nabla^2 w = q \quad \text{in } \Omega, \tag{11.99}$$

$$w = f \quad \text{and} \quad \partial w/\partial n = g \quad \text{on } \partial\Omega, \tag{11.99a}$$

where Ω is a plane region in the xy-plane with boundary $\partial\Omega$. Concretely, the function $w = w(x, y)$ may be interpreted as the deflection of a thin elastic plate acted upon by a transverse load of intensity $p(x, y) = qD$, where D is the flexural rigidity of the plate.

We introduce five auxiliary functions[†] h_0, h_1, h_2, h_3, and h_4 such that:

(a) $$\nabla^2\nabla^2 h_0 = 0 \quad \text{in } \Omega, \tag{11.100}$$

$$h_0 = -r^2 \ln r \quad \text{and} \quad \frac{\partial h_0}{\partial n} = -\frac{\partial}{\partial n}(r^2 \ln r) \quad \text{on } \partial\Omega; \tag{11.100a}$$

here, r is the magnitude of the radius emanating from a fixed point P as the origin and h_0 is the regular part of the Green's function

$$G(x, y; P) = r^2 \ln r + h_0(x, y), \tag{11.101}$$

$r^2 \ln r$ being a fundamental solution of the biharmonic equation $\nabla^2\nabla^2\phi = 0$.

(b) $h_1 = h_1(x, y)$ is a function satisfying the boundary conditions (11.99a),

$$h_1 = f \quad \text{and} \quad \frac{\partial h_1}{\partial n} = g \quad \text{on } \partial\Omega. \tag{11.102}$$

(c) $h_2 = h_2(x, y)$ is a function satisfying equation (11.99),

$$\nabla^2\nabla^2 h_2 = q \quad \text{in } \Omega. \tag{11.103}$$

(d) As a counterpart to h_1 and h_2, the function h_3 obeys the boundary conditions (11.100a),

$$h_3 = -r^2 \ln r \quad \text{and} \quad \frac{\partial h_3}{\partial n} \equiv -\frac{\partial}{\partial n}(r^2 \ln r) \quad \text{on } \partial\Omega. \tag{11.104}$$

(e) Finally, the function h_4 is a solution of the equation (11.100),

$$\nabla^2\nabla^2 h_4 = 0 \quad \text{in } \Omega. \tag{11.105}$$

[†] See Ref. 54 and Ref. 75. Our discussion is patterned after the first of these papers. Compare also the remarks concerning the Washizu method, infra.

In order to proceed further, we first recall the second Green identity in the form

$$\int_{\Omega} \nabla^2 u \nabla^2 v \, dx \, dy = \int_{\Omega} u \nabla^2 \nabla^2 v \, dx \, dy + \int_{\partial\Omega} \left(\nabla^2 v \frac{\partial u}{\partial n} - u \frac{\partial(\nabla^2 v)}{\partial n} \right) ds.$$

$$(11.106)$$

We now interchange u and v on the right-hand side of the preceding equation and equate the results. We next surround—in the well-known manner—the fixed point P in Ω by a circle C_δ of radius δ and remove the interior of this circle from Ω to obtain the domain Ω^1. Finally, we take $u = w$ and $v = r^2 \ln r$, apply in Ω^1 the equation obtained from Green's identity, and, after passing to the limit with $\delta \to 0$, arrive at the central equation

$$8\pi w(P) = \int_{\Omega} S \nabla^2 \nabla^2 w \, dx \, dy + \int_{\partial\Omega} \left(w \frac{\partial \nabla^2 S}{\partial n} - \nabla^2 S \frac{\partial w}{\partial n} \right) ds$$

$$+ \int_{\partial\Omega} \left(\nabla^2 w \frac{\partial S}{\partial n} - S \frac{\partial \nabla^2 w}{\partial n} \right) ds, \qquad (11.107)$$

where $S \equiv r^2 \ln r$.

Following Diaz and Greenberg, we introduce a positive semi-definite inner product in the form

$$(u, v) = \int_{\Omega} \nabla^2 u \nabla^2 v \, dx \, dy; \qquad (11.108)$$

we write

$$(h_1 - h_2, h_1 - h_2) = (h_1 - w + w - h_2, h_1 - w + w - h_2)$$

$$= (h_1 - w, h_1 - w) + (w - h_2, w - h_2),$$

since $(h_1 - w, w - h_2) = 0$ by Green's identity, after considering the properties of the functions h_1 and h_2. The preceding identity implies that

$$(h_1 - h_2, h_1 - h_2) \geq \begin{cases} (h_1 - w, h_1 - w) \\ (h_2 - w, h_2 - w). \end{cases} \qquad (11.109)$$

By similar reasoning, there is

$$(h_3 - h_4, h_3 - h_4) \geq \begin{cases} (h_3 - h_0, h_3 - h_0) \\ (h_4 - h_0, h_4 - h_0). \end{cases} \qquad (11.110)$$

We now observe that, by the Cauchy–Schwarz inequality,

$$(h_2 - w, h_4 - h_0)^2 \leq (h_2 - w, h_2 - w)(h_4 - h_0, h_4 - h_0), \quad (11.111)$$

so that, in view of the inequalities (11.109) and (11.111),

$$(h_2 - w, h_4 - h_0)^2 \leq (h_1 - h_2, h_1 - h_2)(h_3 - h_4, h_3 - h_4). \quad (11.112)$$

Turning our attention to Green's identity (11.106), we conclude that

$$(h_2 - w, h_4 - h_0) = (h_2, h_4) + (h_0, w - h_2) - (w, h_4)$$

$$= (h_2, h_4) + \int_{\partial\Omega} \left[\nabla^2(w - h_2)\frac{\partial h_0}{\partial n} - h_0 \frac{\partial \nabla^2(w - h_2)}{\partial n} \right] ds$$

$$- \int_{\partial\Omega} \left(\nabla^2 h_4 \frac{\partial w}{\partial n} - w \frac{\partial \nabla^2 h_4}{\partial n} \right) ds. \quad (11.113)$$

We observe that the only integral on the right-hand side of the foregoing equation involving unknown quantities is

$$I \equiv \int_{\partial\Omega} \left(\nabla^2 w \frac{\partial h_0}{\partial n} - h_0 \frac{\partial \nabla^2 w}{\partial n} \right) ds. \quad (11.114)$$

Since $h_0 = -S$ on $\partial\Omega$, this is the same integral as the last one in equation (11.107), save for a minus sign [cf. equations (1.100a)]. Adding and subtracting the two remaining line integrals on the right-hand side of (11.113) from the right-hand side of (11.107) and combining with the inequality (11.112), we easily arrive at the desired formula

$$[8\pi w(P) - \alpha]^2 \leq \beta\gamma, \quad (11.115)$$

where

$$\alpha = (h_2, h_4) - \int_{\partial\Omega} \left(\nabla^2 h_2 \frac{\partial h_0}{\partial n} - h_0 \frac{\partial \nabla^2 h_2}{\partial n} \right) ds - \int_{\partial\Omega} \left(\nabla^2 h_4 g - f \frac{\partial \nabla^2 h_4}{\partial n} \right) ds$$

$$+ \int_{\partial\Omega} \left(f \frac{\partial \nabla^2(r^2 \ln r)}{\partial n} - \nabla^2(r^2 \ln r)g \right) ds + \int_{\Omega} qr^2 \ln r \, dx \, dy, \quad (11.116)$$

$$\beta = (h_1 - h_2, h_1 - h_2),$$

$$\gamma = (h_3 - h_4, h_3 - h_4).$$

We can consider α to be an approximate value of $8\pi w(P)$ and $\beta\gamma$ as the "error" of this approximation.

To give an idea of the effectiveness of the Diaz–Greenberg procedure, consider the following simple example.

Example 11.3. A circular plate of radius a and bending rigidity D^* is built-in at its boundary ($f = g = 0$ on $\partial\Omega$) and subjected to a uniform load of intensity p. It is required to find an approximate value α of the deflection of the plate at its center, at which we set $r = 0$. Bearing in mind that the third and fourth integrals on the right-hand side of equation (11.116) vanish, we select the remaining auxiliary functions h_0, h_2, and h_4 as follows: In view of the conditions (11.100) and (11.100a), we set

$$h_0 = -r^2 \ln r \quad \text{in} \quad \Omega,$$

$$\frac{\partial h_0}{\partial n} = -\frac{\partial}{\partial r}(r^2 \ln r) \quad \text{on} \quad \partial\Omega. \tag{11.117}$$

Referring to equation (11.103), we make

$$h_2 = \frac{q}{64} r^4 + C_1 r^2 \quad \text{in} \quad \Omega, \tag{11.118}$$

where C_1 is a coefficient to be determined later. Finally, in accordance with equations (11.102) and (11.105) we simply set

$$h_1 = 0, \quad h_4 = 0 \quad \text{in} \quad \Omega. \tag{11.119}$$

In order to determine the coefficient C_1, we minimize the inner product β; this, of course, can only sharpen the bound. Requiring

$$2\pi \int_0^a \left(\frac{qr^2}{4} + 4C_1\right)^2 r \, dr = \min, \tag{11.120}$$

we find

$$C_1 = -\frac{qa^2}{32}. \tag{11.120a}$$

A straightforward calculation then yields

$$\alpha = \frac{\pi qa^4}{8} \quad \text{and} \quad w(0) = \frac{pa^4}{64D^*}. \tag{11.121}$$

We were thus fortunate enough to select the trial functions h_i, $i = 0, 1, 2$, and 4, in such forms as to obtain the exact value of the deflection at the center of the plate (Ref. 32, p. 60).

In general, there will be a need to improve the bounds given by the inequality (11.115). In this case, an iteration procedure is applied, consisting of the addition of finite linear combinations of functions to some of the functions h_i and minimization of the right-hand side of (11.115) in order to determine the coefficients involved in these combinations (as done above with regard to the function h_2).

11.1.3. The Washizu Procedure[76]

The Washizu technique is somewhat reminiscent of the procedure of Diaz and Greenberg, in that it makes use of fundamental solutions of differential equations, containing the basic singularities of Green's functions. It has been applied for the derivation of bounds for solutions of problems of elasticity in the general two- and three-dimensional cases, bending of plates, and torsion of bars. We confine ourselves below to an examination of the application of the method in the theory of bending of thin plates, governed by the system composed of the following equations:

(a) Equations of equilibrium in the presence of a transverse load $p = p(x_1, x_2)$,

$$m_{ij,ij} + p = 0, \tag{11.122}$$

where m_{ij}, $i, j = 1, 2$, denote the bending and twisting moments, and $x_1 \equiv x$, $x_2 \equiv y$ are Cartesian rectangular coordinates parametrizing the middle plane.

(b) Compatibility equations,

$$\kappa_{ij} = -w_{,ij}, \tag{11.123}$$

where κ_{oj}, $i, j = 1, 2$ are the curvatures of the middle plane of the plate and w its deflection.

(c) Constitutive equations,

$$m_{ij} = D^*[(1 - v)\kappa_{ij} + v\kappa_{kk}\,\delta_{ij}], \tag{11.124}$$

with D^* as the bending rigidity and v as Poisson's ratio.

(a_1) Substitution of (11.123) and (11.124) into (11.122) yields the second form of the equilibrium condition,

$$D^*w_{,ijij} = p. \tag{11.125}$$

(d) With the notation

$$m_n = m_{11}l^2 + m_{22}m^2 + 2m_{12}lm,$$

$$m_{ns} = m_{12}(l^2 - m^2) + (m_{22} - m_{11})lm, \tag{11.125a}$$

$$q_n = \left(\frac{\partial m_{11}}{\partial x_1} + \frac{\partial m_{12}}{\partial x_2}\right) + \left(\frac{\partial m_{22}}{\partial x_2} + \frac{\partial m_{12}}{\partial x_1}\right)m$$

(l, m are the direction cosines of the direction n, perpendicular to the direction s), the boundary conditions become

$$m_n = f_1(s), \qquad v_n \equiv q_n + \frac{\partial m_{ns}}{\partial s} = f_2(s) \tag{11.126}$$

on a certain portion $\partial\Omega_1$ of the contour $\partial\Omega = \partial\Omega\Omega_1 + \partial\Omega_2$ of the plate and

$$w = g_1(s), \qquad \partial w/\partial n = g_2(s) \tag{11.127}$$

on the remaining portion $\partial\Omega_2$ of the contour. Clearly, n and s denote directions, respectively, normal and tangential to the contour. We introduce the following state vectors:

S, denoting the actual solution of the problem,

S^m, obeying equations (11.122) $[\equiv(11.125)]$ and (11.126) on $\partial\Omega_1$,

S^c, obeying equations (11.123) and (11.127) on $\partial\Omega_2$,

Φ, denoting a (singular) fundamental state satisfying equation (11.123) and either of equations (11.122) or (11.125) in which p is replaced by the delta function with the singularity† at $P = (\xi_1, \xi_2)$:

$$D^* w_{,ijij} = 8\pi D^* \, \delta(x_1 - \xi_1, x_2 - \xi_2). \tag{11.127a}$$

U^m, satisfying the homogeneous version of equation (11.122) $[\equiv(11.125)]$ and equating $-\Phi$ on $\partial\Omega_1$,

U^c, satisfying equations (11.123) and equating $-\Phi$ on $\partial\Omega_2$,

W, satisfying equations (11.123), the homogeneous version of equations (11.122) $[\equiv(11.125)]$, and equal to $-\Phi$ on both $\partial\Omega_1$ and $\partial\Omega_2$.

In order to avoid formal difficulties resulting from the presence of the pathological delta function in equation (11.127a), it is convenient to imagine that this function is replaced by some well-behaved function such as $\delta_k(r) = ke^{-kr^2}/\pi$, $r^2 = (x_1 - \xi_1)^2 + (x_2 - \xi_2)^2$, belonging to a so-called δ-sequence‡ and tending to δ for $k \to \infty$. All required operations are then performed under this assumption, and the passage to the limit for $k \to \infty$ is executed at the end of the pertinent calculations.

Now let S' and S'' denote vectors representing two arbitrary elastic states. We introduce an inner product in the space of these vectors in the form

$$(S', S'') = \int_\Omega m'_{ij} k''_{ij} \, dx \, dy$$

$$= \int_\Omega m''_{ij} k'_{ij} \, dx \, dy, \tag{11.128}$$

where $x = x_1$, $y = x_2$, and Ω is the region occupied by the plate. Note that in

---

† Note that a solution of the equation $w_{,ijij} = \delta(x_1 - \xi_1, x_2 - \xi_2)$ is $(1/8\pi)r^2 \ln r$; see, e.g., Greenberg.[72] Below we shall use the fundamental solution of the biharmonic equation in the form $r^2 \ln r$ satisfying equation (11.127a).

‡ See, e.g., Greenberg (Ref. 72, pp. 11 and 60). A simple calculation shows that, for $k \to \infty$,

$$\int_0^\infty 2\pi k \frac{e^{-kr^2}}{\pi} r \, dr \to 1, \qquad \delta_k \to 0 \quad \text{if } r \neq 0, \quad \text{and} \quad \delta_k \to \infty \quad \text{if } r = 0.$$

(11.128) the reciprocal theorem of elasticity has been used. Replacing in the equation above vector S'' by the vector U^c, we obtain

$$(U^c, S') = \int_\Omega m'_{ij} k^c_{ij} \, dx \, dy$$

$$= - \int_\Omega m'_{ij} w^c_{,ij} \, dx \, dy$$

$$= - \int_\Omega w^c m'_{ij,ij} \, dx \, dy + \oint_{\partial\Omega} m'_{ij,j} w^c n_i \, ds - \oint_{\partial\Omega} m'_{ij} w^c_{,i} n_j \, ds.$$

$$(11.129)$$

A straightforward calculation, using the formulas (11.125a), yields[†] the central relation of Washizu's procedure,

$$(U^c, S') = \oint_{\partial\Omega} \left[v_n' w^c - \frac{\partial w^c}{\partial s} m'_{ns} \right] ds - \int_\Omega w^c m'_{ij,ij} \, dx \, dy. \qquad (11.130)$$

Bearing in mind the symmetry of the inner product, we replace U^c by S' and S' by Φ and write the resulting form of the preceding equation in two ways, namely,

$$(S, \Phi) = (\Phi, S) = [S, \Phi]_{\partial\Omega_1} + [S, \Phi]_{\partial\Omega_2} + [S, \Phi]_\Omega,$$

$$(\Phi, S) = [\Phi, S]_{\partial\Omega_1} + [\Phi, S]_{\partial\Omega_2} + [\Phi, S]_\Omega, \qquad (11.131)$$

where the following notation has been used:

$$[S, \Phi]_{\partial\Omega_i} = \oint_{\partial\Omega_i} \left[w v_n^\Phi - \frac{\partial w}{\partial s} m^\Phi_{ns} \right] ds, \qquad i = 1, 2, \qquad (11.132a)$$

$$[S, \Phi]_\Omega = - \int_\Omega w m^\Phi_{ij,ij} \, dx \, dy, \qquad (11.132b)$$

with obvious changes regarding the brackets $[\Phi, S]$.

A lengthy, but not particularly instructive, computation[76] taking into account the properties of the functions S^m, S^c, U^m, U^c, and W leads to the following relation:

$$[S, \Phi]_\Omega = [\Phi + U^c, S]_{\partial\Omega_1} - [S, \Phi + U^m]_{\partial\Omega_2}$$

$$+ \int_\Omega p(\Phi + U^c) \, dx \, dy + (S, U^m - U^c). \qquad (11.133)$$

[†] Note that $- \oint_{\partial\Omega} m'_{ns}(\partial w^c/\partial s) \, ds = \oint_{\partial\Omega} (\partial m'_{ns}/\partial s) w^c \, ds$, by the continuity of $m'_{ns} w^c$ as a function of the coordinate s measured along the contour.

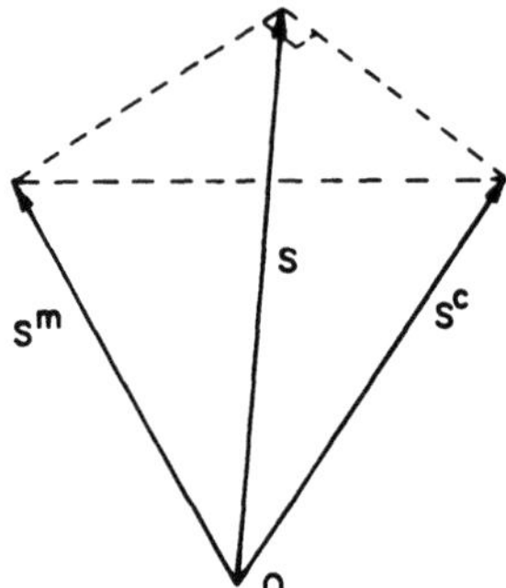

Figure 11.10. Orthogonality of the vectors $S - S^m$ and $S - S^c$.

We can write the last term in the preceding equation in two forms:

$$(S, U^m - U^c) = \begin{cases} (S^m, U^m - U^c) + (S - S^m, U^m - U^c), \\ (S^c, U^m - U^c) + (S - S^c, U^m - U^c). \end{cases} \tag{11.134}$$

But the state $S - S^m$ satisfies the homogeneous forms of equations (11.122) and (11.126), and the state $(S - S^c)$ satisfies equations (11.123) and the homogeneous forms of equations (11.127). Consequently, these states correspond to the states Y_p^τ and Y_q^ε in Chapter 12.† Since this is so, they are orthogonal,

$$(S - S^m, S - S^c) = 0. \tag{11.135}$$

From Figure 11.10, we thus infer that

$$\left.\begin{array}{c} \|S - S^m\| \\ \|S - S^c\| \end{array}\right\} \le \|S^c - S^m\|. \tag{11.136}$$

It follows by the Cauchy–Schwarz inequality that equation (11.133) can be used to write the two inequalities

$$\left| [S, \Phi]_\Omega - \left\{ [\Phi + U^c, S]_{\partial\Omega_1} - [S, \Phi + U^m]_{\partial\Omega_2} \right.\right.$$
$$\left.\left. - \int_\Omega p(\Phi + U^c)\, dx\, dy + (S^c, U^m - U^c) \right\} \right| \le \|S^m - S^c\| \, \|U^m - U^c\| \tag{11.137}$$

and

$$\left| [S, \Phi]_\Omega - \left\{ [\Phi + U^c, S]\partial_{\Omega_1} - [S, \Phi + U^m]\partial_{\Omega_2} \right.\right.$$
$$\left.\left. - \int_\Omega p(\Phi + U^c)\, dx\, dy + (S^m, U^m - U^c) \right\} \right| \le \|S^m - S^c\| \, \|U^m - U^c\|. \tag{11.138}$$

† See the text referring to Figure 12.5.

Suppose that as the fundamental function Φ we take the singular solution of the equation (11.127a), that is,[†] $w = \Phi = r^2 \ln r$. Then equation (11.132b) becomes

$$[S, \Phi]_\Omega = D^* \int_\Omega w\Phi_{,ijij}\, dx\, dy$$

$$= D^* \int_\Omega w 8\pi\, \delta(x - \xi_1, y - \xi_2)\, dx\, dy \qquad (11.139)$$

$$= 8\pi D^* w(\xi_1, \xi_2),$$

where we have used equation (11.127a) and the property of the Dirac function.

Example 11.4. Circular Plate (Numerical Example). As a simple illustration, consider the example already discussed in the preceding section, that is, a circular plate of radius a built-in along its contour and subject to a uniform load p. In this case, $\Omega = \Omega_2$ and Ω_1 is absent. We take U^c in the form $-a^2 \ln a$ and $S^c \equiv 0$, each satisfying the conditions imposed on these functions. Bearing in mind that the state S satisfies the homogeneous boundary conditions obtained from (11.127), we find from the inequality (11.137) the "approximate" value of the deflection at the center of the plate:

$$8\pi D^* w(0) \approx -2\pi \int_0^l p(r^2 \ln r - a^2 \ln a) r\, dr$$

or

$$w(0) \approx \frac{pa^4}{64D^*}(1 + 4\ln a). \qquad (11.140)$$

For the radius a set equal to 1, this result coincides with the exact solution (11.121).

Problems

1. Expand the function

$$f(t) = \begin{cases} 0 & \text{for } -\pi < t < 0, \\ t & \text{for } \quad 0 \le t < \pi \end{cases}$$

into a trigonometric Fourier series in the interval $-\pi < t < \pi$ after normalizing the trigonometric functions. Verify Bessel's inequality (11.2) for $n = 1, 2, 3, 4, 5$. Also verify the inequalities (11.1) and (11.3) for $n = 1$ and $n = 5$, respectively.

[†] Various modifications of this expression may also turn out to be serviceable. See Washizu (Ref. 76, Table 4).

2. Expand the function $f(t) = t$ into a trigonometric Fourier series in the interval $0 \leq t \leq \pi$ by considering its graph as a segment of that of an even sawtooth-like function of period 2π. Verify the Parseval equality (cf. p. 69) after normalizing the trigonometric functions.

3. Prove the Cauchy–Schwarz inequality for the Hilbert inner product by reducing the expression $\|f\|^2\|g\|^2 - (f, g)^2$ to a non-negative-valued integral.

4. Examine the proof of the basic lemma of the calculus of variations (see, e.g., R. Weinstock, *Calculus of Variations*, McGraw-Hill, New York, 1952, p. 16).

5. Modeling the discussion on the example solved in the text, perform the simplified calculations for the torsion of an isotropic bar.

6. Find an upper bound for the function $f(t)$ from Problem 1 by making use of the inequality (11.9) and a single term i^1.

7. Referring to the bounds given by inequality (11.34) for a function f in terms of a function c and a constant r [cf. equation (11.25)], supply a verification by setting $f(t) = t$ and $c(t) = \alpha t (\alpha = a$ constant) for $-\pi \leq t \leq \pi$ and $0 < \alpha \leq 1$.

8. Verify inequality (11.40) in the special case obtained by setting $f = t$, $g = t$, and $h = \alpha t$, $-\pi \leq t \leq \pi$.

9. Find the formal adjoint L^* of the operator $L \equiv e^{\alpha x}(u'' + \alpha u')$, $0 \leq x \leq 1$, with the associated boundary conditions $u(1) = 0$, $u'(0) = 0$. Determine the boundary conditions associated with the operator L^* in order for L^* to be the strict adjoint of L.

12

The Method of the Hypercircle

The method of the hypercircle, initiated by Prager and Synge in 1947[77] for approximating solution of boundary value problems of mathematical physics, translates the analytical content of a problem into the language of function space, thereafter studying the problem in geometric terms. Although the fundamental relations of the method turn out to follow almost directly from the Cauchy–Schwarz and Bessel inequalities, the remarkable pictorial merits of the method make it a useful instrument for the approximate solution of concrete problems. Our exposition is based on Refs. 40 and 77.

The starting point of the hypercircle method for application to problems of elastostatics—in which we are primarily interested—is the representation of the *elastic states* of material continua by vectors in vector space. A vector then stands for the state of stress and deformation of the *entire* body, the general stress–strain relations for an anisotropic material being (using the summation convention involving repeated indices)

$$e_{ij} = d_{ijkl}\tau_{kl}, \tag{12.1}$$

where d_{ijkl} denotes the tensor of elastic compliance. If the material is isotropic, we have

$$e_{ij} = \frac{1}{2\mu}\left(\tau_{ij} - \frac{\lambda}{2\mu + 3\lambda}\tau_{kk}\delta_{ij}\right), \tag{12.2}$$

where λ and μ are the Lamé coefficients and δ_{ij} designates the Kronecker delta.

As is well known, in order for the body to remain continuous and at rest, it is required that the deformations obey the equations of compatibility, while the stresses must satisfy the equations of equilibrium. We shall see later that the hypercircle method involves the relaxation of these requirements, and operates with elastic states which may violate the demands of either of these two systems of equations. The validity of the constitutive equations, however, is maintained.

We recall in this connection that the standard (fundamental) boundary value problems of elasticity involve a prescribed distribution of either displacements, u_i, or external forces, $t_{(n)i}$, on the surface of the body.[†] In view of this, it is usually convenient to formulate the entire problem in terms of displacements or in terms of stresses, respectively. In the first case, the strains are constructed from the strain-displacement relations (8.15),

$$e_{ij} = \tfrac{1}{2}(u_{i,\,j} + u_{j,\,i}), \tag{12.3}$$

and the stresses from the constitutive equations (12.1) or (12.2). The compatibility equations,

$$e_{ij,\,kl} + e_{kl,\,ij} - e_{ik,\,jl} - e_{jl,\,ik} = 0, \tag{12.4}$$

are then satisfied automatically, provided the displacements are sufficiently regular functions of position. This fact is verified directly by substituting from equations (12.3) into equations (12.4). The sole static conditions remaining to be satisfied in this case are the equilibrium equations,

$$\tau_{ij,\,j} + F_i = 0, \tag{12.5}$$

where F_i denotes the body force.

Inasmuch as, in the alternative considered, the only unknown functions are the displacements, it is convenient to cast the preceding equations into the form (8.22),

$$\mu u_{i,\,kk} + (\lambda + \mu)u_{kk,\,i} + F_i = 0. \tag{12.6}$$

The resulting boundary value problem consists, therefore, of the field equations (12.6) and the boundary conditions imposed on the displacements.

In solving problems of the second type (external tractions prescribed on the boundary), we must satisfy the equilibrium equations in the form (12.5) and—via Hooke's relations—the compatibility equations (12.4). On the surface of the body, the stresses are to obey the equilibrium equations,

$$t_{(n)i} = \tau_{ij}n_j, \tag{12.7}$$

where n_i denotes the external normal.

---

† As already noted in Chapter 8 [cf. equations (8.23c)], the third (or mixed) boundary value problem involves the prescription of displacements on a part of the surface and the distribution of external tractions over the remaining portion. There may also be combinations in which, at the same points of the surface, some components of the displacement and some components of the external traction are prescribed. In such cases, the specified components of the surface displacement must be orthogonal to the specified components of the surface traction. See, e.g., Pearson (Ref. 78, Sec. 6.5).

It is worth keeping in mind that the fulfillment of the equilibrium equations does *not* automatically guarantee the compatibility of the deformations associated with the given state of stress. On the contrary, it is quite conceivable that, despite the equilibrium of stresses, the partial differential equations for the displacements, (12.3), have no solution, whence the deformed elements of the body cannot be brought together to form a coherent whole.

In what follows, we consider a fairly general boundary value problem of linear elasticity, involving a body occupying the region V bounded by the surface Ω. The equations to be satisfied by the exact solution S are:

(a) compatibility equations (12.4),

(b) equations of equilibrium (12.5),

$$\tau_{ij,j} + F_i = 0 \quad \text{in } V, \tag{12.7a}$$

(c) traction boundary conditions,

$$\tau_{ij} n_j = t_{(n)i} = f_i \quad \text{on } \Omega_\tau, \tag{12.7b}$$

where f_i is a function of position prescribed on the portion Ω_τ of Ω, and

(d) displacement boundary conditions,

$$u_i = g_i \quad \text{on } \Omega_u, \tag{12.7c}$$

where g_i is a function of position prescribed on the portion Ω_u of Ω. We require that $\Omega_\tau + \Omega_u = \Omega$.

The fundamental feature of the Prager–Synge hypercircle method consists of approaching the exact solution of an elastic problem by means of a pair of incomplete (in a sense, defective) solutions.† Both of these solutions are to obey the constitutive equations relating stress to strain.

One of these pseudosolutions, distinguished by the index τ, is supposed also to satisfy the equilibrium equations,

$$\tau^\tau_{ij,j} + F_i = 0 \quad \text{in } V, \tag{12.7d}$$

and the traction boundary conditions,

$$t^\tau_{(n)i} = \tau^\tau_{ij} n_j = f_i \quad \text{on } \Omega_\tau. \tag{12.7e}$$

The elastic state so defined, denoted by S^τ, is expected to violate the compatibility equations (12.4) via the constitutive equations. This violation implies that there exists no (compatible) system of displacement associated with the system of stress τ^τ_{ij}, inasmuch as the relations (12.3) are now meaningless.

---

† More precisely, a pair of tensor fields τ^τ_{ij} and e^ε_{ij} which approximate the tensor fields τ_{ij} and e_{ij} corresponding to the true solution u_i.

The second incomplete solution, designated by S^ε, is supposed to satisfy the equations (12.4),

$$e^\varepsilon_{ij,\,kl} + e^\varepsilon_{kl,\,ij} - e^\varepsilon_{ik,\,jl} - e^\varepsilon_{jl,\,ik}, \tag{12.7f$'$}$$

so that the strain-displacement relations (12.3),

$$u^\varepsilon_{i,\,j} + u^\varepsilon_{j,\,i} = 2e^\varepsilon_{ij} \quad \text{in } V, \tag{12.7f$''$}$$

have a solution $u_i{}^\varepsilon$ which satisfies the displacement boundary conditions,

$$u_i{}^\varepsilon = g_i \quad \text{on } \Omega_u. \tag{12.7g}$$

The state S^ε is permitted to violate the equilibrium equations (in the form (12.6), for example) and consequently does not guarantee the equilibrium of the body.

We now introduce a vector space of elastic states endowed with an *inner product* discussed previously [cf. (8.13)],

$$(S_1, S_2) = \int_V \tau^1_{ij} e^2_{ij} \, dV, \tag{12.8}$$

where the indices 1 and 2 denote (perhaps different) elastic states. The distance in this space, then, is defined in terms of the strain energy.†

An appeal to Hooke's law now gives

$$\int_V (2\mu e^1_{ij} + \lambda e^1_{kk}\delta_{ij})e^2_{ij} \, dV = \int_V (2\mu e^2_{ij} + \lambda e^2_{kk}\delta_{ij})e^1_{ij} \, dV$$

or, more concisely,

$$\int_V \tau^1_{ij} e^2_{ij} \, dV = \int_V \tau^2_{ij} e^1_{ij} \, dV, \tag{12.9}$$

the last form expressing the familiar Betti–Rayleigh reciprocity relation.

An important conclusion from the latter equation is that the inner product is symmetric, $(S_1, S_2) = (S_2, S_1)$, as it must be. Furthermore, since the elastic strain energy is assumed to be a positive definite quantity, so is the metric obtained from

$$(S, S) = \int_V \tau_{ij} e_{ij} \, dV. \tag{12.10}$$

It is clear that each of the classes of vectors $\{S^\tau\}$ and $\{S^\varepsilon\}$ constitutes a subspace in the function space of states. We designate these subspaces by $\mathscr{S}_*{}^\tau \equiv \{S^\tau\}$ and $\mathscr{S}_*{}^\varepsilon \equiv \{S^\varepsilon\}$, respectively, and assume, for the time being, that

---

† The connection between the two energy metrics (10.13) and (12.10) needs no explanation if one merely refers to the definition (8.14).

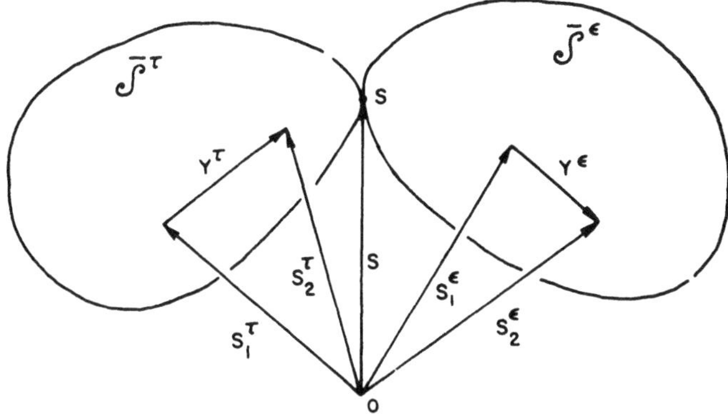

Figure 12.1. The translated subspaces $\bar{\mathscr{S}}^\tau$ and $\bar{\mathscr{S}}^\varepsilon$.

both are *translated* subspaces, i.e., neither contains the space origin. Since the space origin corresponds to the natural (stress- and deformation-free) state of the body, this assumption implies that there are no states corresponding to $F_i \equiv 0, f_i \equiv 0$, and $g_i \equiv 0$ in the subspaces $\bar{\mathscr{S}}_*^\tau$ and $\bar{\mathscr{S}}_*^\varepsilon$, respectively. We note that, in this case, the vectors S^τ and S^ε play the roles of position vectors of points in $\bar{\mathscr{S}}_*^\tau$ and $\bar{\mathscr{S}}_*^\varepsilon$, respectively, and constitute, in the earlier terminology, extrinsic vectors.[†] Vectors having as their extremities the tips of state vectors of type S^τ or S^ε, respectively, are denoted by Y^τ and Y^ε, respectively, and the corresponding subspaces by

$$\bar{\mathscr{S}}^\tau \equiv \{Y^\tau\}, \qquad \bar{\mathscr{S}}^\varepsilon \equiv \{Y^\varepsilon\} \tag{12.10a}$$

(see Figure 12.1). Inasmuch as Y^τ and Y^ε, being intrinsic vectors, constitute differences of position vectors in $\{S^\tau\}$ and $\{S^\varepsilon\}$, respectively, they satisfy homogeneous governing and boundary conditions. Thus, each element of $\{Y^\tau\}$ satisfies the equations

$$\tau^\tau_{ij,\,j} = 0 \quad \text{in } V, \tag{12.10b}$$

$$\tau^\tau_{ij} n_j = 0 \quad \text{on } \Omega_\tau; \tag{12.10c}$$

each element of $\{Y^\varepsilon\}$ satisfies the compatibility equation (12.4) and

$$u_i{}^\varepsilon = 0 \quad \text{on } \Omega_u, \tag{12.10d}$$

where, recall, $\Omega = \Omega_\tau + \Omega_u$.

[†] Compare the discussion of intrinsic and extrinsic vectors in Chapter 9 in connection with Figure 9.1. For conciseness, we denote a point (state) and position vector of the point by the same letter S; we also use interchangeably the expressions "state (point) S" and "position vector S."

An important property of the translated subspaces $\mathscr{S}^\tau$ and $\mathscr{S}^\varepsilon$, essential for making use of the hypercircle method,[†] is that they are *mutually orthogonal*, the orthogonality concerning, of course, vectors Y^τ and Y^ε *lying in* the subspaces $\mathscr{S}^\tau$ and $\mathscr{S}^\varepsilon$, respectively.

In fact, there is

$$(Y^\tau, Y^\varepsilon) = \int_V \tau^\tau_{ij} e^\varepsilon_{ij}\, dV$$

$$= \frac{1}{2} \int_V \tau^\tau_{ij}(u^\varepsilon_{i,j} + u^\varepsilon_{j,i})\, dV$$

$$= \int_V \tau^\tau_{ij} u^\varepsilon_{i,j}\, dV, \qquad \text{by symmetry of the strain tensor,}$$

$$= \int_V (\tau^\tau_{ij} u^\varepsilon_i)_{,j}\, dV - \int_V \tau^i_{ij,j} u^\varepsilon_i\, dV$$

$$= \int_\Omega \tau^\tau_{ij} n_j u^\varepsilon_i\, d\Omega - \int_V \tau^\tau_{ij,j} u^\varepsilon_i\, dV = 0$$

$$= \int_{\Omega_\tau} \tau^\tau_{ij} n_j u^\varepsilon_i\, d\Omega + \int_{\Omega_u} \tau^\tau_{ij} n_j u^\varepsilon_i - \int_V \tau^\tau_{ij,j} u^\varepsilon_i\, dV = 0, \qquad (12.11)$$

the last result by virtue of the conditions (12.10b)–(12.10d).

It is almost self-evident that the objective of the procedure is the determination of the point, or points, of intersection (S in Figure 12.1) of the translated *subspaces* $\mathscr{S}^\tau$ and $\mathscr{S}^\varepsilon$, inasmuch as any such point represents an elastic state satisfying all conditions of, and, consequently, furnishing the exact solution to, the problem. At this stage, the important question arises concerning the number of elements contained in any conceivable set of intersections of $\mathscr{S}^\tau$ and $\mathscr{S}^\varepsilon$. To the rescue comes a *uniqueness theorem* of the function space, which states that two orthogonal translated subspaces, such as $\mathscr{S}^\tau$ and $\mathscr{S}^\varepsilon$, intersect in at most one point. A simple proof of the theorem proceeds as follows.

Suppose that there exist two points, S_1 and S_2, common to the orthogonal translated subspaces $\mathscr{S}^\tau$ and $\mathscr{S}^\varepsilon$. The vector $S_1 - S_2$ then lies in both subspaces $\mathscr{S}^\tau$ and $\mathscr{S}^\varepsilon$.[‡] In view of the *orthogonality* of the subspaces, however, every vector lying in one subspace is orthogonal to every vector lying in the other. Thus, $S_1 - S_2$ is orthogonal to itself, so that

$$(S_1 - S_2, S_1 - S_2) = 0, \qquad (12.12)$$

---

† This was linked by McConnell with the *possibility* of deriving the differential problems from variational principles. Compare Ref. 40, p. 292.

‡ Compare the remarks on the convexity of a linear manifold in Chapters 6 and 9.

and, by the positive definiteness of the metric, $S_1 - S_2$ is the zero vector. It follows that $S_1 = S_2 = S$, say, and the uniqueness of the point of intersection of $\mathscr{S}^\tau$ and $\mathscr{S}^\varepsilon$ is established.

We now make two important assumptions. (1) We suppose that the elastic boundary value problem of interest has a solution and that this solution is unique.† We note here that, although for our physical intuition the existence of a unique solution (whose representation vector lies in a prescribed space) seems, in most practical situations, to be an undeniable fact, this is not necessarily so from the standpoint of a rigorous mathematical analysis. Actually, for each class of problems, existence and uniqueness of solutions must be proved, and such proofs often turn out to be the hardest points of the theory.‡ (2) Our second assumption is that the subspaces $\mathscr{S}^\tau$ and $\mathscr{S}^\varepsilon$ include all imaginable (admissible) states Y^τ and Y^ε. Practically, there is little chance that such a welcome situation occurs. First, the subspaces $\mathscr{S}^\tau$ and $\mathscr{S}^\varepsilon$ cannot include all admissible functions one can imagine. Second, it is doubtful whether one could be fortunate enough to select, even from such "complete" subspaces, just the right pair of functions whose combination satisfies all conditions of the problem at hand. Instead of pursuing this fairly delusive course, therefore, it seems more logical to satisfy oneself with a more modest procedure: to introduce two finite-dimensional subspaces which do not necessarily intersect and to simply expose the points of their *closest approach*. It should be intuitively clear (an aspect to be examined later in more detail) that information about the points of closest approach—called *vertices* and denoted by V^τ and V^ε, respectively—enables one to establish bounds for the exact solution.

Thus, we shall replace the infinite-dimensional translated subspaces $\mathscr{S}^\tau$ and $\mathscr{S}^\varepsilon$ by certain translated subspaces, $\mathscr{S}_m^{\ \tau}$ and $\mathscr{S}_n^{\ \varepsilon}$, of finite dimensionali-

---

† Here a solution means a pair of tensor fields e_{ij} and τ_{ij} satisfying the constitutive relations, the equilibrium equations, the compatibility equations, and the boundary conditions. The displacement field u_i corresponding to e_{ij} and τ_{ij} need not be unique.

‡ It is not our intention to discuss these questions in any detail. We wish only to note that, according to classical elastostatics (cf. Love, Sokolnikoff), which makes appeal to the positive-definiteness of the strain energy function, there is *at most one* displacement field in the case of the second (displacement) and the third (mixed), but not of the first (traction) boundary value problem concerning a bounded isotropic body. Modern approaches use different criteria and are more selective. For example, it is found that necessary and sufficient conditions for the unique solution of the displacement boundary value problem are $\mu \neq 0$ and $-\infty \leq \nu \leq \frac{1}{2}$, while for uniqueness of the solution of the traction boundary value problem, sufficient conditions are $-1 < \nu < 1$ and a star-shaped form of the body (necessary conditions still being uncertain). Compare Love (Ref. 110, p. 170), Sokolnikoff (Ref. 31, p. 87), Knops and Payne,[111] Wang and Truesdell,[112] Fichera,[113] and Gurtin.[114] We remind the reader that in the present chapter, indeed, throughout the entire book, our discussions concern elastic systems assumed to be in *stable* equilibrium.

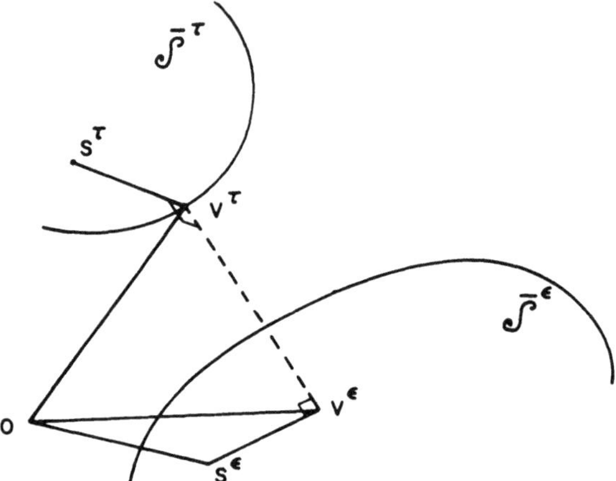

Figure 12.2. Orthogonality of $V^\tau - V^\iota$ and $\mathscr{P}^\tau$ and $\mathscr{P}^\iota$.

ties m and n, respectively. While this decision considerably impoverishes the class of admissible functions, it makes our goal more realistic.

A first question demanding resolution concerns the uniqueness of vertices. The appropriate proof is reminiscent of that of the uniqueness of intersection of $\bar{\mathscr{P}}^\tau$ and $\bar{\mathscr{P}}^\varepsilon$. Then let V^τ and V^ε be vertices, and let S^τ and S^ε be two arbitrary points in $\mathscr{P}^\tau$ and $\bar{\mathscr{P}}^\varepsilon$, respectively (Figure 12.2).† The vector $V^\tau - V^\varepsilon$ is orthogonal to both subspaces $\mathscr{P}^\tau$ and $\mathscr{P}^\varepsilon$, inasmuch as $\| V^\tau - V^\varepsilon \|$ is the shortest distance from the point V^τ to $\mathscr{P}^\varepsilon$ and from the point V^ε to $\mathscr{P}^\tau$; this property is possessed by a normal to each subspace alone.‡ Consequently,

$$(V^\tau - V^\varepsilon, V^\tau - S^\tau) = 0,$$
$$(V^\tau - V^\varepsilon, V^\varepsilon - S^\varepsilon) = 0. \tag{12.13}$$

Likewise,

$$(V^\tau - S^\tau, V^\varepsilon - S^\varepsilon) = 0. \tag{12.14}$$

We now write the identity

$$S^\tau - S^\varepsilon = (S^\tau - V^\tau) + (V^\tau - V^\varepsilon) + (V^\varepsilon - S^\varepsilon), \tag{12.15}$$

---

† Clearly, a similar proof will hold with $\mathscr{P}^\tau$ and $\mathscr{P}^\iota$ replaced by $\mathscr{P}_m^\iota$ and $\mathscr{P}_n^\iota$, respectively.
‡ Compare the statement preceding equation (9.30).

and take the inner product of each side with itself. In view of equations (12.13) and (12.14), the cross terms vanish and there remains

$$(S^\tau - S^\varepsilon, S^\tau - S^\varepsilon) = (S^\tau - V^\tau, S^\tau - V^\tau)$$
$$+ (S^\varepsilon - V^\varepsilon, S^\varepsilon - V^\varepsilon) + (V^\tau - V^\varepsilon, V^\tau - V^\varepsilon). \quad (12.16)$$

Suppose now that there exists another pair of vertices, say, $\bar{V}^\tau$ and $\bar{V}^\varepsilon$. If this is so, then the square of the distance between the new vertices must be equal to the minimum of the right-hand side of equation (12.16), which is $(V^\tau - V^\varepsilon, V^\tau - V^\varepsilon)$. Thus,

$$(\bar{V}^\tau - \bar{V}^\varepsilon, \bar{V}^\tau - \bar{V}^\varepsilon) = (V^\tau - V^\varepsilon, V^\tau - V^\varepsilon), \quad (12.17)$$

and, in view of the vanishing of the first two terms in equation (12.16), that $\bar{V}^\tau = V^\tau$ and $\bar{V}^\varepsilon = V^\varepsilon$. This proves the assertion

Let us now select a single state S^τ from the translated subspace $\mathscr{S}^\tau$ and evaluate the product (S, S^τ), where S is the exact solution of the given problem corresponding to the intersection point S of the translated subspaces $\mathscr{S}^\tau$ and $\bar{\mathscr{S}}^\varepsilon$ (provided such an intersection exists).

At this stage, we wish to derive some formulas which will be of use in future discussions. For this purpose, we select two states S^τ and S^ε, and evaluate the product $(S - S^\varepsilon, S - S^\tau)$, where S is the exact solution of the given problem, corresponding to the intersection point S of the translated subspaces $\mathscr{S}^\tau$ and $\bar{\mathscr{S}}^\varepsilon$ (provided such an intersection exists). We have

$$(S - S^\varepsilon, S - S^\tau) = \int_V (e_{ij} - e_{ij}^\varepsilon)(\tau_{ij} - \tau_{ij}^\tau)\, dV$$

$$= \int_V (u_{i,j} - u_{i,j}^\varepsilon)(\tau_{ij} - \tau_{ij}^\tau)\, dV$$

$$= \int_V [(u_i - u_i^\varepsilon)(\tau_{ij} - \tau_{ij}^\tau)]_{,j}\, dV$$

$$- \int_V (u_i - u_i^\varepsilon)(\tau_{ij,j} - \tau_{ij,j}^\tau)\, dV$$

$$= \int_{\Omega_\tau} (u_i - u_i^\varepsilon)(t_{(n)i} - t_{(n)i}^\tau)\, d\Omega$$

$$+ \int_{\Omega_u} (u_i - u_i^\varepsilon)(\tau_{ij} - \tau_{ij}^\tau)n_j\, d\Omega$$

$$- \int_V (u_i - u_i^\varepsilon)(\tau_{ij,j} - \tau_{ij,j}^\tau)\, dV. \quad (12.18)$$

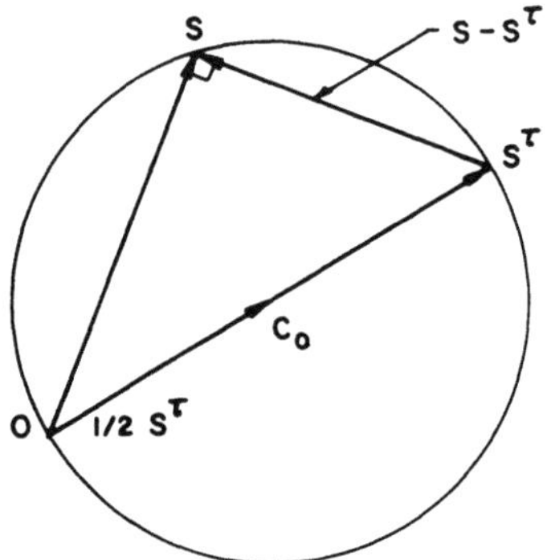

Figure 12.3. The hypersphere, equation (12.20).

All of the above integrals vanish: the first and second on account of (12.7b) and (12.7e) as well as (12.7c) and (12.7g), respectively; the third one in view of (12.7a) and (12.7d). Thus, finally,

$$(S - S^\varepsilon, S - S^\tau) = 0 \tag{12.19}$$

or, if we assume that the translated subspace $\bar{\mathscr{S}}^\varepsilon$ includes the space origin, O (so that $\bar{\mathscr{S}}^\varepsilon$, in fact, coincides with $\mathscr{S}^\varepsilon$), and choose the state $S^\varepsilon \equiv \theta$,

$$(S, S - S^\tau) = 0, \tag{12.20}$$

which implies, in turn, that

$$\|S - \tfrac{1}{2}S^\tau\|^2 = (\tfrac{1}{2}\|S^\tau\|)^2. \tag{12.21}$$

A comparison of the last equation with equation (9.37) indicates that equation (12.21) represents a *hypersphere* with center C_o at the point $\tfrac{1}{2}S^\tau$ and the radius $\tfrac{1}{2}\|S^\tau\|$ (Figure 12.3); in the present case, the origin O lies on the hypersphere. Equation (12.20) now implies that the vectors S and $S - S^\tau$, subtending the diameter $S^\tau - \theta$, are orthogonal. The presence of the origin on the hypersphere indicates the possibility of the natural (stress-free) state of the body associated with the trivial solution, $S \equiv \theta$.

Retaining the assumption that the space origin is located in the subspace $\mathscr{S}^\varepsilon$, let us select two states, S^τ from $\mathscr{S}^\tau$ and S^ε from $\mathscr{S}^\varepsilon$, the latter perhaps different from zero. Subtracting the relation (12.20) from (12.19) gives

$$(S - S^\tau, S^\varepsilon) = 0 \tag{12.22}$$

or, explicitly,

$$(S, S^\varepsilon) = (S^\tau, S^\varepsilon). \tag{12.23}$$

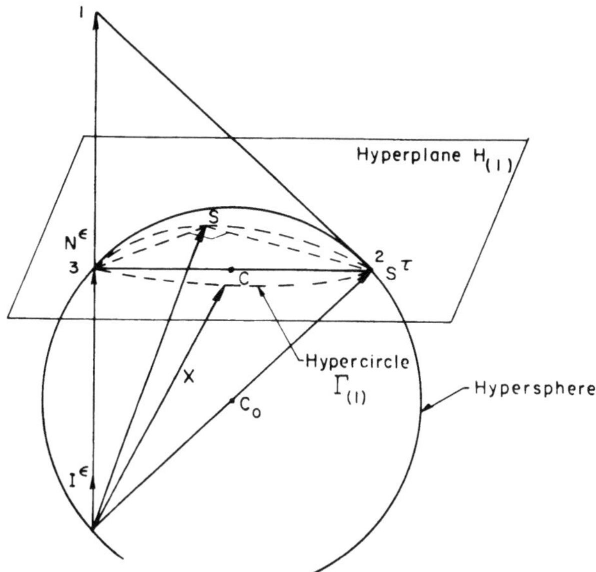

Figure 12.4. The hypercircle $\Gamma_{(1)}$ [equations (12.21) and (12.23)].

Imagine now that we select two particular admissible states S^τ and S^ε, different from zero, so that $(S^\tau, S^\varepsilon) = \alpha_1$, say, is a given fixed scalar. Since S^ε is a fixed nonzero vector, comparison of equations (12.23) and (9.11) shows that the point S is on a certain *hyperplane* of the first class, $H_{(1)}$, whose equation is $(X, S^\varepsilon) = \alpha_1$, with X as the variable point on $H_{(1)}$ (Figure 12.4). On the other hand, on account of equation (12.21), the point S is on a *hypersphere* with center at $\frac{1}{2}S^\tau$ and radius $\frac{1}{2}\|S^\tau\|$. It follows that the point S, representing the exact solution, lies on the intersection of a hyperplane and a hypersphere, that is, on a *hypercircle* $\Gamma_{(1)}$, the latter being of class one because one of the intersecting partners is of this class. It is helpful to keep in mind here that, despite its pictorial name, a hypercircle is, in fact, a set of infinite dimensionality. Figure 12.4, in which $I^\varepsilon = S^\varepsilon/\|S^\varepsilon\|$ is a unit vector in the S^ε-direction, illustrates the situation. Writing the equation of the hypercircle in the form

$$(X, I^\varepsilon) = (S^\tau, I^\varepsilon), \tag{12.24}$$

we conclude that the projection on S^ε of the variable vector X, as its tip moves along $\Gamma_{(1)}$, is constant and equal to (S^τ, I^ε). Thus, $S^\varepsilon \perp H_{(1)}$, and the normal N^ε from the origin to $H_{(1)}$ is

$$N^\varepsilon = (S^\tau, I^\varepsilon)I^\varepsilon. \tag{12.25}$$

Now, since the product

$$(S - N^\varepsilon, S - S^\tau) = (S, S - S^\tau) - \|N^\varepsilon\|(I^\varepsilon, S - S^\tau) \qquad (12.26)$$

vanishes on account of equations (12.20) and (12.24), it follows that the points N^ε and S^τ are the extremities of a diameter of $\Gamma_{(1)}$ passing through the tip of N^ε. Furthermore, Figure 12.4 indicates that the center of the hypercircle is at the point

$$C = S^\tau + \tfrac{1}{2}(N^\varepsilon - S^\tau)$$
$$= \tfrac{1}{2}(N^\varepsilon + S^\tau). \qquad (12.27)$$

Again, from the right triangle $ON^\varepsilon S^\tau$, we infer that the radius of the hypercircle is

$$R = \tfrac{1}{2}[\|S^\tau\|^2 - \|N^\varepsilon\|^2]^{1/2}. \qquad (12.27a)$$

12.1. Bounds on an Elastic State

Let us again select two particular states S^τ and S^ε. Then the vector N^ε in (12.25) is to be considered as given, since S^τ is known and $I^\varepsilon = S^\varepsilon/\|S^\varepsilon\|$. Figure 12.4 illustrates the fact (easily corroborated analytically) that, among all states represented by the points on the hypercircle $\Gamma_{(1)}$, the state represented by the vector N^ε is the closest to, and the vector S^τ the most distant from, the zero vector of the natural state 0. This observation provides the following bounds for the unknown exact solution,

$$\|N^\varepsilon\| \leq \|S\| \leq \|S^\tau\| \qquad (12.28)$$

or, more explicitly,

$$(S^\tau, I^\varepsilon)^2 \leq (S, S) \leq (S^\tau, S^\tau). \qquad (12.28a)$$

A weaker lower bound for (S, S) is obtained by applying the cosine theorem to the triangle 012. We find

$$2(S^\tau, S^\varepsilon) - (S^\varepsilon, S^\varepsilon) = (S^\tau, S^\tau) - (S^\tau - S^\varepsilon, S^\tau - S^\varepsilon). \qquad (12.28b)$$

But, from the right triangle 123,

$$\|S^\tau - (S^\tau, I^\varepsilon)I^\varepsilon\| \leq \|S^\tau - S^\varepsilon\|, \qquad (12.28c)$$

so that, from the last two equations,

$$2(S^\tau, S^\varepsilon) - (S^\varepsilon, S^\varepsilon) \leq (S^\tau, I^\varepsilon)^2. \qquad (12.28d)$$

Combining this inequality with the inequality (12.28a) yields one of the *central* inequalities of the hypercircle method,

$$2(S^\tau, S^\varepsilon) - (S^\varepsilon, S^\varepsilon) \le (S^\tau, I^\varepsilon)^2 \le (S, S) \le (S^\tau, S^\tau). \qquad (12.28\text{e})$$

It should be noted that the preceding formula is derived for a hypercircle of *class one*, and that—unless we are lucky—the bounds may be far off and, consequently, of limited usefulness. Another knotty point is that the bounds do not involve directly the stresses and displacements actually sought, but only implicitly *via* the total strain energy stored in the body.

In order to improve the closeness of the bounds, it is mandatory to introduce into the competition more than a single vector from each of the subspaces $\mathscr{S}^\tau$ and $\mathscr{S}^\varepsilon$. Such additional (*auxiliary*) states are usually required to satisfy the homogeneous equations of equilibrium, or compatibility, and homogeneous boundary conditions. Denoting them by $Y_p{}^\tau$ and $Y_q{}^\varepsilon$, respectively, where p and q are to take values in certain index sets, we have, following (12.10b)–(12.10d),

$$\tau^{\tau p}_{ij,\,j} = 0 \quad \text{in } V, \qquad (12.28\text{f})$$

$$\tau^{\tau p}_{ij} n_j = 0 \quad \text{on } \Omega_\tau, \qquad (12.28\text{g})$$

and

$$u^{\varepsilon q}_i = 0 \quad \text{on } \Omega_u, \qquad (12.28\text{h})$$

where $\tau^{\tau p}_{ij}$ and $u^{\varepsilon q}_i$ are the stresses and displacements associated with the states $Y_p{}^\tau$ and $Y_q{}^\varepsilon$, respectively.

The actual equilibrium equations and conditions on the boundary are accounted for by the states denoted as $S_0{}^\tau$ and $S_0{}^\varepsilon$ and are called *fundamental states*. By analogy to (12.7d), (12.7e), we thus have

$$\tau^{\tau 0}_{ij,\,j} + F_i = 0 \quad \text{in } V, \qquad (12.28\text{i})$$

$$\tau^{\tau 0}_{ij} n_j = f_i \quad \text{on } \Omega_\tau, \qquad (12.28\text{j})$$

and

$$u^{\varepsilon 0}_i = g_i \quad \text{on } \Omega_u, \qquad (12.28\text{k})$$

where $\tau^{\tau 0}_{ij}$ and $u^{\varepsilon 0}_i$ are the stresses and displacements associated with the states $S_0{}^\tau$ and $S_0{}^\varepsilon$, respectively.

We assume for the time being that the space origin lies outside the translated subspaces denoted earlier by $\mathscr{S}^\tau$ and $\mathscr{S}^\varepsilon$. Later, we shall shift the space origin into one of these subspaces for convenience.†

---

† The location of the origin in both translated subspaces would, of course, indicate that the only solution to the problem is the trivial zero vector.

Summarizing our remarks, we have decided to select in the translated subspaces $\mathscr{S}^\tau$ and $\mathscr{S}^\varepsilon$ some finite-dimensional translated subspaces $\mathscr{S}_m{}^\tau$ and $\mathscr{S}_n{}^\varepsilon$, respectively, and to approximate the exact solution to the problem, S, by means of the following systems of elastic states:

(a) State $S_0{}^\tau$ (denoted earlier by S^τ) called the *fundamental state in* $\mathscr{S}_m{}^\tau$. This state satisfies the equilibrium equations (including the body force) and the stress boundary conditions, that is, equations (12.28i) and (12.28j).

(b) State $S_0{}^\varepsilon$ (denoted earlier by S^ε) called the *fundamental state in* $\mathscr{S}_n{}^\varepsilon$. This state satisfies the compatibility equations and the displacement boundary conditions, that is, equations (12.4) and (12.28k).

(a_1) States $Y_p{}^\tau$, $p = 1, 2, \ldots, m$; these are states augmenting the fundamental state $S_0{}^\tau$. They satisfy the homogeneous equations of equilibrium and homogeneous stress boundary conditions, that is, equations (12.28f) and (12.28g).

(b_1) States $Y_q{}^\varepsilon$, $q = 1, 2, \ldots, n$; these are states *auxiliary* to the fundamental state $S_0{}^\varepsilon$. They satisfy the compatibility equations and homogeneous displacement boundary conditions, that is, equations (12.4) and (12.28h).

We now define the translated subspaces $\mathscr{S}_m{}^\tau$ and $\mathscr{S}_n{}^\varepsilon$ by the following equations, respectively:

$$S^\tau = S_0{}^\tau + \sum_{p=1}^{m} \alpha_p Y_p{}^\tau,$$

$$S^\varepsilon = S_0{}^\varepsilon + \sum_{q=1}^{n} \beta_q Y_q{}^\varepsilon,$$

$$(12.29)$$

where S^τ and S^ε represent current position vectors emanating from the space origin O (Figure 12.5a), $S_0{}^\tau$ and $S_0{}^\varepsilon$ position vectors of certain points in $\mathscr{S}_m{}^\tau$ and $\mathscr{S}_n{}^\varepsilon$, respectively, and $\{Y_p{}^\tau\}$ and $\{Y_q{}^\varepsilon\}$ vectors lying in the corresponding subspaces; $\{\alpha_p\}$ and $\{\beta_q\}$ are arbitrary coefficients. As shown in Chapter 5, it is always possible to replace a set of independent vectors (and both $\{Y_p{}^\tau\}$ and $\{Y_q{}^\varepsilon\}$ are considered to be such sets) by an *orthonormal* set. With no loss in generality, therefore, we assume that

$$(Y_p{}^\tau, Y_r{}^\tau) = \delta_{pr}, \qquad p, r = 1, 2, \ldots, m,$$

$$(Y_q{}^\varepsilon, Y_s{}^\varepsilon) = \delta_{qs}, \qquad q, s = 1, 2, \ldots, n.$$

$$(12.30)$$

Since the numbers of orthonormal vectors in $\{Y_p{}^\tau\}$ and $\{Y_q{}^\varepsilon\}$ are equal to the dimensions of the corresponding subspaces, $\mathscr{S}_m{}^\tau$ and $\mathscr{S}_n{}^\varepsilon$, respectively, the sets $\{Y_p{}^\tau\}$ and $\{Y_q{}^\varepsilon\}$ form bases for these subspaces.

By virtue of the orthogonality of $\mathscr{S}_m{}^\tau$ and $\mathscr{S}_n{}^\varepsilon$, we have

$$(Y_p{}^\tau, Y_q{}^\varepsilon) = 0 \qquad (12.31)$$

for each $p = 1, 2, \ldots, m$ and each $q = 1, 2, \ldots, n$.

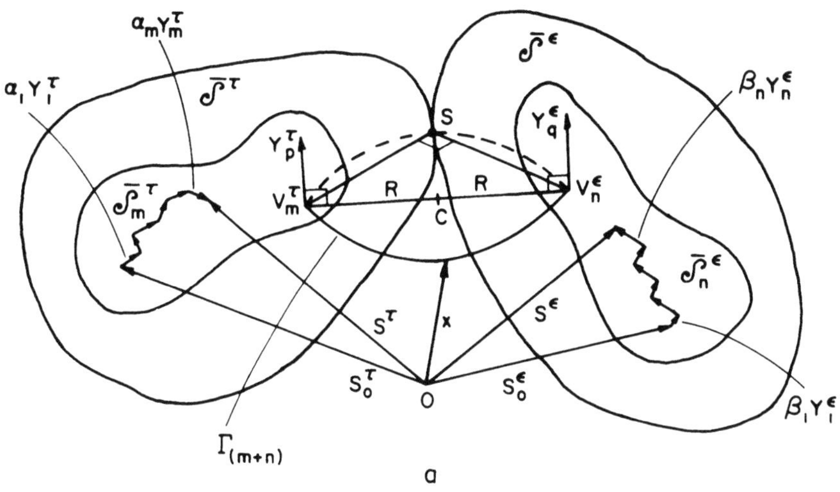

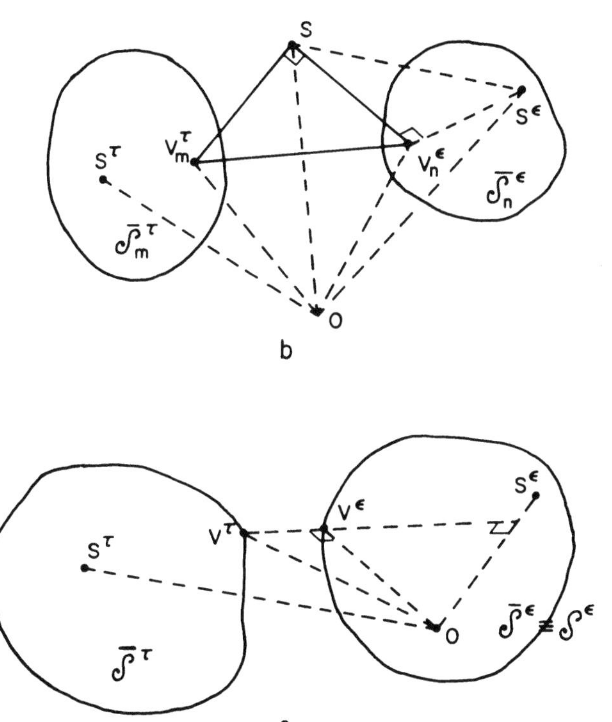

Figure 12.5. Subspaces $\bar{\mathscr{P}}^\tau$ and $\bar{\mathscr{P}}^\epsilon$.

The diagram in Figure 12.5a displays the state of affairs. The translated subspaces $\mathscr{S}^\tau$ and $\mathscr{S}^\varepsilon$ meet at the point marking the exact (unknown) solution to the problem. To make the situation more general, the space origin O is located (for the time being) exterior to both translated subspaces $\mathscr{S}^\tau$ and $\mathscr{S}^\varepsilon$. The specialized translated subspaces $\bar{\mathscr{S}}_m{}^\tau$ and $\bar{\mathscr{S}}_n{}^\varepsilon$, embedded correspondingly in $\mathscr{S}^\tau$ and $\mathscr{S}^\varepsilon$, display points of closest mutual approach marked by the tips of the vectors $V_m{}^\tau$ and $V_n{}^\varepsilon$, respectively. Their separation $\|V_m{}^\tau - V_n{}^\varepsilon\|$ is, according to the definition of vertices, the least distance between $\bar{\mathscr{S}}_m{}^\tau$ and $\bar{\mathscr{S}}_n{}^\varepsilon$. By appeal to equations (12.29) and (12.31), it is not difficult to verify that

$$
\begin{aligned}
(S^\tau - S^\varepsilon, S^\tau - S^\varepsilon) &= \left(S_0{}^\tau - S_0{}^\varepsilon + \sum_p \alpha_p Y_p{}^\tau, \; S_0{}^\tau - S_0{}^\varepsilon + \sum_p \alpha_p Y_p{}^\tau \right) \\
&\quad + \left(S_0{}^\tau - S_0{}^\varepsilon - \sum_q \beta_q Y_q{}^\varepsilon, \; S_0{}^\tau - S_0{}^\varepsilon - \sum_q \beta_q Y_q{}^\varepsilon \right) \\
&\quad - (S_0{}^\tau - S_0{}^\varepsilon, \; S_0{}^\tau - S_0{}^\varepsilon).
\end{aligned}
\tag{12.32}
$$

In order to minimize the right-hand side of this equation, one need only minimize the first two products separately, inasmuch as the parameters α_p and β_q are independent of each other. Carrying out the differentiations, with equation (12.31) in mind, we find

$$
\begin{aligned}
\alpha_p &= -(S_0{}^\tau - S_0{}^\varepsilon, Y_p{}^\tau), & p &= 1, 2, \ldots, m, \\
\beta_q &= (S_0{}^\tau - S_0{}^\varepsilon, Y_q{}^\varepsilon), & q &= 1, 2, \ldots, n.
\end{aligned}
\tag{12.33}
$$

Substitution into equations (12.29) yields the locations of the vertices as

$$
\begin{aligned}
V_m{}^\tau &= S_0{}^\tau - \sum_{p=1}^{m} (S_0{}^\tau - S_0{}^\varepsilon, Y_p{}^\tau) S_p{}^\tau, \\
V_n{}^\varepsilon &= S_0{}^\varepsilon + \sum_{q=1}^{n} (S_0{}^\tau - S_0{}^\varepsilon, Y_q{}^\varepsilon) S_q{}^\varepsilon.
\end{aligned}
\tag{12.34}
$$

As already noted, the segment joining the vertices is orthogonal to the vectors lying in the subspaces (for example, to $Y_p{}^\tau$ and $Y_q{}^\varepsilon$); this is designated in Figure 12.5a by the conventional orthogonality sign.

One of the conclusions of our earlier discussion was that the extremity of the exact solution vector, S, lies on the hypercircle, $\Gamma_{(1)}$, produced by the intersection of the hypersphere (12.21) and the hyperplane (12.23) of class one (Figure 12.4). It is now of importance to find out how the situation changes if we select the combinations (12.29) and replace the translated subspaces $\mathscr{S}^\tau$ and $\mathscr{S}^\varepsilon$ by the finite-dimensional translated subspaces $\mathscr{S}_m{}^\tau$ and $\mathscr{S}_n{}^\varepsilon$.

At the outset, we observe that, in view of the orthogonality of these subspaces, the vectors $S - V_m^\tau$ and $S - V_n^\varepsilon$ (lying in the respective subspaces) are mutually orthogonal (Figure 12.5a). By similar reasoning, every vector Y_q^ε is orthogonal to $S - V_m^\tau$ and every vector Y_p^τ to $S - V_n^\varepsilon$. Symbolically,

$$(S - V_m^\tau, S - V_n^\varepsilon) = 0,$$

$$(S - V_m^\tau, Y_q^\varepsilon) = 0 \qquad \text{for } q = 1, 2, \ldots, n, \tag{12.35}$$

$$(S - V_n^\varepsilon, Y_p^\tau) = 0 \qquad \text{for } p = 1, 2, \ldots, m.$$

The first of the preceding relations can be cast into the form

$$(S - \tfrac{1}{2}[V_m^\tau + V_n^\varepsilon] - \tfrac{1}{2}[V_m^\tau - V_n^\varepsilon], \; S - \tfrac{1}{2}[V_m^\tau + V_n^\varepsilon] + \tfrac{1}{2}[V_m^\tau - V_n^\varepsilon]) = 0, \tag{12.35a}$$

or, using the notation

$$\begin{aligned} C &= \tfrac{1}{2}(V_m^\tau + V_n^\varepsilon), \\ R &= \tfrac{1}{2}\|V_m^\tau - V_n^\varepsilon\|, \end{aligned} \tag{12.35b}$$

into either of the equivalent forms ($J = $ arbitrarily directed unit vector from point C),

$$(S - C, S - C) = (RJ, RJ) \tag{12.36a}$$

or

$$\|S - C\|^2 = R^2. \tag{12.36b}$$

If one thinks of S as a current position vector and compares the latter relation with relation (12.21) (inspect Figure 12.4a), one concludes that the tip of S lies on a *hypersphere*, a diameter of which is $V_m^\tau - V_n^\varepsilon$. Consequently, the position vector of the center of the hypersphere and the radius of the latter are given by the two equations (12.35b), respectively; here, V_m^τ and V_n^ε are known vectors defined by equations (12.34).

The still remaining equations (12.35), when compared with equation (12.23), imply that the point S lies on a *hyperplane* $H_{(m+n)}$ of class $m + n$. It follows that the point S representing the exact solution lies on the intersection of a hypersphere and a hyperplane of class $m + n$, that is, on a *hypercircle* $\Gamma_{(m+n)}$ of class $m + n$.

Now let X be the position vector of a point on the hypercircle $\Gamma_{(m+n)}$. In view of equations (12.35), the hypercircle is defined by

$$(X - V_m^\tau, X - V_n^\varepsilon) = 0,$$

$$(X - V_m^\tau, Y_q^\varepsilon) = 0, \qquad q = 1, 2, \ldots, n, \tag{12.37}$$

$$(X - V_n^\varepsilon, Y_p^\tau) = 0, \qquad p = 1, 2, \ldots, m.$$

From the last of the preceding equations, we have

$$(X - \tfrac{1}{2}[V_m{}^{\tau} + V_n{}^{\varepsilon}] + \tfrac{1}{2}[V_m{}^{\tau} - V_n{}^{\varepsilon}], \; Y_p{}^{\tau}) = (X - C, \; Y_p{}^{\tau})$$

$$= 0, \qquad p = 1, 2, \ldots, m, \qquad (12.38)$$

where reference was made to equations (12.35), as well as to the orthogonality of the vector $V_m{}^{\tau} - V_n{}^{\varepsilon}$ and the subspace $\mathscr{S}_m{}^{\tau}$. Similarly, from the second of equations (12.37)

$$(X - C, \; Y_q{}^{\varepsilon}) = 0, \qquad q = 1, 2, \ldots, n. \qquad (12.39)$$

Accordingly, a vector joining a generic point on the hypercircle and the point C is orthogonal to all $Y_p{}^{\tau}$'s and all $Y_q{}^{\varepsilon}$'s (Figure 12.5a).

Let us now return to equations (9.14) and (9.15a) in Chapter 9, identifying the vectors $Y_p{}^{\tau}$ and $Y_q{}^{\varepsilon}$ with those denoted there by i^{ν}. We conclude that the sets $\{Y_p{}^{\tau}\}$ and $\{Y_q{}^{\varepsilon}\}$ are orthogonal to the hyperplane, and, consequently, that the point C lies on the hyperplane.

Combining equations (12.38) and (12.39) with equations (12.35b) and (12.36b), we conclude that the *hypercircle* is defined by the system of the following equations:

$$\|X - C\| = R^2,$$

$$(X - C, \; Y_p{}^{\tau}) = 0, \qquad p = 1, 2, \ldots, m, \qquad (12.40)$$

$$(X - C, \; Y_q{}^{\varepsilon}) = 0, \qquad q = 1, 2, \ldots, n.$$

New bounds for the exact solution are now found easily by inspecting the triangle $SV_n{}^{\varepsilon}S^{\varepsilon}$ in Figure 12.5b. Here, S^{ε} is an arbitrary point in $\mathscr{S}_n{}^{\varepsilon}$ and the angle at $V_n{}^{\varepsilon}$ is a right angle, since the vector $S - V_n{}^{\varepsilon}$ is orthogonal to every vector $S^{\varepsilon} - V_n{}^{\varepsilon}$ in the subspace $\mathscr{S}_n{}^{\varepsilon}$. Thus,

$$(S - S^{\varepsilon}, S - S^{\varepsilon}) = (S - V_n{}^{\varepsilon}, S - V_n{}^{\varepsilon}) + (S^{\varepsilon} - V_n{}^{\varepsilon}, S^{\varepsilon} - V_n{}^{\varepsilon}). \qquad (12.41)$$

Likewise, from the right triangle $V_m{}^{\tau}SV_n{}^{\varepsilon}$,

$$(S - V_n{}^{\varepsilon}, S - V_n{}^{\varepsilon}) \le (V_n{}^{\varepsilon} - V_m{}^{\tau}, V_n{}^{\varepsilon} - V_m{}^{\tau}), \qquad (12.42)$$

and the last two formulas give the bounds for $\|S - S^{\varepsilon}\|$ contained in

$$(S^{\varepsilon} - V_n{}^{\varepsilon}, S^{\varepsilon} - V_n{}^{\varepsilon}) \le (S - S^{\varepsilon}, S - S^{\varepsilon}) \le (S^{\varepsilon} - V_n{}^{\varepsilon}, S^{\varepsilon} - V_n{}^{\varepsilon})$$

$$+ (V_n{}^{\varepsilon} - V_m{}^{\tau}, V_n{}^{\varepsilon} - V_m{}^{\tau}). \qquad (12.43)$$

By a similar argument applied to the translated subspace $\mathscr{S}_m{}^{\tau}$, we easily derive a second inequality, symmetric to (12.43),

$$(S^{\tau} - V_m{}^{\tau}, S^{\tau} - V_m{}^{\tau}) \le (S - S^{\tau}, S - S^{\tau}) \le (S^{\tau} - V_m{}^{\tau}, S^{\tau} - V_m{}^{\tau})$$

$$+ (V_n{}^{\varepsilon} - V_m{}^{\tau}, V_n{}^{\varepsilon} - V_m{}^{\tau}), \qquad (12.44)$$

where again S^{τ} is an arbitrary point in $\mathscr{S}_m{}^{\tau}$.

An inspection of these two inequalities implies that the error committed by selecting any vector S^ε or S^τ in place of the exact solution S does not exceed $\| V_n{}^\varepsilon - V_m{}^\tau \|$. This result is of major importance in practical applications of the hypercircle method.

Let us now reflect upon our earlier remark to the effect that, in order to account for the natural state, it is necessary to place the space origin in either of the translated subspaces $\mathscr{S}^\tau$ or $\mathscr{S}^\varepsilon$ (or $\mathscr{S}_m{}^\tau$ or $\mathscr{S}_n{}^\varepsilon$, respectively): see Figure 12.5c. Assume then that the space origin is in $\mathscr{S}^\varepsilon$. A decision of this sort simplifies certain of the formulas derived earlier, at the cost, however, of stripping them of the symmetry they possess. Equations (12.13) and (12.14) reduce to

$$(V^\tau - V^\varepsilon, V^\tau - S^\tau) = 0,$$

$$(V^\tau - V^\varepsilon, V^\varepsilon) = 0, \tag{12.45}$$

$$(V^\tau - S^\tau, V^\varepsilon) = 0$$

(compare Figure 12.5c). The second of the preceding equations implies that the segment connecting the vertices is orthogonal to the vector V^ε. This, of course, should be clear since, in the present case, the last-named vector lies in the subspace $\mathscr{S}^\varepsilon$.

A glance at the equation (12.16) leads now to the following inequalities (after setting $S^\varepsilon \equiv 0$):

$$(S^\tau, S^\tau) \geq (V^\tau - V^\varepsilon, V^\tau - V^\varepsilon), \tag{12.46}$$

$$(S^\tau, S^\tau) \geq (S^\tau - V^\tau, S^\tau - V^\tau), \tag{12.47}$$

$$(S^\tau, S^\tau) \geq (V^\varepsilon, V^\varepsilon), \tag{12.48}$$

where, as before, S^τ is an arbitrary point in $\mathscr{S}^\tau$. From Figure 12.5c the geometric sense of the foregoing inequalities is clearly expressed in the statement that the distance $\| S^\tau - \theta \|$ is greater than any of the distances $\| V^\tau - V^\varepsilon \|$, $\| S^\tau - V^\tau \|$, or $\| V^\varepsilon - \theta \|$.

The assumption that the translated subspace $\mathscr{S}^\varepsilon$ (respectively, $\mathscr{S}_n{}^\varepsilon$) includes the origin implies similar simplifications when the class of the hypercircle is greater than one. In this case, we set $S_0{}^\varepsilon = 0$ in the second group of equations (12.29) and, in Figure 12.5a, place the origin in the subspace $\mathscr{S}_n{}^\varepsilon$. In order to derive an upper bound for (S, S), we now replace S^τ in the inequality (12.28a) by $V_m{}^\tau$ and obtain

$$(S, S) \leq (V_m{}^\tau, V_m{}^\tau). \tag{12.49}$$

Likewise, we replace S^ε by $V_n{}^\varepsilon$ and S^τ by $V_m{}^\tau$ in the left terminal expression in the formula (12.28e). This gives

$$2(V_m{}^\tau, V_n{}^\varepsilon) - (V_n{}^\varepsilon, V_n{}^\varepsilon) \leq (S, S), \tag{12.50}$$

and, combined with (12.49) and the second of equations (12.45), yields finally the symmetric inequality

$$(V_n^{\,\varepsilon}, V_n^{\,\varepsilon}) \le (S, S) \le (V_m^{\,\tau}, V_m^{\,\tau}). \tag{12.51}$$

We note, bearing in mind that the location of the space origin is in the translated subspace $\mathcal{S}_n^{\,\varepsilon}$, that we can also infer this two-sided inequality directly from Figure 12.5a.

Inequalities (12.51) and (12.28e) constitute *two central formulas* of the hypercircle method. The first of these is given a more serviceable form if one refers to the relations (12.34) with $S_0^{\,\varepsilon}$ set equal to the zero vector. With the aid of the orthogonality relations (12.30), we then arrive at the equations

$$(V_n^{\,\varepsilon}, V_n^{\,\varepsilon}) = \sum_{q=1}^{n} (S_0^{\,\tau}, Y_q^{\,\varepsilon})^2,$$

$$(V_m^{\,\tau}, V_m^{\,\tau}) = (S_0^{\,\tau}, S_0^{\,\tau}) - \sum_{p=1}^{m} (S_0^{\,\tau}, Y_p^{\,\tau}), \tag{12.52}$$

and finally at the inequality,

$$\sum_{q=1}^{n} (S_0^{\,\tau}, Y_q^{\,\varepsilon})^2 \le (S, S) \le (S_0^{\,\tau}, S_0^{\,\tau}) - \sum_{p=1}^{m} (S_0^{\,\tau}, Y_p^{\,\tau})^2. \tag{12.53}$$

This completes the discussion of the general aspects of the hypercircle method. As an illustration, we concentrate on the following rather specific problem.[79]

Example 12.1. Elastic Cylinder in Gravity Field (Numerical Example). An elastic, heavy, solid, circular cylinder of mass density ρ, Young's modulus E, diameter $2a$, and height $2l$ is enclosed in a perfectly rigid case (Figure 12.6). The direction of the gravity field g is parallel to the axis of the cylinder, the latter being referred to a cylindrical coordinate system R, θ, Z with origin at the mass center and the Z-axis pointing upwards. To avoid carrying nonessential constants, we put $\kappa = \rho g a / E$ and introduce the non-dimensional variables $r = R/a$, $z = Z/a$, $\tau_{ij} = \bar{\tau}_{ij}/E$, $u = (1/\kappa a)u_R$, and $w = (1/\kappa a)u_Z$, with u_R and u_Z as the radial and axial components of the displacement, respectively, and $\bar{\tau}_{ij}$ as the actual stress. In the present case of axial symmetry, the equilibrium equations are

$$\frac{\partial \tau_{rr}}{\partial r} + \frac{\partial \tau_{rz}}{\partial z} + \frac{\tau_{rr} - \tau_{\theta\theta}}{r} = 0,$$

$$\frac{\partial \tau_{rz}}{\partial r} + \frac{\partial \tau_{zz}}{\partial z} + \frac{\tau_{rz}}{r} - \kappa = 0. \tag{12.54}$$

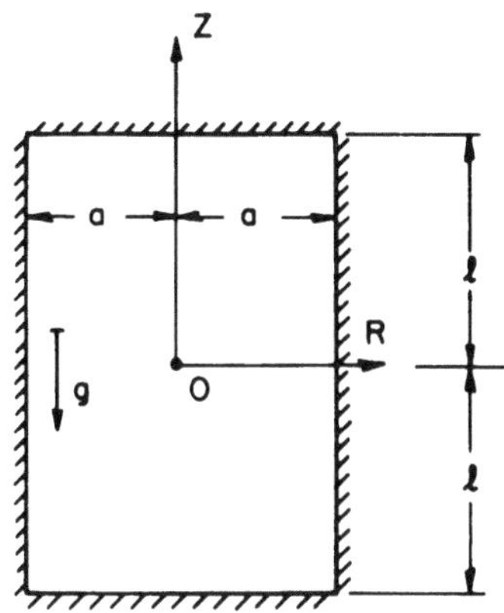

Figure 12.6. Cylinder enclosed in a rigid case.

The strain components and Hooke's law are defined in the customary manner. In conformity with the assumption of rigid enclosure, we adopt the following boundary conditions[†]:

$$u = w = 0 \qquad \text{for } z = \pm 1 \quad \text{and} \quad 0 \le r \le 1,$$
$$u = w = 0 \qquad \text{for } -1 \le z \le 1 \quad \text{and} \quad r = 1. \tag{12.55}$$

Since the work of the gravity field is equal to the work of the weight of the body located at the center of mass, we have

$$(S, S) = 4\pi\delta, \tag{12.56}$$

where $\delta = \Delta/a$ is the nondimensionalized sagging of the mass center,[40, 80] while the inner product of two state vectors and the metric are defined by equations (12.8) and (12.10), respectively. Equation (12.56) implies that any bounds for (S, S) will automatically produce bounds for δ.

We now approach the exact solution of the problem at hand, S, by means of two sets of approximations: the set $\{S^\tau\}$ of states satisfying the equilibrium equations, and the set $\{S^\varepsilon\}$ of states satisfying the compatibility equations and the displacement boundary conditions. We recall that the translated subspaces $\mathscr{S}^\tau \equiv \{S^\tau\}$ and $\mathscr{S}^\varepsilon \equiv \{S^\varepsilon\}$ are mutually orthogonal, and that the tip of the exact solution vector coincides with the point of intersection of $\mathscr{S}^\tau$ and $\mathscr{S}^\varepsilon$. We now confine our attention to portions $\mathscr{S}_m^\tau$ and $\mathscr{S}_n^\varepsilon$ of the translated subspaces $\mathscr{S}^\tau$ and $\mathscr{S}^\varepsilon$, respectively, and expose the vertices V_m^τ

---

† We note that, since the displacements are prescribed on the entire boundary, the problem under discussion is a displacement boundary value problem, in which $\Omega_u = \Omega$ (and Ω_τ is absent).

and V_n^ε as the points of closest approach of $\bar{\mathscr{S}}_m^\tau$ and $\bar{\mathscr{S}}_n^\varepsilon$. As already explained before, the terminal points of the vectors S, V_m^τ, and V_n^ε lie on a hypercircle with center at $C = (V_m^\tau + V_n^\varepsilon)/2$ and radius $R = \|V_m^\tau - V_n^\varepsilon\|/2$, while the vertices coincide with the end points of a diameter of the hypercircle (Figure 12.7).

The relation of fundamental importance now is the inequality (12.51),

$$(V_n^\varepsilon, V_n^\varepsilon) \le (S, S) \le (V_m^\tau, V_m^\tau). \tag{12.57}$$

To make use of the latter, we represent the approximate solutions in the forms (12.29),

$$S^\tau = S_0^\tau + \sum_{k=1}^{m} \alpha_k Y_k^\tau,$$

$$S^\varepsilon = \sum_{l=1}^{n} \beta_l Y_l^\varepsilon, \tag{12.58}$$

where it is assumed, as before, that the space origin lies in $\mathscr{S}_n^\varepsilon$ (i.e., $S_0^\varepsilon = 0$). We repeat that the auxiliary states Y_k^τ satisfy the homogeneous equilibrium equations, while the states Y_l^ε obey the compatibility equations and homogeneous displacement boundary conditions. Again, the α_k's and β_l's are those coefficients determined from the condition that the distance $\|S^\tau - S^\varepsilon\|$ be minimal. The vectors Y_k^τ and Y_l^ε being, in general, *not orthonormal*, we find the location of the vertices from the equations

$$(V_m^\tau, V_n^\tau) = (S_0^\tau, S_0^\tau) + \sum_{k=1}^{m} \alpha_k (S_0^\tau, Y_k^\tau),$$

$$(V_n^\varepsilon, V_n^\varepsilon) = \sum_{l=1}^{n} \beta_l (S_0^\tau, Y_l^\varepsilon), \tag{12.59}$$

where the coefficients α_k and β_l are determined from the equations (Ref. 40,

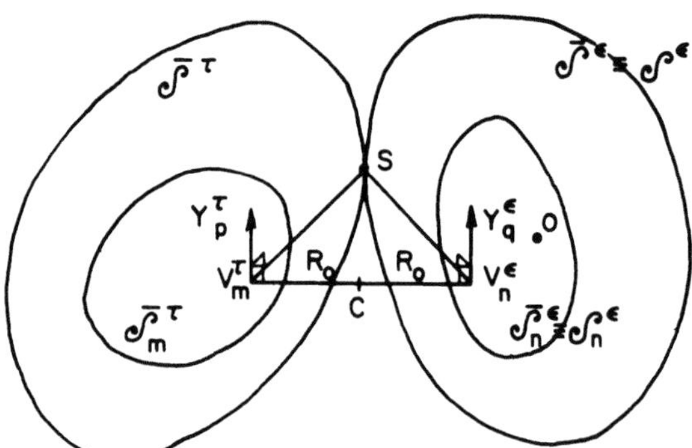

Figure 12.7. Translated subspaces $\mathscr{S}_m^\tau$ and $\bar{\mathscr{S}}_n^\tau$.

pp. 119 and 341)

$$\sum_{p=1}^{m} \alpha_p(Y_p^{\tau}, Y_r^{\tau}) + (S_0^{\tau}, Y_r^{\tau}) = 0, \qquad r = 1, 2, \ldots, m,$$

$$\sum_{q=1}^{n} \beta_q(Y_q^{\varepsilon}, Y_s^{\varepsilon}) - (S_0^{\tau}, Y_s^{\varepsilon}) = 0, \qquad s = 1, 2, \ldots, n.$$

(12.60)

We next select the following three sets of displacement components from the translated subspace $\bar{\mathscr{S}}_n^{\varepsilon}$:

$$(1) \quad u = 0, \qquad w = (r^2 - 1)(1 - z^2);$$

$$(2) \quad u = 0, \qquad w = (r^2 - 1)(1 - z^2)z^2;$$

$$(3) \quad u = r(r^2 - 1)(1 - z^2)z, \qquad w = 0;$$

(12.61)

these satisfy the displacement boundary conditions (12.55), the conditions of symmetry, and they possess certain intuitively conceived characteristics of the resulting deformation. With these in mind, we construct three vectors in the subspace $\bar{\mathscr{S}}_n^{\varepsilon}$ with λ and μ as the Lamé constants.

Vector Y_1^{ε}:

$$e_{rr} = e_{\theta\theta} = 0, \qquad e_{zz} = -2(r^2 - 1)z, \qquad e_{rz} = r(1 - z^2);$$

$$\tau_{rr} = \tau_{\theta\theta} = -2\lambda\kappa(r^2 - 1)z/E,$$

$$\tau_{zz} = -2\kappa(2\mu + \lambda)(r^2 - 1)z/E, \qquad \tau_{rz} = 2\mu\kappa r(1 - z^2)/E.$$

(12.62)

Vector Y_2^{ε}:

$$e_{rr} = e_{\theta\theta} = 0, \qquad e_{zz} = 2(r^2 - 1)(1 - 2z^2)z, \qquad e_{rz} = r(1 - z^2)z^2;$$

$$\tau_{rr} = \tau_{\theta\theta} = \frac{2\lambda\kappa}{E}(r^2 - 1)(1 - 2z^2)z,$$

$$\tau_{zz} = \frac{2\kappa}{E}(2\mu + \lambda)(r^2 - 1)(1 - 2z^2)z, \qquad \tau_{rz} = \frac{2\mu\kappa}{E}r(1 - z^2)z^2.$$

(12.63)

Vector Y_3^{ε}:

$$e_{rr} = (3r^2 - 1)(1 - z^2)z, \qquad e_{\theta\theta} = (r^2 - 1)(1 - z^2)z,$$

$$e_{rz} = \tfrac{1}{2}r(r^2 - 1)(1 - 3z^2);$$

$$\tau_{rr} = \frac{2\kappa}{E}[\mu(3r^2 - 1) + \lambda(2r^2 - 1)](1 - z^2)z,$$

$$\tau_{\theta\theta} = \frac{2\kappa}{E}[\mu(r^2 - 1) + \lambda(2r^2 - 1)](1 - z^2)z,$$

$$\tau_{zz} = \frac{2\lambda\kappa}{E}(2r^2 - 1)(1 - z^2)z, \qquad \tau_{rz} = \frac{\mu\kappa}{E}r(r^2 - 1)(1 - 3z^2).$$

(12.64)

In like manner, we select four vectors in the subspace $\mathscr{S}_m{}^{\tau}$ (with v as Poisson's ratio):

Vector $Y_1{}^{\tau}$:

$$\tau_{rr} = \tau_{\theta\theta} = r^2 z, \qquad \tau_{zz} = \tfrac{2}{3}z^3, \qquad \tau_{rz} = -rz^2;$$

$$e_{rr} = e_{\theta\theta} = \frac{1}{\kappa}[r^2 z - v(r^2 z + \tfrac{2}{3}z^3)], \tag{12.65}$$

$$e_{zz} = \frac{2}{\kappa}(\tfrac{1}{3}z^2 - vr^2)z, \qquad e_{rz} = -\frac{E}{2\mu\kappa}rz^2, \qquad e_{zz} = -\frac{2v}{\kappa}z.$$

Vector $Y_2{}^{\tau}$:

$$\tau_{rr} = \tau_{\theta\theta} = z; \qquad e_{rr} = e_{\theta\theta} = \frac{z}{\kappa}(1-v), \qquad e_{zz} = -\frac{2v}{\kappa}z. \tag{12.66}$$

Vector $Y_3{}^{\tau}$:

$$\tau_{zz} = 2z, \qquad \tau_{rz} = r;$$

$$e_{rr} = e_{\theta\theta} = \frac{2v}{\kappa}z, \qquad e_{zz} = -\frac{2z}{\kappa}, \qquad e_{rz} = \frac{rE}{2\mu\kappa}. \tag{12.67}$$

Vector $Y_4{}^{\tau}$:

$$\tau_{zz} = -4r^2 z, \qquad \tau_{rz} = r^3;$$

$$e_{rr} = e_{\theta\theta} = \frac{4v}{\kappa}r^2 z, \qquad e_{zz} = -\frac{4}{\kappa}r^2 z, \qquad e_{rz} = \frac{r^3 E}{2\mu\kappa}. \tag{12.68}$$

As the vector $S_0{}^{\tau}$, we choose† $\tau_{zz} = \kappa z, e_{rr} = e_{\theta\theta} = -vz, e_{zz} = z$. For the sake of brevity, we now choose to set $v = 0$.

A lengthy calculation gives the inner products appearing in the relations (12.60) as follows:

$$(Y_1{}^{\varepsilon})^2 = 88\pi\kappa/45, \qquad (Y_2{}^{\varepsilon})^2 = (Y_1{}^{\varepsilon}, Y_2{}^{\varepsilon}) = 104\pi\kappa/315,$$

$$(Y_3{}^{\varepsilon})^2 = 85\pi\kappa/315, \qquad (Y_1{}^{\varepsilon}, Y_3{}^{\varepsilon}) = -4\pi\kappa/45, \tag{12.69}$$

$$(Y_2{}^{\varepsilon}, Y_3{}^{\varepsilon}) = 4\pi\kappa/315.$$

---

† We repeat that, since the stresses on the boundary of the body are not prescribed, it is only hoped that those associated with the selected stress systems provide acceptable approximations to the actual stresses on Ω (associated with the state S).

Similarly, for the pertinent vectors in the subspace $\mathscr{S}_m^{\tau}$, we find

$$(Y_1^{\tau})^2 = 102\pi/105\kappa, \qquad (Y_2^{\tau})^2 = 4\pi/3\kappa,$$

$$(Y_3^{\tau})^2 = 14\pi/3\kappa, \qquad (Y_4^{\tau})^2 = 41\pi/9\kappa, \qquad (Y_1^{\tau}, Y_2^{\tau}) = 2\pi/3\kappa,$$

$$(Y_1^{\tau}, Y_3^{\tau}) = -6\pi/5\kappa, \qquad (Y_1^{\tau}, Y_4^{\tau}) = -44\pi/45\kappa, \tag{12.70}$$

$$(Y_2^{\tau}, Y_3^{\tau}) = (Y_2^{\tau}, Y_4^{\tau}) = 0, \qquad (Y_3^{\tau}, Y_4^{\tau}) = 4\pi/\kappa.$$

Moreover,

$$(S_0^{\tau}, Y_1^{\varepsilon}) = 2\pi\kappa/3, \qquad (S_0^{\tau}, Y_2^{\varepsilon}) = 2\pi\kappa/15, \qquad (S_0^{\tau}, Y_3^{\varepsilon}) = 0,$$

$$(S_0^{\tau})^2 = 2\pi\kappa/3, \qquad (S_0^{\tau}, Y_1^{\tau}) = 4\pi/15, \qquad (S_0^{\tau}, Y_2^{\tau}) = 0, \tag{12.71}$$

$$(S_0^{\tau}, Y_3^{\tau}) = (S_0^{\tau}, Y_4^{\tau}) = -4\pi/3.$$

Substitution of the above results into equations (12.60) gives the values of the coefficients: $\alpha_1 = 0.2066\kappa$, $\alpha_2 = -0.1033\kappa$, $\alpha_3 = 0.2020\kappa$, $\alpha_4 = 0.1597\kappa$; $\beta_1 = 0.3348$, $\beta_2 = 0.0649$, $\beta_3 = 0.1072$.

It is now a straightforward matter to find from (12.59) that

$$(V_n^{\varepsilon}, V_n^{\varepsilon}) = 0.2319\pi\kappa, \qquad (V_m^{\tau}, V_m^{\tau}) = 0.2399\pi\kappa. \tag{12.72}$$

Consequently, the bounds for the squared norm of the exact solution S become

$$0.2319\pi\kappa \leq (S, S) \leq 0.2399\pi\kappa. \tag{12.73}$$

With the arithmetic average of the values in (12.72) equal to $0.2359\pi\kappa$, the deviation of (V_m^{τ}, V_m^{τ}) from $(V_n^{\varepsilon}, V_n^{\varepsilon})$ evaluated with respect to their arithmetic average amounts to 3.39%. This seems to be an acceptable difference, considering the small number of vectors Y_p^{ε} and Y_q^{τ} adopted in the calculations.

The last formula, together with (12.56), implies that the non-dimensionalized sagging of the mass center below the geometric center of the cylinder is contained between the bounds given in

$$0.0580\kappa \leq \delta \leq 0.0600\kappa, \tag{12.74}$$

or, in a dimensional form, $0.0580 \leq \Delta E/\rho g a^2 \leq 0.0600$. By the very construction of the hypercircle, the position vector of its center, C, constitutes a *good approximation* to the exact solution (Figure 12.8). In the present case, the squared norm $\|C\|^2 = (C, C) = 0.2339\pi\kappa$. Remaining, of course, between the bounds of (12.73), it differs insignificantly from the arithmetic mean of (12.72), equal to $0.2359\pi\kappa$.

As already noted in connection with the fundamental inequality (12.28e), the bounds derived above by the hypercircle method (compare

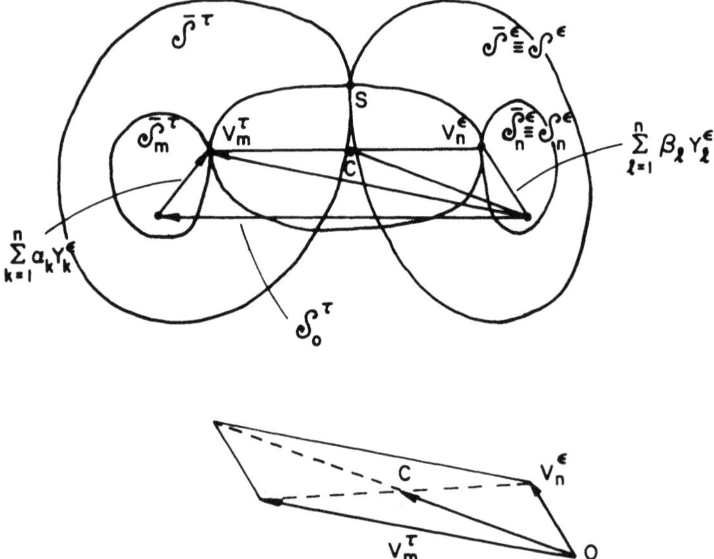

Figure 12.8. Sketch for equation (12.58).

(12.73), for example) bracket the quantities of interest (such as the deformation, strain, or stress) in an *integral* manner, that is, indirectly via the strain energy stored in the body. In order to obtain what may be called *pointwise* bounds for the unknown quantities, modified procedures must be applied.†
We shall discuss this point in some detail a little later in this chapter, in Section 12.2.

At this stage, it is attractive to look at the problem, not so much from an axiomatic, but rather from a plainly utilitarian point of view. We observe that equations (12.61)–(12.68) contain a great amount of information about the stress and strain distributions in the body which would be good to utilize. The difficulty, however, is that each constituent distribution is associated with a particular representative vector $(S_0{}^\tau, Y_k{}^\tau, \text{ or } Y_k{}^\varepsilon, k = 1, 2, 3)$, and the rule for combining them into a single, resultant distribution is unclear. Actually, what is needed is the knowledge of a scaling factor for each particular distribution, that is, its "weight" in a linear combination of the available distributions. Inasmuch as each function vector is associated with the corresponding stress tensor, there is some justification for theorizing that the scaling factors of the particular stress tensors are the same as those of the corresponding function vectors. With this hypothesis in mind, we turn to

---

† In the present case, the only pointwise bounds obtained involve the vertical component of the displacement vector at the origin of coordinates [see the inequality (12.74)].

equations (12.58), in which the values of the coefficients are borrowed from equations (12.60). We then write the equations

$$\tau_{ij}^{(\tau)} = \tau_{ij}^{(\tau)0} + \sum_{k=1}^{m} \alpha_k \tau_{ij}^{(\tau)k},$$

$$\tau_{ij}^{(\varepsilon)} = \sum_{l=1}^{n} \beta_l \tau_{ij}^{(\varepsilon)l}, \tag{12.74a}$$

in which the notation should be self-explanatory. As the resultant of the partial sums $\tau_{ij}^{(\tau)}$ and $\tau_{ij}^{(\varepsilon)}$, we take the arithmetic mean reminiscent of the mean of function vectors (12.35b). It is shown below that, despite the fact that the suggested procedure lacks strict theoretical motivation, the numerical results obtained compare favorably (at least for the problem under discussion) with those obtained by the approximate method of Galerkin. Oddly enough, Galerkin's method also lacks a sufficiently general substantiation [e.g., Ref. (8)].

Suppose now that by appeal to our heuristic hypothesis, we wish to find, say, the stress component τ_{zz} at two points: $r = 0$, $z = -1$, and $r = \frac{1}{2}$, $z = -\frac{1}{2}$. From equations (12.74a) we obtain

$$\tau_{zz} = \tfrac{1}{2}[\kappa z + \tfrac{2}{3}\alpha_1 z^3 - 2\alpha_3 z - 4\alpha_4 r^2 z - 2\kappa\beta_1(r^2 - 1)z$$

$$+ 2\kappa\beta_2(r^2 - 1)(1 - 2z^2)z], \tag{12.75}$$

so that

$$\tau_{zz}(0, -1) = -0.7666\kappa, \qquad \tau_{zz}(\tfrac{1}{2}, -\tfrac{1}{2}) = -0.2311\kappa. \tag{12.76}$$

The exact values of the stress components are, of course, unknown, but we might conclude that the accuracy of the results (12.76) is sufficiently good by recalling the closeness of the bounds in inequality (12.73).

Such a conjecture is confirmed by invoking a different approximating method, such as the well-known *Galerkin procedure* which, more often than not, leads to adequate results.

We thus assume that $u = \sum_m A_m \phi_m(r, z)$ and $w = \sum_m B_m \psi_m(r, z)$, where A_m and B_m are constant coefficients to be determined later, and the functions ϕ_m and ψ_m satisfy the homogeneous boundary conditions imposed on the displacements.

The Galerkin equations then become (Ref. 74, p. 159)

$$2\pi \int_0^1 r\, dr \int_{-1}^1 \phi_m(r, z)\left[\frac{\partial \tau_{rr}}{\partial r} + \frac{\partial \tau_{rz}}{\partial z} + \frac{\tau_{rr} - \tau_{\theta\theta}}{r}\right] dz = 0,$$

$$2\pi \int_0^1 r\, dr \int_{-1}^1 \psi_m(r, z)\left[\frac{\partial \tau_{rz}}{\partial r} + \frac{\partial \tau_{zz}}{\partial z} + \frac{\tau_{rz}}{r} - \kappa\right] dz = 0, \tag{12.77}$$

for $v = 0$, reducing to

$$2\pi\kappa \int_0^1 r \, dr \int_{-1}^1 \left| \frac{\partial^2 u}{\partial r^2} + \frac{1}{r}\frac{\partial u}{\partial r} + \frac{1}{z}\frac{\partial^2 u}{\partial z^2} - \frac{u}{r} + \frac{1}{2}\frac{\partial^2 w}{\partial r \, \partial z} \right| \phi_m(r, z) \, dz = 0,$$

$$2\pi\kappa \int_0^1 r \, dr \int_{-1}^1 \left[\frac{\partial^2 w}{\partial z^2} + \frac{1}{2}\frac{1}{r}\frac{\partial w}{\partial r} + \frac{1}{2}\frac{\partial^2 w}{\partial r^2} + \frac{1}{2}\frac{\partial^2 u}{\partial r \, \partial z} + \frac{1}{2}\frac{1}{r}\frac{\partial u}{\partial z} - 1 \right] \tag{12.78}$$

$$\times \, \psi_m(r, z) \, dz = 0.$$

We take the displacement components in the form

$$u = A_1 r(r^2 - 1)(1 - z^2)z + A_2 r^3(r^2 - 1)(1 - z^2)z^3,$$
$$w = B_1(r^2 - 1)(1 - z^2) + B_2(r^2 - 1)(1 - z^2)z^2, \tag{12.79}$$

and arrive easily at the following system of four simultaneous linear equations for the coefficients A_1, A_2, B_1, and B_2:

$$-\frac{17}{1261}A_1 - \frac{107}{4725}A_2 + \frac{2}{45}B_1 - \frac{2}{315}B_2 = 0,$$

$$-\frac{107}{4725}A_1 - \frac{893}{103950}A_2 + \frac{1}{105}B_1 + \frac{1}{945}B_2 = 0,$$

$$\frac{2}{45}A_1 + \frac{1}{105}A_2 - \frac{44}{45}B_1 - \frac{52}{315}B_2 = -\frac{1}{3}, \tag{12.80}$$

$$\frac{2}{315}A_1 + \frac{1}{945}A_2 - \frac{52}{315}B_1 - \frac{52}{315}B_2 = -\frac{1}{15}.$$

A straightforward calculation gives, for example, $B_1 = 0.3184$ and $B_2 = 0.0701$, so that, via equations (12.79), the normal stress component

$$\tau_{zz}(r, z) = -2\kappa(r^2 - 1)r[B_1 + (2z^2 - 1)B_2] \tag{12.81}$$

at the points $(0, -1)$ and $(\tfrac{1}{2}, -\tfrac{1}{2})$ becomes

$$\tau_{zz}(0, -1) = -0.7770\kappa, \qquad \tau_{zz}(\tfrac{1}{2}, -\tfrac{1}{2}) = -0.2125\kappa. \tag{12.82}$$

A comparison of the preceding results with those of (12.76) seems to confirm the anticipated accuracy of the hypercircle solution and the acceptability of our working hypothesis.

12.2. Bounds for a Solution at a Point

In discussing, in this section, the application of the hypercircle method to the derivation of bounds for a solution (and possibly its derivatives) at a point of a physical domain, we take a slightly different standpoint from that which was appropriate in the examination of problems of a particular elastic class in the preceding section. There is little need to emphasize that, for practical purposes, pointwise bounding is considerably more important than the estimation of a solution in an integral manner via the elastic strain energy, for example, as was done before.

Let us then consider a pair of intersecting orthogonal translated subspaces $\bar{\mathscr{S}}'$ and $\bar{\mathscr{S}}''$ with the space origin in $\bar{\mathscr{S}}''$ say (Figure 12.9a). We select two auxiliary translated subspaces $\bar{\mathscr{S}}_r'$ and $\bar{\mathscr{S}}_s''$ of dimensions r and s, immersed in $\bar{\mathscr{S}}'$ and $\bar{\mathscr{S}}''$, respectively. S is the point of intersection of $\bar{\mathscr{S}}'$ and $\bar{\mathscr{S}}''$ and corresponds to the exact solution of a problem, V' and V'' are vertices of $\bar{\mathscr{S}}_r'$ and $\bar{\mathscr{S}}_s''$, respectively, and C is the center of the hypercircle $\Gamma_{(r+s)}$ of class $r + s$; as explained before (cf., e.g., Figure 12.5), the hypercircle passes

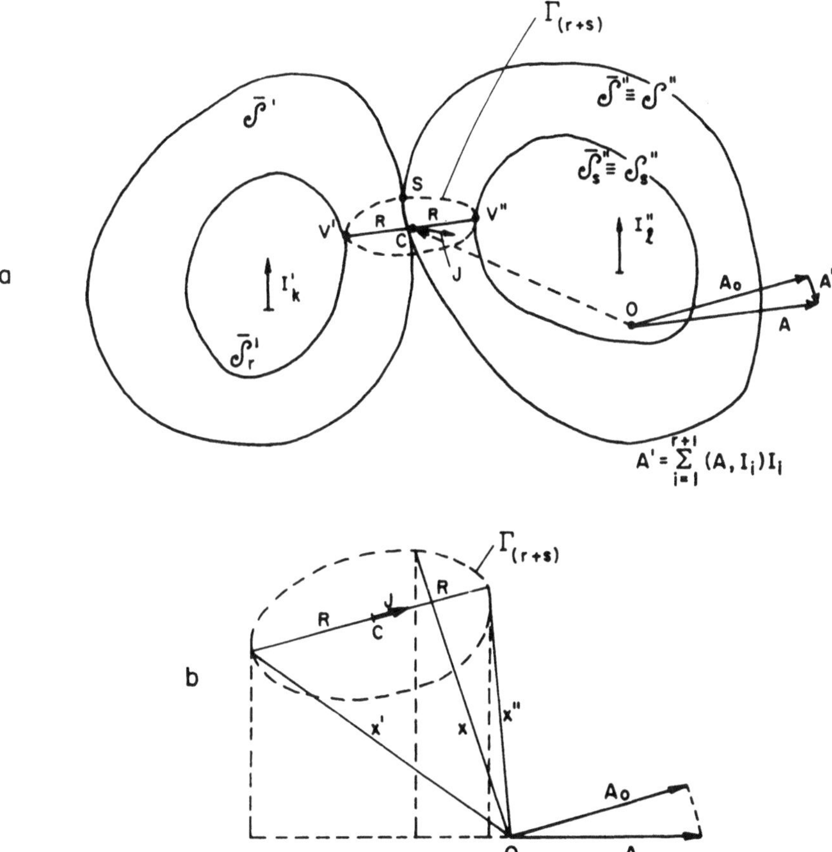

Figure 12.9. Intersection of orthogonal subspaces.

through the vertices and includes the point S. As bases for the subspaces $\mathscr{S}_r{}'$ and $\mathscr{S}_s{}''$, we adopt sets of orthonormal vectors $I_k{}'$, $k = 1, 2, \ldots, r$ and $I_l{}''$, $l = 1, 2, \ldots, s$ contained in these subspaces, respectively. If in equations (12.40) $Y_p{}^\tau$ is replaced by $I_k{}'$, $Y_q{}^\varepsilon$ by $I_l{}''$, and $X - C$ by J, where J is a unit vector, then the equations defining the *hypercircle* $\Gamma_{(r+s)}$ become

$$(X - C, X - C) = R^2,$$

$$(J, I_k{}') = 0, \qquad k = 1, 2, \ldots, r, \tag{12.83}$$

$$(J, I_l{}'') = 0, \qquad l = 1, 2, \ldots, s,$$

where C is the center of the hypercircle and R is its radius. Combining equations (12.35b) and (12.34), setting $S_0{}^\varepsilon = 0$, and changing the notation, we have

$$V' = S_0 - \sum_{k=1}^{r} (S_0, I_k{}')I_k{}'$$

$$\tag{12.84}$$

$$V'' = \sum_{l=1}^{s} (S_0, I_l{}'')I_l{}''$$

and

$$C = \tfrac{1}{2}(V' + V''), \qquad R = \tfrac{1}{2}\|V' - V''\|, \tag{12.84a}$$

S_0 being a preassigned vector in the translated subspace $\bar{\mathscr{S}}_r{}'$.

Equations (12.83) can be cast into the parametric form [compare the passage from (9.37)–(9.38)]

$$X = C + RJ,$$

$$\|J\| = 1,$$

$$(J, I_k{}') = 0, \qquad k = 1, 2, \ldots, r, \tag{12.85}$$

$$(J, I_l{}'') = 0, \qquad l = 1, 2, \ldots, s.$$

One can now imagine, for convenience, that $\mathscr{S}_r{}'$ and $\mathscr{S}_s{}''$ are independent, i.e., their intersection is the zero vector. Then the last two of equations (12.85) become simply

$$(J, I_i) = 0, \qquad i = 1, 2, \ldots, r + s. \tag{12.86}$$

Since the preceding relations are true for any direction J of the radius of the hypercircle and for any I_i, $i = 1, 2, \ldots, r + s$, in $\mathscr{S}_{r+s} = \mathscr{S}_r{}' \oplus \mathscr{S}_s{}''$, one can express this fact by saying that the subspace $\mathscr{S}_{r+s}$ of dimension $(r + s)$ is orthogonal to the hypercircle $\Gamma_{(r+s)}$.

Consider now an arbitrary vector A in a space including the subspace

$\mathscr{S}_{r+s}$. We resolve A into two components: A' lying in $\mathscr{S}_{r+s}$ and A_0 orthogonal to $\mathscr{S}_{r+s}$ (Figure 12.9a). Accordingly,

$$A = A_0 + \sum_{i=1}^{r+s} (A, I_i)I_i, \tag{12.87}$$

$$(A_0, I_i) = 0, \qquad i = 1, 2, \ldots, r + s. \tag{12.87a}$$

The inner product of the first of equations (12.85) with A gives

$$(X, A) = (C, A) \pm R \,|(A, J)|. \tag{12.88}$$

By (12.87a), however,

$$|(A, J)| = |(A_0, J)| \leq \|A_0\|, \tag{12.89}$$

which yields the important inequality

$$(C, A) - R\|A_0\| \leq (X, A) \leq (C, A) + R\|A_0\|, \tag{12.90}$$

where A_0 is determined from (12.87).

By appealing to our space intuition, we draw in Figure 12.9b a diameter[†] of the hypercircle "parallel" to the vector A_0 and construct the orthogonal projections of the position vectors X' and X'' of the end points of this diameter on A. It is easily verified that the lower and upper bounds in (12.90) are attained for

$$X' = C - RA_0/\|A_0\|$$

and
$$\tag{12.91}$$
$$X'' = C + RA_0/\|A_0\|,$$

respectively.

We now turn our attention to the Dirichlet problem,

$$\nabla^2\psi = 0, \qquad \psi\Big|_{S_2} = f \tag{12.92}$$

in a three-dimensional domain $V = V_1 + V_2$ with bounding surface S_2 (Figure 12.10). At a point P of V, the latter being referred to a Cartesian rectangular coordinate system ξ_i, $i = 1, 2, 3$, a second Cartesian rectangular coordinate system x_i, $i = 1, 2, 3$, is constructed; it is required to find pointwise bounds for the function $\psi(\xi)$, $\xi = (\xi_1, \xi_2, \xi_3)$ at the arbitrarily selected point P within V.

As is well known, such a Dirichlet problem for the Laplace equation is encountered in many areas of mathematical physics: electrostatics, heat conduction, fluid dynamics, elasticity, and others.

---

† Note that for this diameter, $J = \pm A_0/\|A_0\|$.

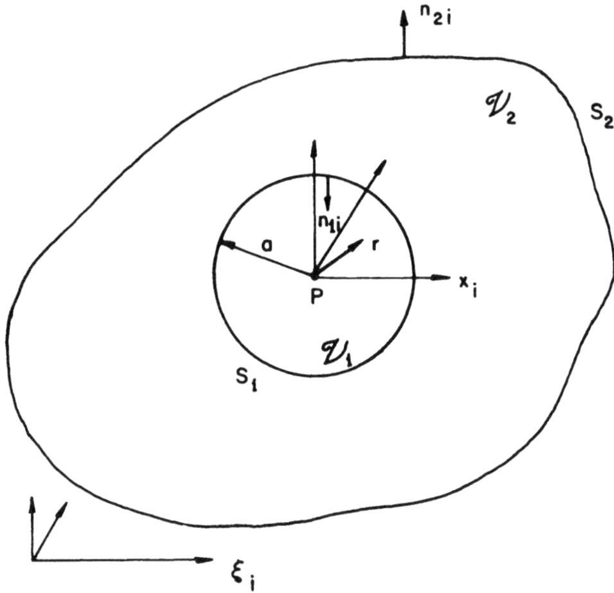

Figure 12.10. Sketch for the derivation of pointwise bounds.

We start the discussion by recalling that the fundamental solution of Laplace's equation in three dimensions (containing the basic singularity of the *Green's function*) is

$$\psi_{\text{fund}} = \frac{1}{r}. \tag{12.93}$$

Following, e.g., Maple[81], we introduce the vector

$$m_i = \begin{cases} -\dfrac{x_i}{r^3} & \text{for } r \geq a, \\[2mm] 0 & \text{for } r < a, \end{cases} \tag{12.94}$$

where a is the radius of a sphere drawn within the domain V with center at P. We denote by V_1 the domain occupied by the sphere and by Ω_1, the boundary of V_1.

Now let A and S denote vectors in a function space corresponding to the vectors m_i and $\psi_{,i}$, $i = 1, 2, 3$, respectively, where ψ constitutes the solution to the problem (12.92). With this in mind, we cast the inequality (12.90) into the form

$$|(S, A) - (C, A)| \leq R\|A_o\|, \tag{12.95}$$

and calculate the first of the foregoing inner products:

$$(S, A) = \int_V \psi_{,i} m_i \, dV$$

$$= \int_V (\psi m_i)_{,i} \, dV - \int_V \psi m_{i,\,i} \, dV \qquad (12.96)$$

$$= \int_{S_1} \psi m_i n_i \, dS_1 + \int_{S_2} \psi m_i n_i \, dS_2,$$

in which we have used the Gauss–Green theorem and the fact that $m_{i,\,i} = 0$. Noticing that on the boundary S_1 there is $n_i = n_{1i} = -x_i/a$ and $m_i = -x_i/a^3$, we have

$$(S, A) = \frac{1}{a^2} \int_{S_1} \psi \, dS_1 + \int_{S_2} \psi m_i n_i \, dS_2. \qquad (12.97)$$

The function ψ being harmonic, Gauss' theorem of the arithmetic mean implies that the first integral in the preceding equation is equal to $4\pi\psi(P)$, where P is the center of the sphere V_1. Substituting (12.97) into (12.95), we obtain finally

$$\left| 4\pi\psi(P) + \int_{S_2} \psi m_i n_i \, dS_2 - (C, A) \right| \le R \| A_o \|. \qquad (12.98)$$

All quantities appearing herein, except $\psi(P)$, are known[†] [cf. (12.84a) and (12.87a)] provided the associated hypercircle was earlier constructed. (We assume that this was accomplished.) Instead of carrying out the pertinent calculations, however, we shall content ourselves with examining a much simpler situation. Namely, suppose that we were fortunate enough to make the radius of the hypercircle very small, $R \approx 0$. Then $C \approx S$,[‡]

$$(C, A) \approx (S, A) = \int_V m_i \psi_{,i} \, dV, \qquad (12.99)$$

and it is easy to verify that (12.98) implies that, approximately,

$$\psi(P) \approx \frac{1}{4\pi a^2} \int_{S_1} \psi \, dS_1, \qquad (12.100)$$

close to the exact result.

[†] Upon selecting the states S_0, $\{I_k'\}$, and $\{I_l''\}$ at the outset of the calculations (cf. Problem 12.1), we automatically determine V' and V'' by (12.84) and C and R by (12.84a). Likewise, selecting in advance the state A, one determines A_0 by (12.87) and (12.87a).

[‡] Since, by definition, state S (representing the exact solution) lies on the hypercircle, by shrinking the latter to its center, we necessarily accept C as the approximation of S.

Similar reasoning applied to a plane domain yields the formula

$$\psi(P) \approx \frac{1}{2\pi a}\int_{C_1} \psi \, dC_1, \tag{12.101}$$

where a is the radius of a circle whose circumference is C_1. If ψ is interpreted as the conjugate torsion function, for which $\psi = r^2/2$ on the boundary of the cross section of a bar subjected to twist,† and if the cross section V becomes a circular disk of radius a, then

$$\psi(P) \approx \frac{a^2}{2}. \tag{12.102}$$

This is a rigorous result which, incidently, holds at any point of a circular cross section where $\psi = \text{const.}^{(31)}$

By repeating the reasoning applied above to the function ψ, it is a straightforward matter to derive bounds for the first and second partial derivatives of ψ. The associated Maple functions turn out to be $m_{i,\,p}$ and $m_{i,\,pq}$, $p, q = 1, 2$, respectively.

12.3. Hypercircle Method and Function Space Inequalities

A digression seems now to be in order for the purpose of corroborating our earlier remark about the intrinsic connections existing between the hypercircle procedure and the inequalities of Cauchy and Bessel, as was first observed in Ref. (29).

We first wish to recall two *central* formulas of the hypercircle method valid for linear elasticity, namely, the inequality (12.28e),

$$(S^\tau, I^\varepsilon)^2 \le (S, S) \le (S^\tau, S^\tau), \tag{12.103}$$

and the inequality (12.53),

$$\sum_{q=1}^{n} (S_0^\tau, Y_q^\varepsilon)^2 \le (S, S) \le (S_0^\tau, S_0^\tau) - \sum_{p=1}^{m} (S_0^\tau, Y_p^\tau)^2. \tag{12.104}$$

Since‡ $I^\varepsilon = S^\varepsilon/\|S^\varepsilon\|$ and $(S^\tau, S^\varepsilon) = (S, S^\varepsilon)$, we can immediately cast the left-hand member of inequality (12.103) into the form

$$\frac{(S, S^\varepsilon)^2}{(S^\varepsilon, S^\varepsilon)} \le (S, S), \tag{12.105}$$

making manifest its identity with the Cauchy–Schwarz inequality (7.14).

---

† The position of the origin of the coordinates is of no consequence in this problem.
‡ See equation (12.23) and the text following.

Likewise, writing (12.23) for two states $S^{\varepsilon'}$ and $S^{\varepsilon''}$ and considering that, by definition,[†] any auxiliary state Y_q^{ε} is equivalent to a difference between two fundamental states, we obtain, after subtracting, $(S, Y_q^{\varepsilon}) = (S^{\tau}, Y_q^{\varepsilon})$. With reference to (12.3) and (12.29), the left inequality of (12.104) thus becomes

$$\sum_{q=1}^{n} (S, Y_q^{\varepsilon})^2 \le (S, S), \tag{12.106}$$

turning but to coincide with the Bessel inequality (7.21) written for an n-space.

There is no difficulty in arriving at similar conclusions with regard to the right-hand sides of the inequalities (12.103) and (12.104). For this purpose, it is enough to appeal to equation (12.19). In fact, use of the relation $(S, S) = (S, S^{\tau})$ makes it possible to reduce the right-hand side of inequality (12.103) directly to the Cauchy–Schwarz form

$$(S, S^{\tau})^2 \le (S, S)(S^{\tau}, S^{\tau}). \tag{12.107}$$

Similarly, writing equation (12.19) for two states $S^{\tau'}$ and $S^{\tau''}$ and subtracting yields

$$(S, S^{\tau'} - S^{\tau''}) = 0. \tag{12.108}$$

By its very definition, however,[‡] an auxiliary state Y_p^{τ} is equivalent to a difference between two fundamental states S^{τ}. Thus,

$$(S, Y_p^{\tau}) = 0 \tag{12.109}$$

for every value of p between 1 and m. Accordingly, the $m + 1$ vectors $S/\|S\|$ and Y_p^{τ}, $p = 1, 2, \ldots, m$ form an orthonormal set.

Returning to the inequality (12.104), we drop the subscript 0 and transform its right-hand side so as to obtain

$$(S, S) + \sum_{p=1}^{m} (S^{\tau}, Y_p^{\tau})^2 \le (S^{\tau}, S^{\tau})$$

and

$$\left(S^{\tau}, \frac{S}{\|S\|} \right)^2 + \sum_{p=1}^{m} (S^{\tau}, Y_p^{\tau})^2 \le (S^{\tau}, S^{\tau}), \tag{12.110}$$

the last form implied by the relation (12.19) (after squaring). This result coinciding with Bessel's inequality (7.21), our assertion is confirmed.

---

† Compare point (b_1) following equation (12.28k).
‡ Compare point (a_1) following equation (12.28k).

12.4. A Comment

As already noted, a hypercircle of class one, $\Gamma_{(1)}$, is generated by the intersection of a hypersphere [equation (9.37)] centered at C_0, given by

$$\|X - C_0\|^2 = R_0{}^2 \tag{12.111}$$

and a hyperplane $H_{(1)}$ of class one [equation (9.14)], given by

$$(X, I) = \alpha, \tag{12.112}$$

where I is a unit vector normal to the hyperplane (Figure 12.11; see also Figure 12.4). A parametric representation of $\Gamma_{(1)}$ is, of course, given by (9.38),

$$X = C + RJ, \tag{12.113}$$

where $\|J\| = 1$, $(I, J) = 0$, and C is the center of $\Gamma_{(1)}$. In order to find C and R in terms of C_0, I, and R_0, we observe that, from Figure 12.11,

$$C = C_0 + \beta I, \tag{12.114}$$

with β a scalar. But $(X, I) = (C, I) = \alpha$, so that $(C, I) = (C_0, I) + \beta = \alpha$ and

$$C = C_0 + [\alpha - (C_0, I)]I. \tag{12.115}$$

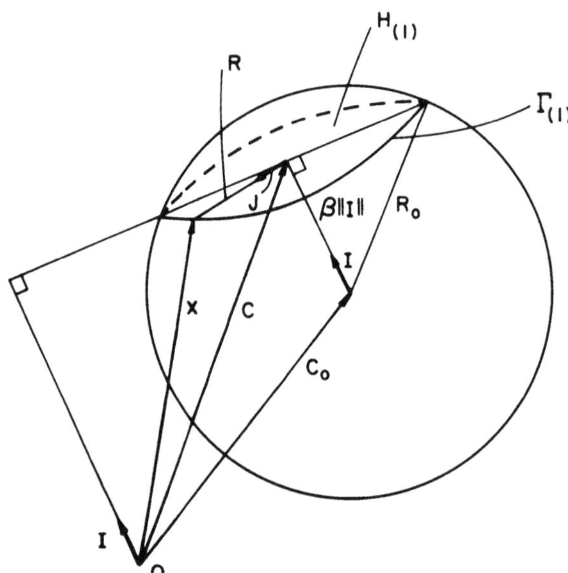

Figure 12.11. A hypercircle of class one.

Now, $(X - C, I) = 0 = (X - C, C - C_0)$. Consequently,

$$(X - C_0, X - C_0) = (X - C + C - C_0, X - C + C - C_0)$$

$$= (X - C, X - C) + (C - C_0, C - C_0). \quad (12.116)$$

By appeal to (12.113) and (12.115), this yields finally

$$R^2 = R_0{}^2 - [\alpha - (C_0, I)]^2. \quad (12.117)$$

Problems

1. Derive the governing equations (12.6).

2. How many dimensions of the function space $\mathcal{L}_2(-\pi, \pi)$, described with respect to the standard trigonometric base vectors $(1, \sin t, \cos t, \sin 2t, \ldots)$, are needed to represent the states: (a) $\sin t$, $\cos t$, $\cos^2 t$; (b) $\sin t$, $\cos t$, $\sin t \cdot \cos t$?

3. Let each of the states $S_1 = \{e_{ij}^{(1)}\}$ and $S_2 = \{e_{ij}^{(2)}\}$ satisfy the compatibility equations (12.4). Does the state $\alpha S_1 + \beta S_2$ (α, β are scalars) satisfy these equations?

4. An (intrinsic) vector $X - Y$ joining any two points X and Y of a one-space has the representation $X - Y = \gamma_1 S_1 + \gamma_2 S_2$, where X, Y, S_1, S_2 are position vectors. Show that $\gamma_1 + \gamma_2 = 0$.

5. Compute bounds for $(I_1{}^\varepsilon, I_2{}^\varepsilon)$ and $(I_1{}^\varepsilon, I_2{}^\varepsilon + I_3{}^\varepsilon)$, where $I_i{}^\varepsilon$, $i = 1, 2, 3$ are unit vectors and $I_2{}^\varepsilon$ and $I_3{}^\varepsilon$ are mutually orthogonal.

6. Let $\mathcal{S}_1$ and $\mathcal{S}_2$ be two mutually orthogonal subspaces in a Hilbert space and let S be any vector in $\mathcal{S}_1 \oplus \mathcal{S}_2$. Show that S has a unique representation $S = S_1 + S_2$, where S_1 is in $\mathcal{S}_1$ and S_2 is in $\mathcal{S}_2$.

7. Consider a hyperplane $H_{(2)}$ of class two including the tips of all vectors S corresponding to functions $f(t)$, $-\pi \le t \le \pi$, which are such that $(S, I^1) = a$, $(S, I^2) = b$, where I^1, I^2 are two orthonormal vectors and a, b are scalars. Find the length of the normal N from the space origin to the hyperplane.

8. Find the representation of a function $f(t)$, $a \le t \le b$ such that $\int_a^b [f(t)]^2 \, dt = \bar{\alpha}$ and $\int_a^b f(t)t \, dt = \bar{\beta}$ by using an analog of the hypercircle method [compare (Ref. 40, p. 95)].

13

The Method of Orthogonal Projections

The concept of the orthogonal projection of a function vector on a subspace was already discussed in Chapter 9. We now wish to proceed with the development of a fruitful method employing this concept, known as the *method of orthogonal projections.*

To introduce the basic ideas, imagine an n-dimensional function space, $\mathscr{V}_n$, a function f in this space, and a k-dimensional subspace $\mathscr{S}_k$ of $\mathscr{V}_n$ such that $k < n$ (Figure 13.1 in which $k = 2$ and $n = 3$).

Let $g_1, g_2, \ldots, g_k$ be an orthonormal basis for $\mathscr{S}_k$, and let $g_1, g_2, \ldots, g_k, \ldots, g_n$ be an orthonormal basis for $\mathscr{V}_n$. By definition of a basis, any function f in $\mathscr{V}_n$ can be represented in the form

$$f = (\alpha_1 g_1 + \cdots + \alpha_k g_k) + (\alpha_{k+1} g_{k+1} + \cdots + \alpha_n g_n)$$
$$= f^* + f^\perp, \tag{13.1}$$

where f^* and $f^\perp$ designate the expressions in parentheses, respectively. Again by definition of a basis, f^* (represented in terms of the basis in $\mathscr{S}_k$) is a vector in $\mathscr{S}_k$. On the other hand, $f^\perp$—as seen from its representation in (13.1)—is not in $\mathscr{S}_k$. It is, in fact, perpendicular to all vectors in $\mathscr{S}_k$, as demonstrated by the following argument. In view of the orthogonality of the basis $g_1, \ldots, g_n$, there is

$$([\alpha_{k+1} g_{k+1} + \cdots + \alpha_n g_n], g_i) = 0, \qquad i = 1, 2, \ldots, k. \tag{13.2}$$

However, any function in $\mathscr{S}_k$ can be represented linearly in terms of the basis $g_1, \ldots, g_k$. Thus, $f^\perp$ is perpendicular to every vector in $\mathscr{S}_k$ and, as a consequence, to $\mathscr{S}_k$ itself. Since f^* is in $\mathscr{S}_k$, then

$$f^* \perp f^\perp \tag{13.3a}$$

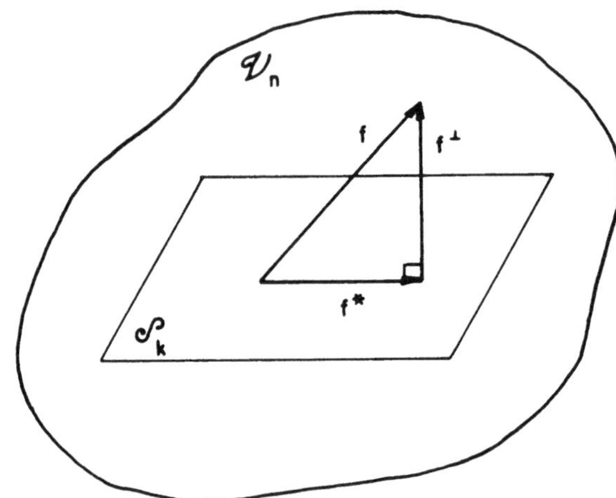

Figure 13.1. Illustration of an orthogonal projection ($n = 3$, $k = 2$).

or

$$(f^*, f^{\perp}) = 0. \tag{13.3b}$$

The so-constructed function f^* is known as the *orthogonal projection* of the function f on the subspace $\mathcal{S}_k$. The reason for its significance will be given a little later on. At this point, we can state that any function in $\mathcal{V}_n$ can be represented as the sum of its orthogonal projection on a subspace $\mathcal{S}_k$ in $\mathcal{V}_n$ and a function orthogonal to $\mathcal{S}_k$. As seen from the symmetry of equation (13.1), there is no essential difference between the parts played by the vectors f^* and $f^{\perp}$ in the decomposition. Accordingly, $f^{\perp}$ can be thought of as the orthogonal projection of f on the subspace $\mathcal{S}_{n-k}$ which is orthogonal to $\mathcal{S}_k$. Denoting the projection of f on $\mathcal{S}_l$ by $\mathrm{Pr}\,\mathcal{S}_l f$, we can write equation (13.1) in the symmetric form

$$f = \mathrm{Pr}\,\mathcal{S}_k f + \mathrm{Pr}\,\mathcal{S}_k^{\perp} f, \tag{13.3c}$$

where $\mathcal{S}_k^{\perp}$ denotes the complementary subspace perpendicular to $\mathcal{S}_k$.

The *method of orthogonal projections* is closely related to the preceding resolution. This is a result of the fact that f^* is a very special function in $\mathcal{S}_k$, namely, the function closest to f. The term "close" is here understood in the sense of the least distance [see Chapter 10], which implies that f^* also gives the best possible approximation to f of all the vectors in $\mathcal{S}_k$.

In order to verify this statement, let us refer to Figure 13.2, in which the subspace $\mathcal{S}_k$ is symbolized by a plane in the three-space $\mathcal{V}_n$, f^* is the orthogonal projection of the given vector f on $\mathcal{S}_k$, and f' is any function in

$\mathscr{S}_k$ different from f^*. Since $f - f^*$ is orthogonal to S_k, then $f - f^*$ is also orthogonal to $f' - f^*$:

$$(f - f^*, f' - f^*) = 0. \tag{13.4}$$

In view of the orthogonality condition (13.4), taking the inner product of the identity

$$f - f' = (f - f^*) + (f^* - f')$$

with itself gives

$$(f - f', f - f') = (f - f^*, f - f^*) + (f^* - f', f^* - f') \tag{13.5}$$

or

$$\|f - f'\|^2 = \|f - f^*\|^2 + \|f^* - f'\|^2. \tag{13.5a}$$

This represents a *Pythagorean theorem* for the symbolic triangle of functions ABC. But $\|f^* - f'\|^2 > 0$, so that

$$\|f - f'\| > \|f - f^*\| \qquad \text{for } f' \neq f^*, \tag{13.6}$$

as asserted.

In geometric language, the foregoing inequality expresses the elementary fact that the hypotenuse $\|f - f'\|$ of the right triangle ABC is greater than a leg of this triangle. Analytically, this means that, of all functions in a

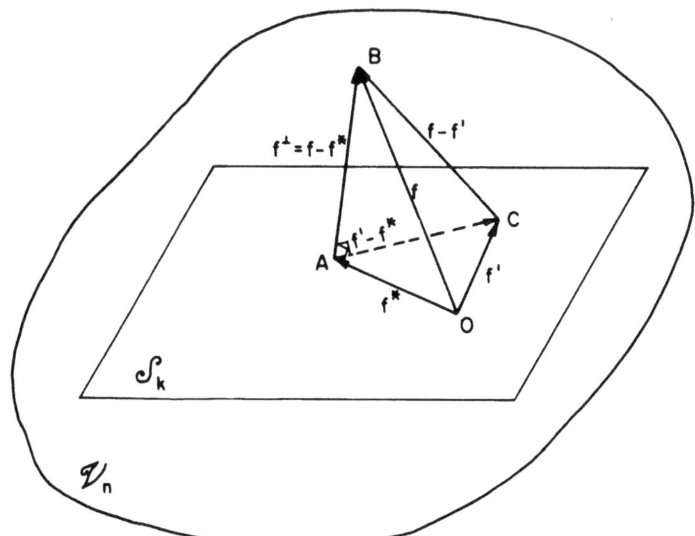

Figure 13.2. Illustration for equation (13.6) ($n = 3$, $k = 2$).

given subspace, the *closest* to a given function is the orthogonal projection of the given function on the subspace.

Since $f - f^*$ is orthogonal to every function h in $\mathscr{S}_k$ (Figure 13.2),

$$(f - f^*, h) = 0 \quad \text{or} \quad (f^*, h) = (f, h) \tag{13.7}$$

for all h in $\mathscr{S}_k$. If $g_1, \ldots, g_k$ is a basis in $\mathscr{S}_k$, then, by definition of a basis,

$$h = \alpha_i g_i, \qquad i = 1, 2, \ldots, k \tag{13.8}$$

(sum over the index i), whence either of equations (13.7) can be replaced by the equivalent system

$$(f^*, g_i) = (f, g_i), \qquad i = 1, 2, \ldots, k. \tag{13.9}$$

It is important to emphasize at this point that the subspace $\mathscr{S}_k$, populated by the vectors f^* [see equation (13.11)], and the subspace $\mathscr{S}_{n-k}$, populated by the vectors $f^{\perp} = f - f^*$, can be used together to construct the entire space $\mathscr{V}_n$. In fact, every vector f in $\mathscr{V}_n$ is represented in terms of a basis that is composed of the bases for $\mathscr{S}_k$ and $\mathscr{S}_{n-k}$ taken together. Symbolically,

$$\mathscr{S}_k \oplus \mathscr{S}_{n-k} = \mathscr{V}_n, \tag{13.10}$$

where the sign $\oplus$ denotes the so-called direct sum, alluded to in Chapter 9 [equation (9.29)].† Evidently, the subspaces are mutually orthogonal,

$$\mathscr{S}_k \perp \mathscr{S}_{n-k}, \tag{13.11}$$

inasmuch as every vector in $\mathscr{S}_k$ is orthogonal to every vector in $\mathscr{S}_{n-k}$ and vice versa.

It is now both informative and instructive to illustrate our somewhat abstract discussion by two concrete problems.

Example 13.1. Arithmetic Progression (Numerical Example). It is required to find the arithmetic progression g closest to the three-element sequence $f = (3, 4, 6)$, where the closeness is understood in the sense of our earlier convention, by which the distance between two functions is measured by the norm of the difference of the functions [see equation (10.6)]. The norm itself is defined to be the square root of the sum of the squares of the components of the function [see equation (5.10)].

---

† This actually follows from the general *projection theorem*, according to which, if $\mathscr{S}$ is a subspace in a Hilbert space $\mathscr{H}$, and $\mathscr{S}^{\perp}$ is its orthogonal complement (cf. Chapter 9 in this book), and $\mathscr{H} = \mathscr{S} \oplus \mathscr{S}^{\perp}$. See, e.g., Lipschutz (Ref. 82, p. 281). In practical applications (leading mostly to approximate solutions) one is often unable to satisfy the requirement of the projection theorem that the linear manifolds involved be closed (cf. p. 53 supra).

Let the vector g represent a function $g(n)$ that takes three discrete values,[†] $g(1) = \alpha$, $g(2) = \alpha + \beta$, $g(3) = \alpha + 2\beta$, so that

$$g = (\alpha, \alpha + \beta, \alpha + 2\beta); \tag{13.12}$$

α and β are certain scalars, to be determined later.

Now let the vector f be represented by a function $f(n)$ taking the following discrete values:

$$f(1) = 3, \qquad f(2) = 4, \qquad f(3) = 6,$$

so that

$$f = (3, 4, 6). \tag{13.13}$$

Represent g in the form

$$g = (\alpha, \alpha, \alpha) + (0, \beta, 2\beta)$$
$$= \alpha g_1 + \beta g_2, \tag{13.14}$$

where $g_1 = (1, 1, 1)$ and $g_2 = (0, 1, 2)$ are two vectors spanning some two-dimensional subspace $\mathscr{S}_2$. We call $\mathscr{S}_2$ the subspace of *arithmetic progressions* and treat it as a subspace of the space $\mathscr{V}$ of *sequences* of three elements. It is thus required to find the progression $g(n)$ in $\mathscr{S}_2$ *closest* to the sequence vector $f = (3, 4, 6)$.

By the property of the orthogonal projection, $g(n)$ is the orthogonal projection of $f(n)$ on $\mathscr{S}_2$ and, consequently, obeys the relations (13.9):

$$(g, g_1) = (f, g_1),$$
$$(g, g_2) = (f, g_2).$$

Upon substituting from equation (13.14), we find

$$\alpha(g_1, g_1) + \beta(g_1, g_2) = (f, g_1),$$
$$\alpha(g_1, g_2) + \beta(g_2, g_2) = (f, g_2).$$

Recalling the definition of the inner product of sequences, and bearing in mind that

$$g_1 = (1, 1, 1), \qquad g_2 = (0, 1, 2), \qquad f = (3, 4, 6),$$

we arrive at the system

$$3\alpha + 3\beta = 13,$$
$$3\alpha + 5\beta = 16, \tag{13.15}$$

[†] We follow here Hausner (Ref. 13, p. 134).

from which, trivially, $\alpha = \frac{17}{6}$ and $\beta = \frac{3}{2}$. Thus, the arithmetic progression *closest* to the sequence (3, 4, 6) is

$$\left(\tfrac{17}{6}, \tfrac{26}{6}, \tfrac{35}{6}\right).$$

We note that the problem at hand may also be approached quite differently by minimizing the expression

$$\| f - g \| = \{(3 - \alpha)^2 + [4 - (\alpha + \beta)]^2 + [6 - (\alpha + 2\beta)]^2\}^{1/2} \quad (13.16)$$

with respect to the parameters α and β [cf. equations (5.10) and (6.24)]. This operation brings us back to equations (13.15) and completes the solution. By proceeding along this line, one looks, in fact, for a function, the deviation of which from the given function is measured according to Gauss' criterion of the mean-square error. This observation discloses an algebraic facet of the problem, invisible in our earlier geometric approach.

Our second illustration of the orthogonal projection method involves a problem in heat conduction.

Example 13.2. A Heated Bar (Numerical Example). A long bar of uniform cross section is heated by an electric current of intensity J. The amount of heat generated in the bar, per unit of volume and unit of time, is

$$h = J^2/\sigma, \quad (13.17)$$

where σ is the electrical conductivity [Ref. 83, equation (23.148)]. Considering the heat supplied by the internal sources, we cast the equation of heat conduction into the form [Ref. 83, equation (6.31a)]

$$\nabla^2 \bar{\theta} = -1 \quad \text{in } \Omega, \quad (13.18)$$

where $\bar{\theta} = k\theta/h$, θ is the absolute temperature, k is the thermal conductivity, Ω is the region occupied by the cross section of the bar, and $\nabla^2 = \partial^2/\partial x^2 + \partial^2/\partial y^2$.

The governing equation (13.18), of Poisson's type, is complemented by the boundary condition

$$\bar{\theta} = 0 \quad \text{on } \partial\Omega \quad (13.19)$$

if we assume (as we do) that the surface of the bar is kept at the temperature zero ($\partial\Omega$ being the contour of Ω).

Following Mikhlin (Ref. 8, pp. 340 and 408), we introduce the vector function†

$$\mathbf{v} = -\operatorname{grad} \bar{\theta}, \quad (13.20)$$

---

† In this problem, we find it convenient to denote vectors by boldface type.

so that equation (13.18) becomes

$$\nabla \cdot \mathbf{v} = 1 \quad \text{in} \quad \Omega, \tag{13.21}$$

where ∇ is the gradient operator.

It is now required to find a vector $\mathbf{v} \equiv (x_x, v_y)$ such that:

(a) $\mathbf{v}$ satisfies equation (13.21) in Ω,
(b) $-\mathbf{v}$ is the gradient of a function $\bar{\theta}$ such that
(c) $\bar{\theta}$ vanishes on $\partial\Omega$.

Symbolically,

$$\frac{\partial v_x}{\partial x} + \frac{\partial v_y}{\partial y} = 1 \quad \text{in } \Omega \tag{13.22a}$$

$$v_x = -\frac{\partial \bar{\theta}}{\partial x}, \qquad v_y = -\frac{\partial \bar{\theta}}{\partial y} \quad \text{in } \Omega \tag{13.22b}$$

$$\bar{\theta} = 0 \quad \text{on } \partial\Omega. \tag{13.22c}$$

We seek the solution vector in a linear space $\mathscr{V}$, in which a Hilbert-type product (8.2) and a norm (7.2) are defined. Specifically, we define

$$\mathscr{V} = \mathscr{S}' \oplus \mathscr{S}'', \tag{13.23}$$

where $\mathscr{S}'$ is the subspace of *irrotational* vectors $\mathbf{v}_1$, such that for some $\overline{\theta^*}$,

$$\mathbf{v}_1 = -\nabla\bar{\theta}^* \quad \text{in } \Omega, \tag{13.24a}$$

where

$$\overline{\theta^*} = 0 \quad \text{on } \partial\Omega, \tag{13.24b}$$

while $\mathscr{S}''$ is the subspace of *solenoidal* vectors $\mathbf{v}_2$, such that

$$\nabla \cdot \mathbf{v}_2 = 0 \quad \text{in } \mathscr{S}''. \tag{13.25}$$

These subspaces satisfy

$$\mathscr{S}' \perp \mathscr{S}''. \tag{13.26}$$

We first prove the assertion (13.26). Using a known relation from vector calculus, we have, for $\mathbf{v}_1$ in $\mathscr{S}'$ and $\mathbf{v}_2$ on $\mathscr{S}''$,

$$\mathbf{v}_1 \cdot \mathbf{v}_2 = -\nabla\overline{\theta^*} \cdot \mathbf{v}_2$$

$$= -\nabla \cdot (\overline{\theta^*}\mathbf{v}_2) + \overline{\theta^*}\nabla \cdot \mathbf{v}_2$$

or, by appeal to the divergence theorem,

$$(\mathbf{v}_1, \mathbf{v}_2) = \int_\Omega \mathbf{v}_1 \cdot \mathbf{v}_2 \, dx \, dy$$

$$= -\int_{\partial\Omega} \mathbf{v}_2 \cdot \mathbf{n}\overline{\theta^*} \, ds + \int_\Omega \overline{\theta^*}\nabla \cdot \mathbf{v}_2 \, dx \, dy. \tag{13.27}$$

The first of the preceding integrals vanishes by virtue of the relation (13.24b); the second integral vanishes because of the condition (13.25). This proves the assertion (13.26).

We are now permitted to represent the solution vector $\mathbf{v}$ as a sum[†]

$$\mathbf{v} = \mathbf{v}_1 + \mathbf{v}_2. \tag{13.28}$$

In order to find the solution explicitly, it is convenient to seek the solution vector in the form

$$\mathbf{v} = \mathbf{V} + \mathbf{W}, \tag{13.29}$$

where $\mathbf{V}$ satisfies the equation

$$\nabla \cdot \mathbf{V} = 1 \quad \text{in } \Omega. \tag{13.30}$$

For this purpose, we select the vector $\mathbf{V}$ in the simple form,

$$V_x = x, \qquad V_y = 0. \tag{13.31}$$

Since the solution vector is to satisfy equation (13.22a), $\mathbf{W}$ must be a solution of the homogeneous equation

$$\frac{\partial W_x}{\partial x} + \frac{\partial W_y}{\partial y} = 0 \quad \text{in } \Omega. \tag{13.32}$$

In view of equation (13.25), the meaning of the equation above is that $\mathbf{W}$ lies in $\mathscr{S}''$: $\mathbf{W}\varepsilon\{\mathbf{v}_2\}$ (compare Figure 13.3, crudely depicting the relation $\mathscr{S}' \perp \mathscr{S}''$). Since the conditions (13.22b) and (13.22c), taken together, coincide with conditions (13.24) defining $\mathscr{S}'$, we conclude that the vector $\mathbf{v}$ lies in $\mathscr{S}'$, i.e., $\mathbf{v}\varepsilon\{\mathbf{v}_1\}$. Moreover, in view of equation (13.29), there is

$$\mathbf{V} = \mathbf{v} - \mathbf{W}, \tag{13.33}$$

so that $\mathbf{v}$ is the orthogonal projection of $\mathbf{V}$ in $\mathscr{S}'$ and $-\mathbf{W}$ is the orthogonal projection of $\mathbf{V}$ in $\mathscr{S}''$. Since $\mathbf{V}$ is a known (or, rather, an easily selected) vector, it is sufficient to determine $\mathbf{W}$ in order to arrive at the solution vector $\mathbf{v}$. Actually, it is easier to determine the solenoidal vector $\mathbf{W}$ than to deal with the projection of $\mathbf{V}$ in $\mathscr{S}'$.

[†] This should come as no surprise if one recalls the Helmholtz theorem of vector calculus. See, e.g., Wills (Ref. 84, p. 121).

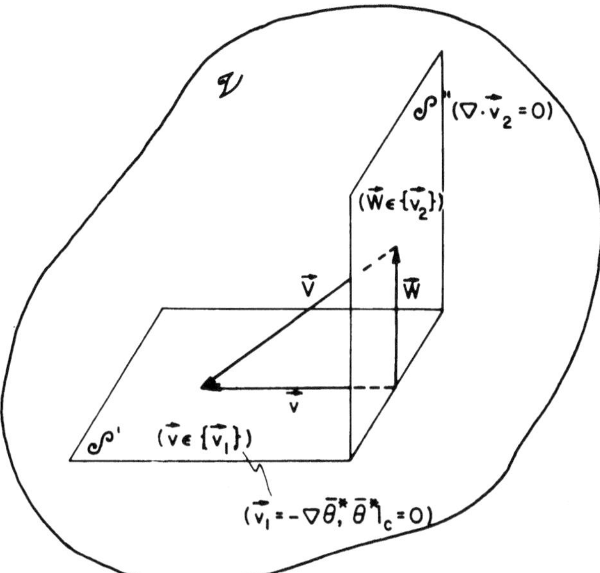

Figure 13.3. Illustration for equations (13.29), $\vec{v} = \vec{V} + \vec{W}$.

We now show the application of the foregoing analysis in the cases involving three types of cross sections: an elliptical, equilateral triangular, and in the form of a circular sector.[85]

A. An Elliptical Cross Section (Figure 13.4)

In view of the symmetry of the cross section and certain physical considerations, the function θ turns out to be an even function of x and y. Hence, $\partial\theta/\partial x$ is odd in x and even in y, while $\partial\theta/\partial y$ is odd in y and even in x.

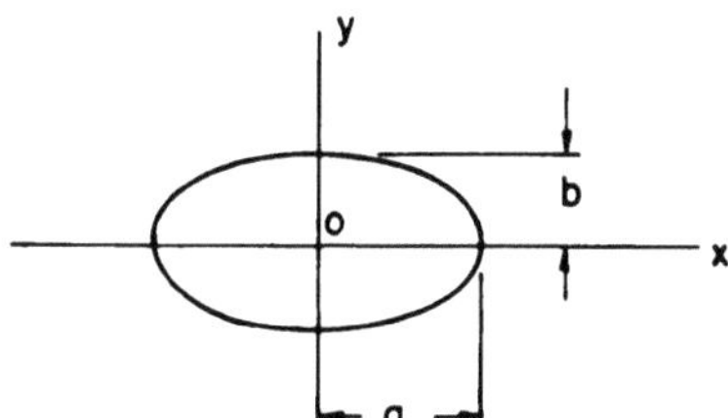

Figure 13.4. An elliptical cross section.

We select the vector $\mathbf{V}$ in the form (13.31), obeying the conditions of symmetry. In finding an approximation to $\mathbf{v}_2 \equiv \mathbf{W}$, it is helpful to recall that, by a generalization of the Weierstrass approximation theorem,[72, 86] any real function of n real variables, continuous on a closed bounded subset of $\mathcal{R}_n$, can be approximated uniformly by polynomials in the n variables. We may thus write

$$-\mathbf{v}_2 \approx \sum_{i=1}^{n} \alpha_i \, \mathbf{v}_{2i}, \tag{13.34}$$

where the α_i's are certain coefficients to be determined later and the $\mathbf{v}_{2i}$'s are monomials in x and y (linearly independent) satisfying equation (13.25). Since our objective is here rather modest, we content ourselves with the value $n = 3$.

Suppose, then, that we take three terms of the representation (13.34), so that the vector giving the approximate solution is

$$\mathbf{v} \equiv (x; 0) - \alpha_1(x; -y) - \alpha_2(x^3; -3x^2y) - \alpha_3(3xy^2; -y^3). \tag{13.35}$$

It is easily verified that the expression above satisfies all the imposed conditions. Now, since $\mathbf{v}_2$ is the orthogonal projection of $\mathbf{V}$ on $\mathscr{S}''$, it is also that vector closest to the vector $\mathbf{v}$ (of all vectors in $\mathscr{S}''$). Symbolically,

$$I(\alpha_i) \equiv \| \mathbf{V} - \mathbf{v}_2 \|^2$$

$$= \int_{\Omega} (\mathbf{V} - \mathbf{v}_2, \mathbf{V} - \mathbf{v}_2)\, dx\, dy = \min. \tag{13.36}$$

From the conditions $\partial I(\alpha_i)/\partial \alpha_i = 0$, $i = 1, 2, 3$, we conclude

$$\alpha_1 = \frac{a^2}{a^2 + b^2}, \qquad \alpha_2 = \alpha_3 = 0. \tag{13.37}$$

It follows that the solution vector is

$$\mathbf{v} \equiv (x; 0) - \frac{a^2}{a^2 + b^2}(x; -y). \tag{13.38}$$

We now refer to equation (13.20) and find the temperature distribution as

$$-\theta = \frac{b^2 x^2 + a^2 y^2}{2(a^2 + b^2)} + C, \tag{13.39}$$

where the integration constant C is evaluated from the boundary condition (13.19). Consequently,

$$\theta = \frac{a^2 b^2}{2(a^2 + b^2)} \left(\frac{x^2}{a^2} + \frac{y^2}{b^2} - 1 \right). \tag{13.40}$$

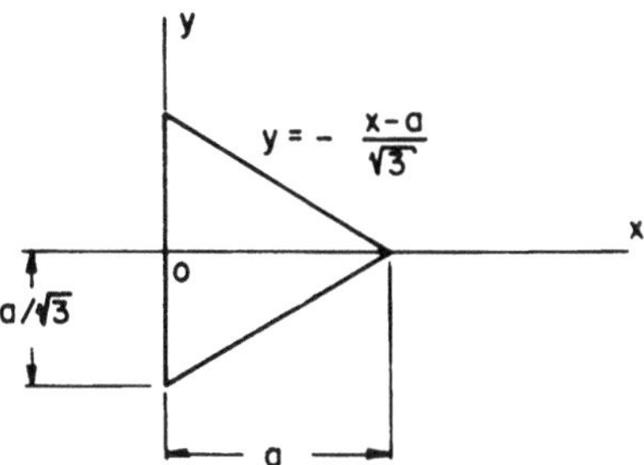

Figure 13.5. An equilateral triangular cross section.

This completes the solution, and it is not difficult to verify that, while look-ing for an approximate solution, we have unexpectedly arrived at the exact solution of the problem at hand (Ref. 74, p. 263).

B. An Equilateral Triangular Cross Section (Figure 13.5)

This cross section has an axis of symmetry coinciding with the x-axis. This implies that $\partial\theta/\partial y$ is odd in y. We select the vector $\mathbf{V}$ in the form (13.31), and approximate $\mathbf{v}_2$ by means of three vectors $\mathbf{v}_{2i}$, $i = 1, 2, 3$, giving

$$\mathbf{v} \equiv (x; 0) - \alpha_1(3x^2; -6xy) - \alpha_2(a^2; 0) - \alpha_3(-3y^2; 0). \tag{13.41}$$

Before proceeding further, we should convince ourselves that the expression above is consistent with the requirement (13.20). Observing that this obtains provided we set $\alpha_3 = \alpha_1$, we proceed as in Case A and find, after some elementary calculations, that $\alpha_1 = \alpha_2 = 1/4a$. Hence, finally, the solution vector is

$$\mathbf{v} \equiv (x; 0) - \frac{1}{4a}(3x^2 - 3y^2; -6xy) - \frac{1}{4a}(a^2; 0). \tag{13.41a}$$

Upon using equation (13.20) once more and setting $\theta|_{x=0,\, y=0} = 0$, as required by the boundary condition, we find the temperature distribution

$$\theta = \frac{1}{4a}(x^3 - 3xy^2 - 2ax^2 + a^2x). \tag{13.42}$$

It is easily verified that we were again fortunate enough to arrive at the exact solution (Ref. 74, p. 266).

C. A Circular Sector (Figure 13.6)

As our last example, let us consider a circular sector of radius a and central angle 2α; it is assumed that $|\alpha| < \pi/4$. Evidently, the temperature

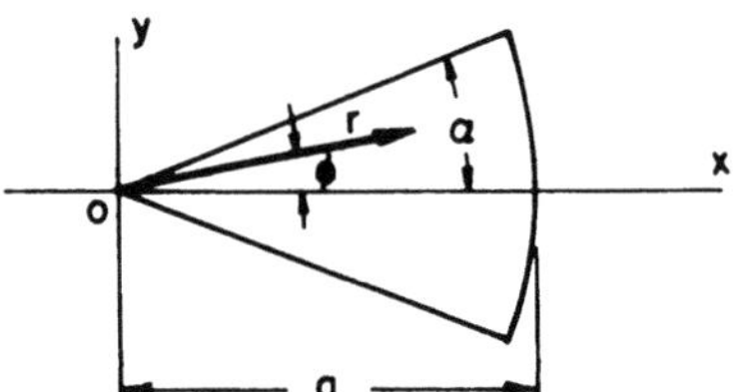

Figure 13.6. A cross section in the form of a circular sector.

field is symmetric with respect to the x-axis. We note that, in polar coordinates r, ϕ,

$$\operatorname{grad} u = \frac{\partial u}{\partial r}\, \mathbf{e}_r + \frac{1}{r}\frac{\partial u}{\partial \phi}\, \mathbf{e}_\phi,$$

$$\operatorname{div} \mathbf{w} = \frac{\partial w_r}{\partial r} + \frac{w_r}{r} + \frac{1}{r}\frac{\partial w_\phi}{\partial \phi},$$

(13.43)

where $\mathbf{e}_r$ and $\mathbf{e}_\phi$ are unit vectors along the radius and in the azimuthal direction, respectively (the remaining notation being self-explanatory). With equations (13.25) and (13.30) and the symmetry of the cross section in mind, we set

$$\mathbf{V} \equiv \left(\frac{r}{2};\, 0\right), \qquad \mathbf{v}_{20} \equiv \left(\frac{r}{2}\cos 2\phi;\, -\frac{r}{2}\sin 2\phi\right),$$

$$\mathbf{v}_{2i} = \left(r^{n\pi/2\alpha - 1}\cos\frac{n\pi}{2\alpha}\phi;\, -r^{n\pi/2\alpha - 1}\sin\frac{n\pi}{2\alpha}\phi\right),$$

(13.44)

$$i = \frac{n+1}{2}, \qquad n = 1,\, 3,\, 5,\, \ldots .$$

A lengthy, but simple, calculation, retracing the earlier argument, leads to the following system of equations:

$$\frac{\partial I}{\partial \alpha_0} = 0,$$

$$\frac{a^4}{16}\sin 2\alpha = \alpha_0 \frac{a^4}{8}\alpha + \sum_{n=1,\,3,\,5,\,\ldots} \alpha_{(n+1)/2}\frac{a^{n\pi/2\alpha + 2}}{(n\pi/2\alpha + 2)(n\pi/2\alpha - 2)}$$

$$\times \sin\left(\frac{n\pi}{2\alpha} - 2\right)\alpha;$$

$$\frac{\partial I}{\partial \alpha_i} = 0, \qquad i = 1, 2, 3, \ldots,$$

$$(-1)^{(m+3)/2} \frac{a^{m\pi/2\alpha + 2}}{(m\pi/\alpha)(m\pi/2\alpha + 2)} = \alpha_0 \frac{a^{m\pi/2\alpha + 2}}{2(m\pi/2\alpha + 2)(m\pi/2\alpha - 2)}$$

$$\times \sin\left(\frac{m\pi}{2\alpha} - 2\right)\alpha + \alpha_{(m+1)/2} \frac{a^{m\pi/\alpha}}{m\pi} \alpha^2,$$

$$m = 1, 3, 5, \ldots . \qquad (13.45)$$

For definiteness, we choose $a = 1$ and $\alpha = \pi/6$, and (with the help of a computer) arrive at the following values for the coefficients:

$$\begin{aligned}
\alpha_0 &\approx +1.998962, \\
\alpha_1 &\approx -0.763349, \\
\alpha_2 &\approx +0.049491, \\
\alpha_3 &\approx -0.017217, \\
\alpha_4 &\approx +0.008693.
\end{aligned} \qquad (13.46)$$

Assuming $\bar{\theta}|_{r=0}$, we thus obtain

$$\bar{\theta} = -\frac{r^2}{4} + \alpha_0 \frac{r^2}{4} \cos 2\phi + \sum_{n=1, 3, 5, \ldots} \alpha_{(n+1)/2} \frac{2\alpha}{n\pi} r^{n\pi/2\alpha} \cos \frac{n\pi}{2\alpha} \phi, \quad (13.47)$$

where the first few coefficients are given by equations (13.46). To verify the satisfaction of the boundary conditions, we check the accuracy of the preceding equation at the points $r = 1$, $\phi = 0$ and $r = 1$, $\phi = \pm\alpha$. It turns out that the values of $\bar{\theta}$ at these points are $+0.000056$ and -0.006032, respectively, instead of zero. This seems to be acceptable accuracy, in view of the fact that the series (13.47) converges rapidly and five terms of the latter were used in the calculations.

13.1. Theory of Approximations

It is quite often demanded that a continuous or discrete analytic formula be selected which approximately represents a function determined by a table, a graph, or an analytic expression too complicated or otherwise unsuitable for the purpose at hand. An example of such an *approximation* procedure is found in Example 13.1, where the method of orthogonal projections in Hilbert space is used. In order to more fully clarify the part played by function space methods in the *theory of approximations*, it is instructive to take recourse to a class of spaces more primitive (general) than the Hilbert spaces, for instance, to the *normed linear spaces*. We recall[†] that a vector space is a normed linear space if there is a real number $\|x\|$, called the norm of x, associated with each vector x in the space; the norm satisfies the requirements imposed by Axiom N [equations (7.13)].

In the following we denote the normed linear space by $\mathscr{N}$ and choose it to be an ensemble of continuous real-valued functions $f(P)$ of n real variables, the common domain of definition of which is $\mathscr{S}$, a closed subset of the n-dimensional point space $\mathscr{R}_n$.[‡]

We shall use a particular type of norm, representing a generalization of the type (5.33),[§] to the present setting of a continuous domain $\mathscr{S}$,

$$\|f\| = \max_{P \in \mathscr{S}} |f(P)|, \tag{13.47a}$$

and known as the *uniform*, or *Chebyshev*, norm.

In the manner of the definition (6.24), we define a *distance* between two vectors $f(P)$ and $g(P)$ by

$$d(f, g) = \|f - g\|$$
$$= \max_{P \in \mathscr{S}} |f(P) - g(P)|. \tag{13.48}$$

Our problem now is: given a function $F(P)$ in $\mathscr{N}$ and a particular linear subspace $\mathscr{S}_m$ spanned by m linearly independent functions f_i in $\mathscr{N}$, it is required to find the coefficient c_i in the sum

$$\bar{f}(P) = \sum_{i=1}^{m} c_i\, f_i(P) \tag{13.48a}$$

such that $\bar{f}(P)$ is the "best" approximation of $F(P)$. In the language of function space (Figure 13.7), the problem requires: (a) an inspection of the *minimal* "error"

$$\bar{d} = \inf_{f \in \mathscr{S}_m} \|F - f\|, \tag{13.49}$$

where $d = \|F - f\|$ is the error (in the selected norm) in approximating F by $f \in \mathscr{S}_m$; (b) to find the set $\bar{\mathscr{S}} \subset \mathscr{S}_m$ of functions $\bar{f}$ for which

$$\|F - \bar{f}\| = \bar{d}. \tag{13.50}$$

Clearly, the number $\bar{d}$ is the shortest distance (in the selected norm) of the vector F from the subspace $\mathscr{S}_m$, and the set $\bar{\mathscr{S}}$ is the set of vectors $\bar{f}$ closest to the vector F (or the set of points "nearest" to the "tip" T of the vector F).

[†] Compare Chapter 1 and Figure 1.1.
[‡] That is, a set of n-tuples of real numbers.
[§] $\|f(P)\| = \max |f(P_1), f(P_2), \ldots, f(P_n)|$.

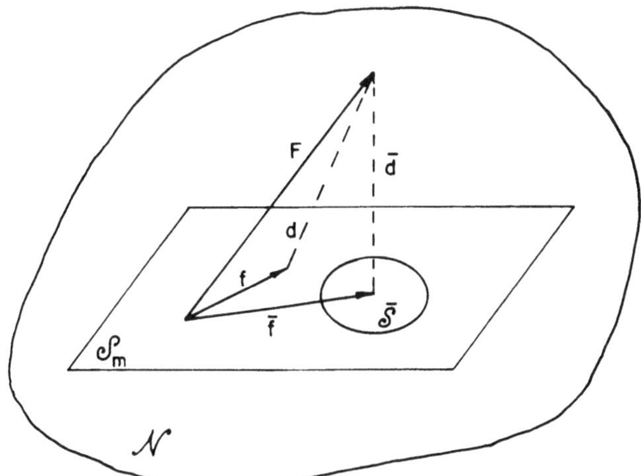

Figure 13.7. Illustration for the formula (13.50).

Inasmuch as the geometry of a normed space does not include the idea of perpendicularity (no concept of the inner product having been introduced), there need not exist a unique point in $\mathscr{S}_m$ nearest to T (or such a point at all[†]). In fact, as will be shown below, there may be infinitely many such points.

Example 13.3. Linear Approximation. (Numerical Example). Buck[88] gives the following elegant example of a *linear approximation* of the vector $F(x, y) = xy$ by a vector

$$f(x, y) = c_1 + c_2(x + y) + c_3(x^2 + y^2) \tag{13.51}$$

from the subspace $\mathscr{S}_3$ with the base 1, $x + y$, and $x^2 + y^2$. As the space V, we select the space of continuous functions of two variables x and y defined in the square $0 \le x \le 1, 0 \le y \le 1$. As the norm, we select the uniform norm (13.47a).

We first require that the error

$$\|F - f\| \le \tfrac{1}{4}. \tag{13.52}$$

This implies that, for $0 \le x \le 1$ and $0 \le y \le 1$,

$$xy - \tfrac{1}{4} \le c_1 + c_2(x + y) + c_3(x^2 + y^2) \le xy + \tfrac{1}{4}. \tag{13.53}$$

---

† It is shown, however, that among the approximating functions of the form $P(x)/Q(x)$, where $P(x)$ and $Q(x)$ are polynomials, there exists at least one function for which $\max_{a \le x \le b} \|P(x)/Q(x) - F(x)\|$ has a minimum, cf., e.g., Achieser (Ref. 87, p. 53).

We take, in turn, $(x, y) = (0, 0)$, $(0, 1)$, $(1, 0)$, and $(1, 1)$, obtaining

$$-\tfrac{1}{4} \le c_1 \le \tfrac{1}{4}, \tag{13.54a}$$

$$-\tfrac{1}{4} \le c_1 + c_2 + c_3 \le \tfrac{1}{4}, \tag{13.54b}$$

$$\tfrac{3}{4} \le c_1 + 2c_2 + 2c_3 \le \tfrac{5}{4}. \tag{13.54c}$$

Adding the first and last of the equations above and comparing with the second gives the necessary conditions

$$c_1 = -\tfrac{1}{4}, \qquad c_2 + c_3 = \tfrac{1}{2}. \tag{13.55}$$

From these and the inequality (13.53), there follows

$$0 \le c_2 \le 1. \tag{13.56}$$

The foregoing results are plotted in Figure 13.8 with reference to a coordinate system c_1, c_2, c_3.

It is seen that the subspace $\mathscr{S}$ of the best approximating functions [in the given norm and obeying the condition (13.52)] is not represented by a single point, but by an entire "line segment" whose terminal points are the functions

$$\bar{f}_1 = -\tfrac{1}{4} + \tfrac{1}{2}(x^2 + y^2),$$

$$\bar{f}_2 = -\tfrac{1}{4} + (x + y) - \tfrac{1}{2}(x^2 + y^2). \tag{13.57}$$

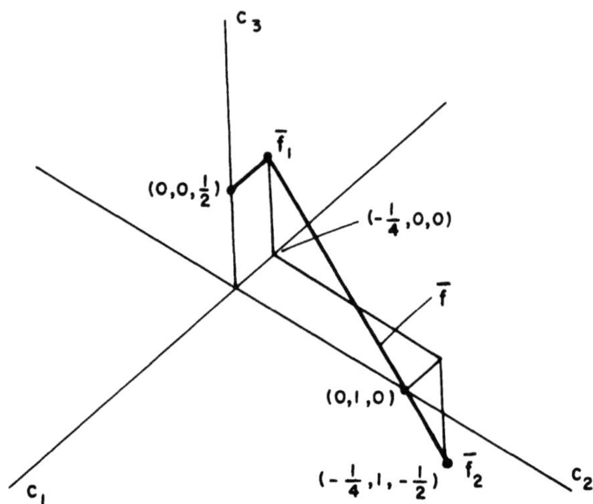

Figure 13.8. Illustration for the formulas (13.55) and (13.56).

For all the functions of this segment,

$$\bar{f} = \alpha \bar{f}_1 + (1 - \alpha)\bar{f}_2, \qquad 0 \le \alpha \le 1, \tag{13.58}$$

the error d in equation (13.50) is equal to $\tfrac{1}{4}$. A different statement of this fact is that the surface of the ball (solid hypersphere) $d \le \tfrac{1}{4}$ centered at T "touches" the "plane" $\mathscr{S}_3$ along a "line segment" $\bar{f}_1 \le \bar{f} \le \bar{f}_2$.

Problems

1. Construct a subspace of the Hilbert space $\mathscr{L}_2(-\pi, \pi)$, of square-integrable functions on $[-\pi, \pi]$, which is orthogonal to the subspace of functions spanned by $\sin kt$, where $k = 1, 2, \ldots$.

2. Determine the orthogonal complements: (1) in $\mathscr{E}_3$, of a subspace $\mathscr{E}_1$ represented by a straight line; (2) in $\mathscr{E}_3$, of a subspace $\mathscr{E}_2$ represented by a plane; (3) in $\mathscr{E}_2$, of a subspace $\mathscr{E}_1$ represented by a straight line.

3. Let $\mathscr{S}$ be a subspace of a Hilbert space $\mathscr{H}$, so that each f in $\mathscr{H}$ possesses a unique decomposition as $f = Pr_{\mathscr{S}} f + Pr_{\mathscr{S}^\perp} f$, where $Pr_{\mathscr{S}} f$ is in $\mathscr{S}$, $Pr_{\mathscr{S}^\perp} f$ is in $\mathscr{S}^\perp$. Verify that $Pr_{\mathscr{S}}(f_1 + f_2) = Pr_{\mathscr{S}} f_1 + Pr_{\mathscr{S}} f_2$.

4. With notation as in Problem 3, show that $Pr_{\mathscr{S}}(cf) = cPr_{\mathscr{S}}(f)$, where c is a scalar.

5. Let $\mathscr{E}_x$, $\mathscr{E}_y$, and $\mathscr{E}_z$ denote three subspaces in $\mathscr{E}_3$ represented by the coordinate axes x, y, and z, respectively, and let $\mathscr{E}_{yz}$ be the subspace represented by the plane yz. (a) Show analytically the correctness of the answers to Problem 2, (1) and (2). (b) What are the orthogonal projections of a vector v in $\mathscr{E}_3$ on $\mathscr{E}_{yz}$ and $\mathscr{E}_x$?

6. In $\mathscr{E}_3$ with an orthogonal basis (g_1, g_2, g_3), the orthogonal projections of a vector u on the subspaces spanned by g_1 and g_2 (straight lines $\mathscr{S}'$ and $\mathscr{S}''$) are u' and u'', respectively. Find the orthogonal projection u''' of u on the subspace spanned by g_1 and g_2 (plane $\mathscr{S}'''$) in terms of u' and u''.

7. If $\mathscr{S}$ is a closed subspace of a Hilbert space $\mathscr{H}$, show that $\mathscr{S} = \mathscr{S}^{\perp\perp}$, where $\mathscr{S}^{\perp\perp}$ is the orthogonal complement of $\mathscr{S}^\perp$.

8. Defining the norm of a continuous function, after Chebyshev, as $\|f\| = \max_{a \le x \le b} |f(x)|$ —see equation (13.47a)—show that the following relation holds: $\|cf\| = |c|\,\|f\|$, where c is any scalar.

9. With notation is in Problem 8, show that $\|f + g\| \le \|f\| + \|g\|$.

14

The Rayleigh–Ritz and Trefftz Methods

14.1. The Rayleigh–Ritz Method

The *Rayleigh–Ritz* method belongs to the so-called *direct methods* of the calculus of variations, inasmuch as it is applied to problems formulated in an integral rather than a conventional, that is, differential, form. More often than not, the procedure involves the minimization of integrals containing unknown functions and their derivatives, without first deriving from these integrals equivalent (the so-called Euler–Lagrange) differential equations.

The method was first proposed by Lord Rayleigh in his celebrated theory of sound, in connection with the vibrations of elastic systems (strings, beams, and plates). Rayleigh, making use of the principle of conservation of energy, was able to show the remarkable accuracy of the approximate values of frequencies and modes of vibrations obtained by means of his method.

It is worth recalling that, in the theory of vibrations, the frequencies and modes represent, respectively, the eigenvalues and eigenfunctions of the boundary value problem describing the particular process of vibration. Specifically, the frequencies, λ_n, are represented through the so-called Rayleigh quotient,

$$\rho(u) = \frac{(Lu, u)}{(u, u)}, \tag{14.1}$$

where L is the differential operator of the problem and u is a function in the domain of L; the inner product is here of the Hilbert type. If $u = u_n$, an eigenfunction of L, then $\rho(u_n) = \lambda_n$. In general u_n is not known, but it is shown that, for $u = u_n$, $\rho(u)$ has a stationary value. Moreover, if we take a function u which is close to u_n, we expect that the value of $\rho(u)$ will be close to λ_n.† It is worth noting that the numerator and denominator in the right-hand member of equation (14.1) are closely related to the potential and kinetic energies of the vibrating system, respectively.

A significant development of Rayleigh's ideas was effected by Ritz, who, in a series of articles published during the years 1908 and 1909, approached the problem from a more general point of view.

Before examining the Rayleigh–Ritz procedure in the context of function space concepts, it is instructive to recall the main points of this procedure.

Suppose, then, that a given integral $I[f]$ is to be *minimized* or, more generally, made stationary over a domain of functions f. To achieve this, we choose a set of functions, say $\{\phi_i\}$, called *coordinate functions*, which, although being to some extent arbitrary, are nonetheless bound to be *admissible*, i.e., to satisfy (ordinarily) the boundary, initial, and some regularity conditions. As a next step, we form a linear combination, f_n, of n of the functions ϕ_i, called the *Ritz expression*,

$$f_n = c_1 \phi_1 + c_2 \phi_2 + \cdots + c_n \phi_n, \tag{14.2}$$

and determine the coefficients c_i from the requirement that the *functional* $I[f_n]$ be stationary over all values of the coefficients. This implies that

$$\frac{\partial I[f_n]}{\partial c_i} = 0, \qquad i = 1, 2, \ldots, n. \tag{14.3}$$

Frequently, the integrand in the functional $I[f]$ turns out to be a quadratic form in f and its derivatives, that is, a homogeneous polynomial of the second degree. Consequently, a solution of the foregoing system of (linear) equations in the c_i's presents no difficulties. A substitution of the so-determined coefficients into equation (14.2) furnishes immediately the required approximate solution.

There are often reasons to believe that the greater the number n of coordinate functions utilized, the smaller the expression for f_n deviates from the exact solution. A proof of the convergence of the Rayleigh–Ritz process, however, is in general difficult (if attainable at all),‡ and one is usually satisfied if one assures that a certain number of the values $I[\bar{f}_k]$ of $I[f]$ provide a nonincreasing sequence,

$$I[\bar{f}_1] \geq I[\bar{f}_2] \geq \cdots . \tag{14.4}$$

---

† See, e.g., Friedman (Ref. 22, p. 207ff), and Timoshenko (Ref. 89, Sec. 61). We recall, however, that since $\rho(u)$ is "flat" at u_n, the value $\rho(u)$ usually provides a better approximation for $\rho(u_n)$ than u does for u_n.

‡ Some questions of convergence of sequences of Rayleigh–Ritz approximations are now successfully treated in the theory of finite elements. See, e.g., Oden (Ref. 90, p. 120).

Here, $\bar{f}_k$, $k = 1, 2, \ldots, n$, denote the functions f_k *minimizing* the functional $I[f]$ over all linear combinations of the ϕ_i, $i = 1, \ldots, k$. The inequality (14.4) follows, naturally, from the fact that each new approximation f_k includes functions of the preceding approximation f_{k-1}; this implies the non-increasing values of successive minima. Since, by hypothesis, $I[f]$ possesses an unqualified minimum, the sequence $I[\bar{f}_k]$ is bounded from below and thus converges. It need not, however, converge to the min $I[f]$. If the coordinate functions are mutually orthogonal, then a general formula for the coefficients c_i can be found directly, without appeal to the whole system (14.3).

It is often convenient to single out a particular coordinate function from the Ritz expression (14.2) by setting

$$f_n = \phi_0 + \sum_{k=1}^{n} c_k \phi_k. \qquad (14.5)$$

In this case, the distinguished function ϕ_0 is often required to satisfy any preassigned nonhomogeneous boundary conditions, while the remaining functions obey boundary conditions of the homogeneous type. Along with that just described, other modifications of Ritz expression are of use. For instance, if a given problem involves more than one independent variable, say three, x, y, and z, then it might be practical to set

$$f_n(x, y, z) = \sum_{k=1}^{n} c_k \phi_k(x) \psi_k(y, z), \qquad (14.6)$$

where the functions ϕ_k are selected in advance and the functions ψ_k are to be determined from other conditions. There are situations in which one should refrain from using linear combinations of the coordinate functions and form the Ritz expression as a transcendental function of the parameters c_k.

While the Ritz expression is usually required to satisfy all, or at least some of, the prescribed auxiliary conditions, it does not, in general, satisfy the governing equation (or equations). In order to minimize this deficiency, the Rayleigh–Ritz method is normally applied in combination with a variational principle, such as that of the minimum potential energy of *Lagrange* or of the minimum complementary energy of *Castigliano*.† The first-named, expressed in terms of displacements, automatically secures the best possible satisfaction of the equations of equilibrium and the stress boundary conditions. The Castigliano principle, on the other hand, is expressed in terms of stresses and automatically guarantees the best possible satisfaction of the compatibility equations and the displacement boundary conditions.

---

† See, e.g., Sokolnikoff (Ref. 31, Secs. 107 and 108). We have here in mind elastic problems.

The inherent connections between the Rayleigh–Ritz method and the geometry of function space are amply illustrated by the following example.

Example 14.1. Bounds for Torsional Rigidity. Suppose that one is required to find the torsional rigidity D of a bar of uniform and simply connected, but otherwise arbitrary, cross section.† It is known that the potential energy of a bar subjected to torsion is

$$U = \frac{\mu\alpha^2}{2} \int_\Omega [(\psi_{,x})^2 + (\psi_{,y})^2] \, dx \, dy + F_1(x, y), \qquad (14.7)$$

where Ω is the region occupied by the cross section, μ is a Lamé constant, α is the angle of twist per unit length, $F_1(x, y)$ is a known function depending on the form of the cross section, and ψ is the so-called *conjugate torsion function*, harmonic in Ω and obeying the boundary condition

$$\psi = \frac{x^2 + y^2}{2} \qquad (14.8)$$

on the contour $\partial\Omega$ of Ω.[31, 74]

The torsional rigidity of the bar is

$$D = \mu\left\{ I_p - \int_\Omega \left[\left(\frac{\partial\psi}{\partial x}\right)^2 + \left(\frac{\partial\psi}{\partial y}\right)^2 \right] dx \, dy \right\}, \qquad (14.9)$$

where I_p is a polar moment of inertia of Ω about the centroid.

By the theorem of minimum potential energy, the functional (14.7) attains its minimum when ψ is the exact solution of the problem. From equation (14.7) it is seen that this happens when the Dirichlet integral, $I[\psi] = \int_\Omega [(\psi_{,x})^2 + (\psi_{,y})^2] \, dx \, dy$ [cf. equation (8.45)] becomes minimum with the stipulation that ψ satisfies condition (14.8).

In the calculus of variations it is demonstrated (Ref. 91, Chap. 4) that the just-described minimization of the Dirichlet integral is equivalent to solving the differential (Dirichlet) boundary-value problem

$$\nabla\psi = 0 \quad \text{in } \Omega, \qquad \psi = \frac{x^2 + y^2}{2} \equiv f \quad \text{on } \partial\Omega. \qquad (14.10)$$

Now let us introduce a Dirichlet-type inner product of functions in the form [cf. equation (8.6)]

$$(\phi_1, \phi_2) = \int_\Omega (\phi_{1,x}\phi_{2,x} + \phi_{1,y}\phi_{2,y}) \, dx \, dy, \qquad (14.11)$$

---

† Weinstein.[50] We review here some of the results obtained in this paper. Compare also the problem in Chapter 8 following equation (8.38).

so that

$$I[\phi] = (\phi, \phi) \quad \text{or} \quad \|\phi\| = [I(\phi)]^{1/2}. \tag{14.11a}$$

Retracing the argument which led to inequality (14.4), we conclude not only that

$$I[\bar{\psi}_1] \geq I[\bar{\psi}_2] \geq \cdots \geq I[\psi], \tag{14.12}$$

but also that every approximate solution $\bar{\psi}$ [satisfying condition (14.8)] furnishes an *upper bound* for the actual minimum of the functional $I[\psi]$. Symbolically,

$$\|\psi\| \leq \|\bar{\psi}\|, \tag{14.12a}$$

$$\bar{\psi} = f \quad \text{on } \partial\Omega. \tag{14.12b}$$

An inspection of equation (14.9) shows that this result furnishes simultaneously a *lower bound* for the torsional rigidity D.

It is easily verified that the inequality (14.12a) is a direct consequence of the Cauchy–Schwarz inequality (3.22). In fact, there is

$$(\psi, \bar{\psi})^2 \leq (\psi, \psi)(\bar{\psi}, \bar{\psi}), \tag{14.13a}$$

and by the first Green identity,

$$(\psi, \bar{\psi} - \psi) = \int_\Omega [\psi_{,x}(\bar{\psi} - \psi)_{,x} + \psi_{,y}(\bar{\psi} - \psi)_{,y}]\, dx\, dy$$

$$= \int_{\partial\Omega} (\bar{\psi} - \psi)\psi_{,n}\, ds - \int_\Omega (\bar{\psi} - \psi)\nabla^2\psi\, dx\, dy$$

$$= 0, \tag{14.13b}$$

in view of the fact that $\psi = \bar{\psi} = f$ on $\partial\Omega$ and $\nabla^2\psi = 0$ in Ω. This implies that

$$(\psi, \bar{\psi} - \psi) = (\psi, \bar{\psi}) - (\psi, \psi) = 0, \tag{14.13c}$$

so that, replacing $(\psi, \bar{\psi})$ by (ψ, ψ) in (14.13a), we recover the inequality (14.12a).

The remarkable geometric sense of the latter inequality is depicted in Figure 14.1, in which relation (14.13b) translates into the perpendicularity of the vectors ψ and $\bar{\psi} - \psi$. Thus, in geometric language, the inequality (14.12a) expresses the theorem that, in a right triangle, the length of a leg is less than that of the hypotenuse.

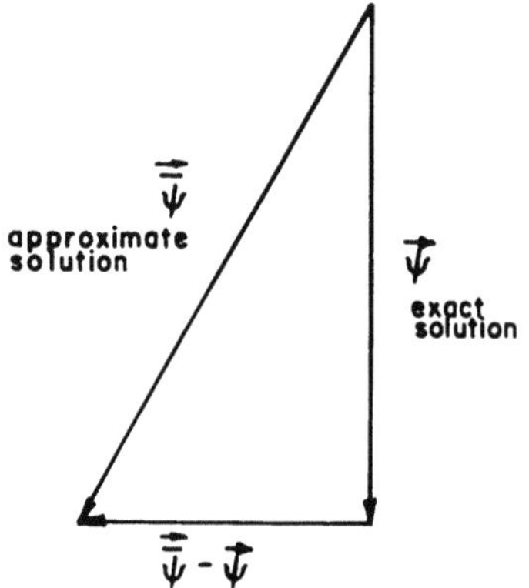

Figure 14.1. Illustration for equation (14.13b).

Example 14.2. Biharmonic Problem. As another illustration, it is informative to examine a geometric interpretation of the Rayleigh–Ritz method when applied to a *biharmonic problem*. Let it then be required to find a function $w = w(x, y)$ satisfying the differential equation

$$\nabla^4 w = q_0/D \quad \text{in } \Omega, \tag{14.14}$$

as well as the boundary conditions

$$w = 0, \quad \frac{\partial w}{\partial n} = 0 \quad \text{on } \partial\Omega, \tag{14.14a}$$

where $\partial\Omega$ is the contour of the region Ω and n is the normal to $\partial\Omega$.

The function sought may be interpreted as the deflection of a thin elastic plate of bending rigidity D, clamped at the contour, and acted upon by a transverse load of uniform intensity q_0. The problem so defined belongs to the class of so-called semi-homogeneous boundary value problems because one defining equation [here (14.14)] is nonhomogeneous, but the remaining two are not. Using the Rayleigh–Ritz procedure, which ensures satisfaction of the boundary conditions, it is convenient to invert the roles by making the governing equation homogeneous[92]; this automatically converts the associated boundary conditions into nonhomogeneous forms. In order to achieve this, we find a particular integral of equation (14.14), denoting the integral and its normal derivative on the contour $\partial\Omega$ by f and g, respectively:

$$w_0 = f, \quad \frac{\partial w_0}{\partial n} = g \quad \text{on} \quad \partial\Omega. \tag{14.15}$$

Let the plate be located in the strip $0 \leq x \leq a$. We select the integral in the form

$$w_0 = \frac{4q_0 a^4}{\pi^5 D} \sum_{k=1, 3, \ldots}^{\infty} \frac{1}{k^5} \sin \frac{k\pi x}{a}, \qquad (14.16)$$

reminiscent of the deflection of an infinite plate strip of width a, uniformly loaded. It can be verified that the foregoing expression satisfies the field equation (14.14) provided the load is represented in the form of a Fourier series,

$$q_0 = \frac{4q_0}{\pi} \sum_{k=1, 3, \ldots}^{\infty} \frac{1}{k} \sin \frac{k\pi x}{a}, \qquad (14.17)$$

odd in x and having period $2a$.

If w_0 obeys the condition (14.15), then the component $\bar{w}$ of the actual solution, taken in the form

$$w = w_0 + \bar{w}, \qquad (14.18)$$

is required to satisfy the homogeneous version of equation (14.14) in Ω, and the conditions

$$\bar{w} = -f, \qquad \frac{\partial \bar{w}}{\partial n} = -g \qquad (14.19)$$

on the contour $\partial\Omega$.

As a consequence, the original problem is reduced to that of finding a function $\bar{w}$ satisfying the homogeneous equation

$$\nabla^4 \bar{w} = 0 \quad \text{in } \Omega \qquad (14.20)$$

and the nonhomogeneous conditions (14.19).

We now employ the Rayleigh–Ritz procedure and select a function $u(x, y)$ in Ritz form satisfying the boundary conditions

$$u = \bar{w}, \qquad \frac{\partial u}{\partial n} = \frac{\partial \bar{w}}{\partial n} \quad \text{on } \partial\Omega. \qquad (14.21)$$

This function is required to minimize the potential energy of the plate,

$$U(u) = \int_{\Omega} W(u)\, d\Omega + \int_{\partial\Omega} M_n(u) \frac{\partial u}{\partial n}\, ds - \int_{\partial\Omega} V_n(u) u\, ds = \min, \quad (14.22)$$

where the first integral is given by equation (8.33) and the remaining two by equation (8.9), in which we set $q = 0$. We assume that u has been determined by the Rayleigh–Ritz procedure, considering it hereafter as known.

By appeal to the Clapeyron theorem (8.9), we have

$$(\bar{w}, u - \bar{w}) = -\int_{\partial\Omega}\left[M_n(\bar{w})\frac{\partial}{\partial n}(u - \bar{w}) - V_n(\bar{w})(u - \bar{w})\right]ds. \quad (14.23)$$

In the preceding equation, the integrand vanishes by (14.21) so that

$$(\bar{w}, u - \bar{w}) = 0. \qquad (14.24)$$

Accordingly,

$$(\bar{w}, u) = (\bar{w}, \bar{w} + (u - \bar{w}))$$
$$= (\bar{w}, \bar{w}) + (\bar{w}, u - \bar{w})$$
$$= (\bar{w}, w). \qquad (14.25)$$

By the Cauchy–Schwarz inequality (3.22),

$$(u, \bar{w})^2 \leq (u, u)(\bar{w}, \bar{w}), \qquad (14.26)$$

and disregarding the trivial case $(\bar{w}, \bar{w}) = 0$, use of (14.25) gives

$$(\bar{w}, \bar{w}) \leq (u, u). \qquad (14.27)$$

This result corroborates our previous finding that from a geometric point of view, the Rayleigh–Ritz procedure leads to an elementary theorem for a right triangle. Figure 14.1 illustrates this fact, provided ψ and $\bar{\psi}$ are replaced by $\bar{w}$ and u, respectively.

It is evident that, in the example just examined, the Rayleigh–Ritz method furnishes an upper bound for the norm (represented by an energy integral) of the solution vector, but not for the solution vector itself. In this connection, we have no information about the deviation of the approximating function from the exact solution $\bar{w}$. The situation is even less certain if one is interested in the derivatives of the function at hand inasmuch as the closeness of functions does not guarantee the closeness of their derivatives. Such a situation usually arises if one applies the Lagrange principle formulated in terms of displacements, and one is concerned with the stress field, that is, with the second derivatives of displacements. As a simple example, consider a cantelever beam of bending rigidity EJ, acted upon by a terminal load. Applying a combination of Lagrange's principle with the Rayleigh–Ritz method, we represent the deflection of the beam in the form

$$w = C\left(1 - \cos\frac{\pi x}{l}\right),$$

where l is the span of the beam. An elementary calculation gives the following values for certain measures of the deviation of the approximating solution from the exact one:

$$\text{maximum deflection, } w(l), \quad -1.0\%,$$

$$\text{slope at the free end, } \left.\frac{dw}{dx}\right|_{x=l}, \quad +3.2\%,$$

$$\text{maximum moment, } -EJ\left.\frac{d^2w}{dx^2}\right|_{x=l}, \quad -19.0\%,$$

$$\text{shear force, } -EJ\left.\frac{d^3w}{dx^3}\right|_{x=l}, \quad +27.5\%.$$

The increase of the error with repeated differentiations, evident in the foregoing list, indicates that the only possibility of diminishing the error is to directly associate the functional to be minimized with the quantity of interest. This was done, for example, in the torsion problem examined earlier [cf. equations (14.9) and (14.11a)].

14.2. The Trefftz Method

We now turn our attention to a second, less popular, method of the calculus of variations, known as the *Trefftz method*. This method is, in a sense, the direct opposite of the Rayleigh–Ritz method, inasmuch as the approximate solution is selected here to satisfy the governing equation, but permitted to violate the boundary conditions. The main attractiveness of the Trefftz method rests in the fact that, occasionally, it is found to furnish bounds complementary to those provided by the Rayleigh–Ritz procedure. As an illustration, consider the following problem. Let us find bounds for the solution of the biharmonic equation

$$\nabla^4 w = 0 \quad \text{in } \Omega, \tag{14.28}$$

satisfying the boundary conditions

$$w = -f \quad \text{and} \quad \frac{\partial w}{\partial n} = -g \quad \text{on } \partial\Omega. \tag{14.28a}$$

Let w be the exact solution of the problem. We select an inner product of vectors in the form (8.9). By the Cauchy–Schwarz inequality,

$$(u, w)^2 \le (u, u)(w, w), \tag{14.29}$$

and, from equation (8.9) (after replacing v by w),

$$(u, w) = \int_{\partial\Omega} [M_n(u)g - V_n(u)f] \, ds, \tag{14.30}$$

where u is the function introduced earlier as that which satisfies equations (14.21) and (14.22).

Combining the last two formulas, we have

$$\frac{\{\int_{\partial\Omega} [M_n(u)g - V_n(u)f] \, ds\}^2}{(u, u)} \leq (w, w). \tag{14.31}$$

All functions in the left-hand member of this inequality are known; thus, the procedure furnishes a *lower bound* for the norm of the exact solution and, in combination with the inequality (14.27), bounds the norm of the solution on both sides.[†] As was the case in the Rayleigh–Ritz method, the function space language here also suggests a pictorial representation of the inequality (14.31). Indeed, introducing the notation

$$\frac{\int_{\partial\Omega} [M_n(u)g - V_n(u)f] \, ds}{(u, u)} = \Lambda, \tag{14.32}$$

we cast equation (14.31) into the compact form

$$(\Lambda u, \Lambda u) \leq (w, w). \tag{14.33}$$

Now,

$$(\Lambda u, w - \Lambda u) = \Lambda(u, w) - \Lambda^2(u, u)$$
$$= 0, \tag{14.34}$$

in view of equations (14.30) and (14.32). A geometric interpretation of this result, not difficult to discover, is that the multiple Λu (of the approximate solution) and the vector difference of the exact solution w and the Λ-multiple of the approximate solution are mutually orthogonal (Figure 14.2). The exact solution represents here the hypotenuse of a right triangle (OAB in Figure 14.2), and the Λ-multiple of the approximate solution is a leg of this triangle. A graphical representation of this result, as well as of the pertinent result of the Rayleigh–Ritz procedure, is given in Figure 14.3.

As another illustration of the Trefftz technique, we wish to examine the *Dirichlet problem*, approached before via the Rayleigh–Ritz method [see

---

† Note that the function $\bar{w}$ satisfying the equations (14.20) and (14.19) is now denoted by w, without an overbar.

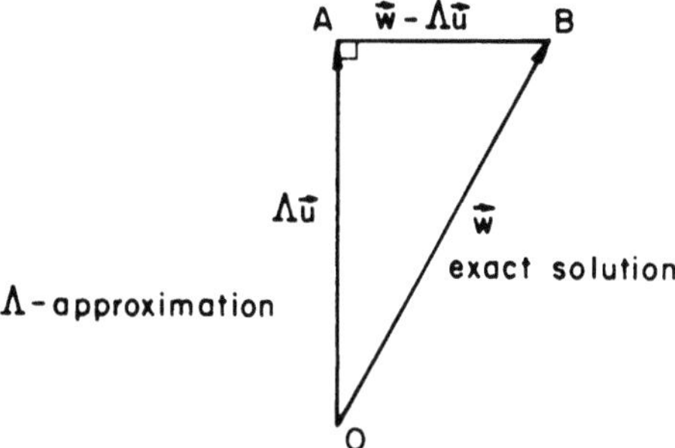

Figure 14.2. Illustration for the Trefftz procedure.

equations (14.10)]. Then, let it be required to find an approximate solution to the boundary value problem,

$$\nabla^2 w = 0 \quad \text{in } \Omega, \tag{14.35}$$

$$w = f \quad \text{on } \partial\Omega. \tag{14.35a}$$

As noted before, this problem is equivalent to minimizing the Dirichlet integral,

$$\int_\Omega [(w_{,x})^2 + (w_{,y})^2] \, dx \, dy = \min, \tag{14.36}$$

under the condition (14.35a). The integrand in the preceding equation evidently represents the square of the magnitude of the gradient of w and, in the Dirichlet notation of (8.46), equation (14.37) takes the compact form

$$(w, w)_D = \min. \tag{14.37}$$

The Trefftz procedure is usually carried out in two consecutive steps. First, a set, $\{\phi_i\}$, of functions satisfying the governing equation—in the present case, equation (14.35)—is selected. The functions are linearly independent, and it

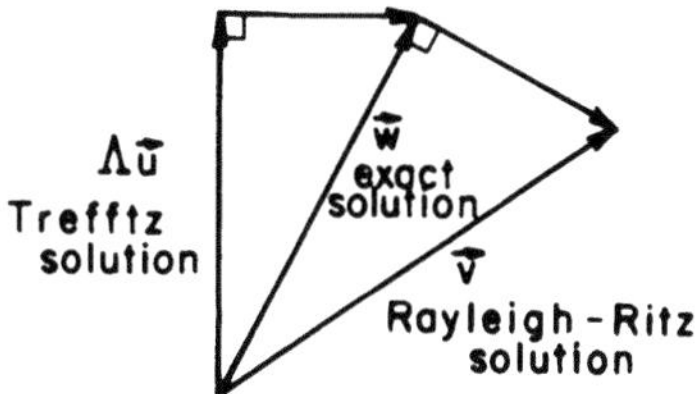

Figure 14.3. Two-sided bounds.

is desirable that they form a complete set according to the following criterion: for any function w harmonic in Ω, and for any positive number ε (however small), there is

$$\int_\Omega \{[(w - w_n)_{,x}]^2 + [(w - w_n)_{,y}]^2\}\, dx\, dy < \varepsilon, \tag{14.38}$$

provided n is sufficiently large, where

$$w_n = \sum_{k=1}^{n} c_k \phi_k \tag{14.39}$$

with the c_k's as some constant coefficients. In order to determine the latter, a simple device is to require the approximate satisfaction of the boundary conditions through

$$\int_{\partial\Omega} [w(s) - w_n(s)]^2\, ds = \min, \tag{14.39a}$$

where w is defined by (14.35a). According to Trefftz himself, however, it is better to go back to equation (14.36) and demand that the functional be minimized with respect to the deviation of w_n from w:

$$(w - w_n, w - w_n)_\Omega = \min. \tag{14.40}$$

Trefftz' idea can be applied to boundary value problems of a considerably more general type than that of (14.35). Instead of the variational equation (14.36), for instance, one can consider the equation

$$\int_\Omega [Aw_{,x}^2 + 2Bw_{,x}w_{,y} + Cw_{,y}^2 + Ew^2 - 2rw)\, dx\, dy = \min, \tag{14.41}$$

with the associated Euler–Lagrange equation

$$L(w) \equiv (Aw_{,x} + Bw_{,y})_{,x} + (Bw_{,x} + Cw_{,y})_{,y} - Ew = -r(x, y), \tag{14.42}$$

where A, B, C, E, and r are given functions of x and y. Suppose, for definiteness, that the function w satisfies the condition

$$w(s) = f(s) \quad \text{on } \partial\Omega, \tag{14.42a}$$

and take the solution of equation (14.42) in the form

$$w_n = \bar{w} + \sum_{k=1}^{n} c_k \phi_k, \tag{14.42b}$$

where $\bar{w}$ is a particular integral of (14.42) and ϕ_k, $k = 1, 2, \ldots, n$, are solutions of the homogeneous version of (14.42). We introduce a "gradient-type" inner product

$$(u, v)_G = \int_\Omega \left[A u_{,x} v_{,x} + B(u_{,x} v_{,y} + u_{,y} v_{,x}) + C u_{,y} v_{,y} + E u v \right] dx\, dy$$

$$(14.42c)$$

and, following Trefftz, require that the square of the distance $\|w - w_n\|_G$ be minimum:

$$(w - w_n,\, w - w_n)_G = \min. \tag{14.42d}$$

It is clear that the condition of the vanishing of the derivatives of the preceding functional with respect to the parameters c_k yields a system of linear equations in these parameters,

$$(w - w_n,\, \phi_k)_G = 0, \qquad k = 1, 2, \ldots, n, \tag{14.43}$$

from which the latter can be determined explicitly provided the unknown function w is eliminated. To achieve this, we apply the Gauss–Green theorem,

$$(u, v)_G + \int_\Omega u L(v)\, dx\, dy = \int_{\partial\Omega} u L^*(v)\, ds, \tag{14.44}$$

where

$$L^*(v) = (A v_{,x} + B v_{,y})\cos(n, x) + (B v_{,x} + C v_{,y})\cos(n, y). \tag{14.44a}$$

In the foregoing identity, we set

$$u = \bar{w} + \sum_{k=1}^{n} c_k \phi_k - w \quad \text{and} \quad v = \phi_l, \qquad l = 1, 2, \ldots, n, \tag{14.44b}$$

obtaining the system of equations

$$(w_n - w,\, \phi_l)_G + \int_\Omega (w_n - w) L(\phi_l)\, dx\, dy - \int_{\partial\Omega} (w_n - w) L^*(\phi_l)\, ds = 0,$$

$$l = 1, 2, \ldots, n. \tag{14.45}$$

At this stage, we note that the condition (14.42d) implies that

$$\frac{\partial}{\partial c_l}(w - w_n,\, w - w_n) = 2(w - w_n,\, \phi_l)$$

$$= 0, \qquad l = 1, 2, \ldots, n, \tag{14.46}$$

so that the first term in equation (14.45) vanishes. Similarly, $L(\phi_l) = 0$ for any $l = 1, 2, \ldots, n$. Thus, after accounting for the boundary condition (14.42a), we arrive at Trefftz' system of equations,

$$\sum_{k=1}^{n} c_k \int_{\partial\Omega} \phi_k L^*(\phi_l) \, ds = \int_{\partial\Omega} (f - \bar{w}) L^*(\phi_l) \, ds, \qquad l = 1, 2, \ldots, n, \quad (14.47)$$

from which the values of the coefficients c_k are readily calculated.

14.3. Remark

As demonstrated, the Rayleigh–Ritz procedure, applied to the semi-homogeneous biharmonic problem (14.19)–(14.20), furnishes an upper bound for the norm of the solution vector [equation (14.27)]. On the other hand, employing the Trefftz method, we are able to determine a corresponding lower bound [equation (14.33)]. An upper bound for the norm of the solution of a Dirichlet problem [equation (14.12a)] was also provided by the Rayleigh–Ritz procedure. A general extrapolation from these examples, to conclude that things always turn out this way would, however, be ill-founded. Actually, there exists no general proof to support the belief that, by utilizing functions which satisfy the boundary conditions, but fail to obey the governing equation, one invariably arrives at an upper bound for the functional in question. Likewise unfounded is the claim that the application of functions satisfying the governing equation, but failing to obey the boundary condition, always furnishes a corresponding lower bound (Ref. 50, p. 153).

14.4. Improvement of Bounds

In practical applications, bounds derived for quantities of interest often turn out to be unacceptably far off, and it is required to take recourse to procedures that enable one to improve the closeness of bounds.

Suppose, then, that by using the Rayleigh–Ritz method, one has reached an approximate solution, w, for, say, a Dirichlet problem [equations (14.10)]. Let the exact solution of this problem be designated by u. By the orthogonality condition (14.13c), we have

$$\begin{aligned}
(w - u, w - u) &= (w - u, w) - (w - u, u) \\
&= (w - u, w) \\
&= (w, w) - (u, u).
\end{aligned} \qquad (14.48)$$

This gives the error of an upper bound in terms of the Dirichlet metric (14.11). In order to make the error smaller, it is helpful to employ a Ritz expression in the form (14.5),

$$w = w_0 - \sum_{k=1}^{n} c_k \phi_k, \tag{14.49}$$

in which w_0 satisfies the nonhomogeneous boundary condition (14.10),

$$w_0 = f \quad \text{on } \partial\Omega \tag{14.50}$$

while we suppose that the functions ϕ_k obey the homogeneous boundary conditions

$$\phi_k = 0 \quad \text{on } \partial\Omega, \qquad k = 1, 2, \ldots, n. \tag{14.51}$$

Evidently, $w = f$ on $\partial\Omega$, as required.

According to the general expression on the left in (14.48), the error of the approximate solution (14.49) is found from

$$\begin{aligned}
(w - u, w - u) &= \left(w_0 - u - \sum_{k=1}^{n} c_k \phi_k, \; w_0 - u - \sum_{k=1}^{n} c_k \phi_k \right) \\
&= (w_0 - u, w_0 - u) + \sum_{\substack{i=1 \\ k=1}}^{n} c_i c_k (\phi_i, \phi_k) \\
&\quad - 2 \sum_{k=1}^{n} c_k (w_0 - u, \phi_k).
\end{aligned} \tag{14.52}$$

If all the coefficients c_k are set equal to zero, then the error of the approximate solution becomes $(w_0 - u, w_0 - u)$, as before. Making use of the representation (14.49), however, we enlarge the family of functions admitted for competition. This circumstance, naturally, decreases (or, at least, does not increase) the error of the approximate solution. Before actually determining the error, it is of interest to recall that, according to our earlier geometric interpretation, the vectors w_0 and u play the parts of the hypotenuse and a leg of a right triangle, respectively (Figure 14.4). Thus,

$$\begin{aligned}
(u, w_0 - u) &= 0 \\
&= \left(u, \; w - u + \sum_{k=1}^{n} c_k \phi_k \right) \\
&= (u, w - u) + \sum_{k=1}^{n} (u, c_k \phi_k).
\end{aligned} \tag{14.53}$$

By the first Green identity (14.13b), however,

$$(u, c_k \phi_k) = \int_{\partial\Omega} c_k \phi_k \frac{\partial u}{\partial n} \, ds - \int_{\Omega} c_k \phi_k \nabla^2 u \, dx \, dy$$

$$= 0, \tag{14.53a}$$

since u is harmonic and the auxiliary functions ϕ_k vanish on the boundary. Therefore,

$$(u, w - u) = 0, \tag{14.54}$$

and w again admits an interpretation as being the hypotenuse of a right triangle, a leg of which is u (Figure 14.4).

The question that now arises is how the hypotenuse w can be made to approach the leg as closely as possible. In geometric language, this is to require that the distance $\|w - u\|$ or its square,

$$\|w - u\|^2 = (w - u, w - u), \tag{14.55}$$

be made as small as possible. To achieve this, we minimize the right-hand member of equation (14.52) with respect to the coefficients c_k, obtaining the system of equations

$$\left(w_0 - \sum_{k=1}^{n} c_k{}^* \phi_k, \phi_i\right) = 0, \qquad i = 1, 2, \ldots, n, \tag{14.56}$$

where use was made of relations (14.53a), and an asterisk is attached to mark the values of the coefficients associated with the minimum.

It is readily verified by inspection that a geometric interpretation of the preceding equations amounts to the following: if the distance represented by

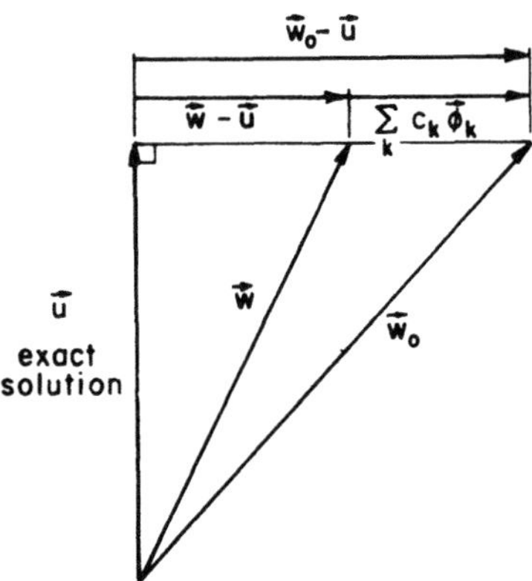

Figure 14.4. Illustration for the Rayleigh–Ritz procedure.

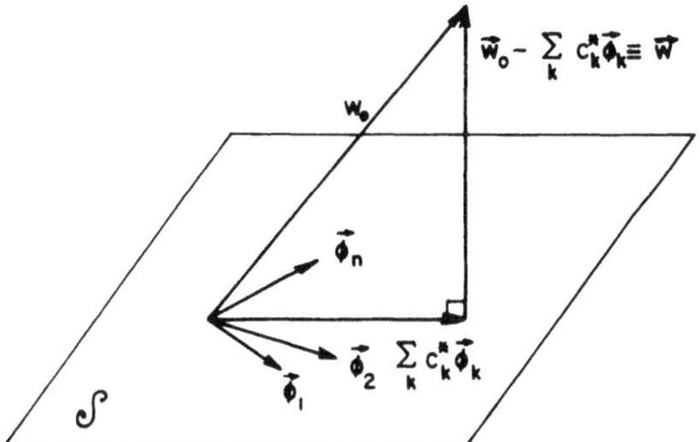

Figure 14.5. Illustration for condition (14.56).

the norm in (14.55) becomes minimum, then the sum $\sum_{k=1}^{n} c_k^* \phi_k$ coincides with the orthogonal projection of the vector w_0 onto the subspace $\mathcal{S}$ spanned by the functions ϕ_i (Figure 14.5). This is equivalent to the statement that the vector w is orthogonal to the subspace $\mathcal{S}$.

Problems

1. Let $\{u_n\}$ be a minimizing sequence for a functional $F[u]$, i.e., such that $F[u_n] \to$ min $F[u]$ for $n \to \infty$. Show that in a Hilbert space $\mathcal{H}$, each minimizing sequence converges to the solution u_0 [see equation (6.3) in Chapter 6†] of the equation $Lu = v$, (1), u, $v \in \mathcal{H}$, where L is a linear symmetric coercive operation and the function $F[u] = (Lu, u) - 2(u, v)$, (2), associated with (1), takes its minimum for $u = u_0$. Note that symmetry implies $(Lu, v) = (u, Lv)$, $u, v \in \mathcal{H}$, while coerciveness signifies that $(Lu, u) \geq c\|u\|^2$ for some $c > 0$, for each $u \in \mathcal{H}$.

2. Select as a basis of the Hilbert space $\mathcal{H}$ a set of coordinate functions φ_i [cf. equation (14.2)], and consider the subspaces $\mathcal{H}_n \subset \mathcal{H}$, a basis for each of which consists of $\varphi_1, \varphi_2, \ldots, \varphi_n$. Let L and $F[u]$ denote the operator and the functional defined in the preceding problem, respectively. Show that the sequence u_n converges to the solution u_0 of $Lu = v$, where $u_n \in \mathcal{H}_n$, $u_n = \min_{u \in \mathcal{H}_n} F[u]$, for each n.

3. Find the error, in the norm, resulting from replacing u_0 by u_n, where u_0 and u_n are as in Problem 1. Assume that the inverse operator L^{-1} exists (note: $L^{-1}Lu = LL^{-1}u = Iu = u$, where I = identity operator) and is bounded, i.e., $\|L^{-1}u\| \leq c\|u\|$, c = a constant.

---

† Convergence "in the norm," $\lim\|u_0 - u_n\| = 0$ for $n \to \infty$, is often known as a *strong* convergence, while convergence of the type $\lim F[u_n] = F[u_0]$ is considered to be *weak*, in case the limiting relation holds for *every* continuous linear functional on $\mathcal{H}$.

4. Considering $\mathcal{H}_n \subset \mathcal{H}$ as in Problem 2 above, show that, if the functional $F[u]$ has its least value at the element $u = u_n$, $u_n \in \mathcal{H}_n$, then the vector $Lu_n - v$ is perpendicular to each of the vectors φ_i, $i = 1, 2, \ldots, n$.

5. Show that for an elastic plate subject to a transverse load q and having built-in edges, the terms associated with the coefficients β_i in the Galerkin equations are symmetric (cf. the solution of the preceding problem). Select the inner product in the Hilbert form.

6. Show that[117] if a function u satisfies Poisson's equation $\nabla^2 u = q$ in Ω, and a function v obeys the boundary conditions $v = \partial v/\partial n = 0$ on $\partial\Omega$, then the biharmonic functional (cf. Problem 1 above) satisfies $F[v] = [(\nabla^4 v, v) - 2(v, q)]_\Omega \geq -(u, u)_\Omega$. Use the inner product in the Hilbert form.

7. Show[118] that the functional $F[u_i] = \sum_{i=1}^{n} (L_i u_i, u_i)$ becomes minimum for $u_i = u_0$, $i = 1, 2, \ldots, n$, where the operators L_i are symmetric and positive definite $[(L_i, u, u) > 0, u \neq 0]$, the operator L is coercive $[(Lu, u) \geq c^2 \|u\|^2]$, $L = \sum_{i=1}^{n} L_i$, $\sum_{i=1}^{n} L_i u_i = f$, and u_0 is the solution of the equation $Lu = f$.

8. Show that, of all functions satisfying Poisson's equation $\nabla^2 u + f = 0$ in Ω, (1), the solution u_0 which satisfies the boundary condition $u = 0$ on $\partial\Omega$ makes the functional $F[u] = \int_\Omega (\nabla u)^2 \, d\Omega$ minimum. Select the Dirichlet-type scalar product for the corresponding bilinear form $F[u, v]$.

15

Function Space and Variational Methods

In the preceding chapters, specifically in Chapters 8 and 14, we were afforded opportunities to make comments on the relations between the concepts of function space and those of the calculus of variations. It seems worthwhile to return to these questions by examining certain of their aspects from somewhat different points of view. We first examine the so-called inverse problem, in which one looks for a functional whose critical points (i.e., extrema and saddle points) are the solutions of the given differential equation.

15.1. The Inverse Method

Looking retrospectively at our exposition, we should recognize the important part played in our derivations by the idea of an inner product. Several different mathematical forms have been given to this concept, such as $\sum_i u_i v_i$ in Euclidean spaces [equation (6.22)], and $\int_\Omega u(P)v(P)\, d\Omega$, $\int_\Omega \sum_i u_i(P)v_i(P)\, d\Omega$, and $\int_V u_{ij} v_{ij}\, dV$ in Hilbert spaces of functions [equations (8.2), (8.6a), and (8.14)]. Even more elaborate expressions proved to be serviceable; recall that given by equation (8.8),

$$(u, v) = \frac{D}{2} \int_\Omega [\nabla^2 u \nabla^2 v - (1 - v)(u_{,xx} v_{,yy} + u_{,yy} v_{,xx} - 2u_{,xy} v_{,xy})]\, dx\, dy \quad (15.1)$$

or that involving convolution of functions,

$$u(t)^* v(t) = \int_0^t u(\tau)v(t - \tau)\, d\tau \quad (15.2)$$

[equation (8.10d)].

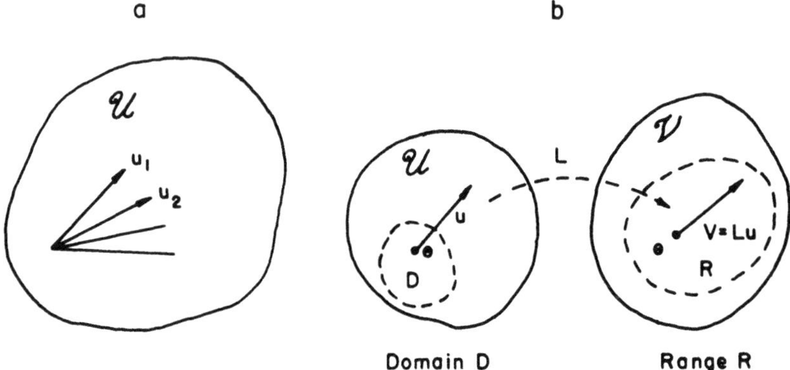

Figure 15.1. Basic concepts.

The just-named, and similar, forms have three characteristic properties in common: they are symmetric, positive definite, and linear in each of the two component function vectors, that is, they are *bilinear*.

In the first four cartesian forms listed above, all vectors belong to a single space, say $\mathscr{U}$ (Figure 15.1a, in which the elements are denoted u_i, $i = 1$, 2, ...). In bilinear forms such as (15.1), in which differential operations appear, the function vectors belong to two spaces: one, call it $\mathscr{U}$, comprising the domain of the operators, and the other, say $\mathscr{V}$, including the range of the operators, and usually different from $\mathscr{U}$ (Figure 15.1b). Actually, it would be more exact, in this latter case, not to apply the standard inner product notation $(\,,\,)$, but a separate one, for example, $[\,,\,]$.

It turns out to be useful to consider the standard inner product as a particular case of a general *bilinear form* associated with two spaces and stripped of the properties ordinarily attributed to the symbol[†] (v, u). We shall denote this form by $\langle v, u \rangle$, where $u \in \mathscr{U}$ and $v \in \mathscr{V}$.

Consider now the equation

$$Lu(P) = f(P) \tag{15.3}$$

[cf. equation (8.16)], where L is a linear differential operator with *domain*,[‡] D, in a linear space $\mathscr{U}$ and *range*, R, in a second linear space $\mathscr{V}$. $u(P)$ is a function to be determined and $f(P)$ is a function preassigned; clearly, $u \in \mathscr{U}$ and $v = Lu \in \mathscr{V}$ (Figure 15.1b).

[†] That is, symmetry and positive definiteness.

[‡] We recall that the *domain D* of an operator, or mapping, L is the set of vectors u for which the expression Lu is defined. The set of vectors $v = Lu$, generated as u varies over D, is called the *range* of the operators. For the concept of mapping, see the footnote preceding equation (8.16).

The *symmetry of L*, defined earlier [equation (8.17)] by

$$(Lu_1, u_2) = (u_1, Lu_2), \tag{15.4}$$

is now replaced by the more general requirement that the operator L be symmetric with respect to the selected bilinear form $\langle u, v \rangle$, that is, that the equality

$$\langle Lu_1, u_2 \rangle = \langle Lu_2, u_1 \rangle \tag{15.5}$$

holds for every pair of elements u_1 and u_2 in D. It is clearly visible that the property of symmetry of an operator is not absolute, but is *relative* to the bilinear form selected[115].†

Our objective now is to establish a variational formulation for equation (15.3), in which the operator L is *symmetric* with respect to the selected bilinear form $\langle v, u \rangle$. We have examined a similar problem in Chapter 8,‡ but there the associated bilinear form had all the properties of a conventional inner product, while, in the present case, the only requirement imposed on the bilinear form is that it is *nondegenerate* on $\mathscr{U}$ and $\mathscr{V}$. This means that, by definition, if

$$\langle v_0, u \rangle = 0 \qquad \text{for } v_0 \in \mathscr{V} \text{ and every } u \in \mathscr{U}, \text{ then } v_0 = \theta, \tag{15.5a}$$

and if

$$\langle v, u_0 \rangle = 0 \qquad \text{for } u_0 \in \mathscr{U} \text{ and every } v \in \mathscr{V}, \text{ then } u_0 = \theta. \tag{15.5b}$$

Let us now introduce the functional [cf. equation (15.3)]

$$F[u] \equiv \tfrac{1}{2}\langle Lu, u \rangle - \langle f, u \rangle. \tag{15.6}$$

We have§

$$\Delta F[u] = F[u + \delta u] - F[u]$$

or, after employing (15.6),

$$\Delta F[u] = \tfrac{1}{2}(\langle Lu, \delta u \rangle + \langle L\, \delta u, u \rangle + \langle L\, \delta u, \delta u \rangle) - \langle f, \delta u \rangle.$$

Thus,

$$\delta F[u] = \langle Lu, \delta u \rangle - \langle f, \delta u \rangle,$$

---

† Clearly, if $\mathscr{U} = \mathscr{V}$ is a Hilbert space and one replaces the bilinear form $\langle v, u \rangle$ by the inner product (v, u) on $\mathscr{U}$, then the symmetry equation (15.5) reduces to the familiar form $(Lu_1, u_2) = (u_1, L_2)$; cf. (8.17).

‡ Compare the text following equation (8.18).

§ We recall that the variation δu of a function vector u is *any* function vector δu (most often small) which, when added to u, gives a new (so-called comparison) function $u + \delta u$.

The variation of $\delta F[u]$ of a functional $F[u]$ is obtained from the total increment of the functional, $\Delta F[u]$, after disregarding terms of order in δu higher than the first.

where we have considered (15.5). Finally,

$$\delta F[u] = \langle Lu - f, \delta u \rangle. \tag{15.7}$$

Now, if u_0 is a solution of (15.3), then $Lu_0 - f = 0$ and $\delta F[u_0] = 0$, so that u_0 makes the functional $F[u]$ *stationary*.†

Conversely, it is easily shown that the stationary points of $F[u]$ are solutions of the equation (15.3). Indeed, if u^0 is a stationary point of $F[u]$, then for every δu in D, we have

$$\delta F[u^0] = \langle Lu^0 - f, \delta u \rangle = 0. \tag{15.8}$$

By the nondegeneracy condition (15.5a), therefore,

$$Lu^0 - f = 0,$$

as claimed. Summing up, we conclude that if a given operator L is symmetric with respect to a selected nondegenerate bilinear form $\langle v, u \rangle$, then it is *always* possible to arrive at a variational formulation for equation (15.3), the corresponding functional being given by (15.6). This result[115] is of great significance, although superficially it may seem somewhat trivial. In fact, a variational formulation for the entire class of initial value problems, for example, cannot be achieved as long as we persist in using the procedures of the classical calculus of variations (as we did in Chapter 8). Even the fundamental principle of Hamilton must artificially be converted into the treatment of motions between two terminal configurations, instead of considering the initial conditions alone[116]. Likewise, such a simple operator as $L = d/dt$, in the field of functions obeying the initial condition $u(0) = 0$ and defined for $0 \leq t \leq 1$, evades variational formulation because (in the classical context) it fails to be symmetric.‡ *On the other hand*, if we take as the bilinear form the convolution integral (15.2), we have§

$$\langle Lu, v \rangle = \int_0^1 \frac{du}{dt} v(1 - t)\, dt$$

$$= u(1)v(0) - u(0)v(1) + \int_0^1 u(1 - t)\frac{dv}{dt}\, dt$$

$$= \langle Lv, u \rangle, \tag{15.9}$$

---

† See the discussion in Chapter 8, where the functional (8.19) is reminiscent of the present functional (15.6) if the operation $\langle , \rangle$ is replaced by the operation $(,)$.

‡ We have $(u, v) = \int_0^1 u(t)v(t)\, dt$ and $(Lu, v) = \int_0^1 (du/dt)v(t)\, dt = u(1)v(1) - (u, Lv)$, so that $(Lu, v) \neq (u, Lv)$ even if we assume that, e.g., $v(1) = 0$.

§ This approach was initiated by M. E. Gurtin,[93] who departed from the custom of using Cartesian forms of the inner product and employed the convolution of two functions.

by virtue of the known symmetry of a convolution integral and the assumption that v is in the given space (that is, $v(0) = 0$). Since, as the matter stands, the possibility of a variational formulation depends, in turn, on the possibility of finding a bilinear form with respect to which the given operator is symmetric, the question arises as to whether the discovery of such a form is always possible. Magri[115] has shown that the answer is in the affirmative and that, moreover, there exists an infinity of such forms.

In the example being discussed, the functional to be subject to variation, (15.6), is

$$F[u] = \frac{1}{2} \int_0^1 u(1 - t)u'(t)\, dt - \int_0^1 u(1 - t) f(t)\, dt, \tag{15.10}$$

so that the variation of $F[u]$ becomes

$$\delta F[u] = \int_0^1 \delta u(1 - t)\frac{du}{dt}\, dt + \tfrac{1}{2}[\delta u(1 - t)u(t)]_0^1 - \int_0^1 \delta u(1 - t) f(t)\, dt$$

$$= \int_0^1 \delta u(1 - t)\left[\frac{du}{dt} - f(t)\right] dt + \tfrac{1}{2}[\delta u(0)u(1) - \delta u(1)u(0)]. \tag{15.11}$$

Inasmuch as $u(0)$ remains unvaried and $\delta u(1 - t)$ and $\delta u(1)$ are arbitrary, one arrives at the Euler–Lagrange equation of the problem in the anticipated form (15.3); in addition, one recovers $u(0) = 0$ as the condition imposed on the space of comparison functions.

15.2. Orthogonal Subspaces

New and interesting aspects of the function space–variational calculus connections come to light if one goes back to Chapter 12 and analyzes certain conclusions which can be drawn from the orthogonality of the subspaces denoted there by $\mathscr{S}^\tau$ and $\mathscr{S}^\varepsilon$ (Figure 15.2). For convenience, let us list the equations obeyed by the states S^τ and S^ε making up the subspaces $\mathscr{S}^\tau$ and $\mathscr{S}^\varepsilon$, respectively.

States S^τ:
(a) equations of equilibrium,

$$\tau'_{ij,j} + F_i = 0, \tag{15.12}$$

(b) stress boundary conditions,

$$\tau'_{ij}n_j = f_i \quad \text{on } \Omega_\tau. \tag{15.13}$$

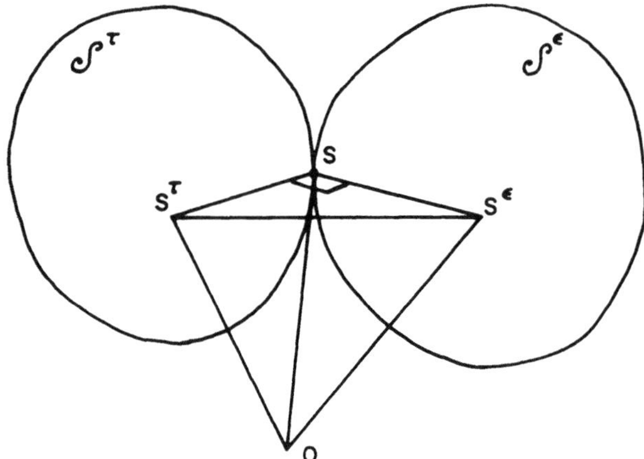

Figure 15.2. Orthogonal subspaces $\mathscr{S}^\tau$ and $\mathscr{S}^\varepsilon$.

States S^ε:

(a) compatibility equations,†

$$u''_{i,j} + u''_{j,i} = 2e''_{ij}, \tag{15.14}$$

(b) displacement boundary conditions,

$$u_i'' = g_i \quad \text{on } \Omega_u. \tag{15.15}$$

In the equations above, f_i and g_i denote functions prescribed on the portions Ω_τ and Ω_u of the surface of the body, respectively, where $\Omega = \Omega_\tau + \Omega_u$.

With the Pythagorean theorem in mind, we conclude, upon inspecting Figure 15.2, that

$$\|S - S^\tau\|^2 + \|S - S^\varepsilon\|^2 = \|S^\tau - S^\varepsilon\|^2 \tag{15.16}$$

or, explicitly,

$$[(S, S) - 2(S, S^\tau)] + (S^\tau, S^\tau) + [(S, S) - 2(S, S^\varepsilon)] + (S^\varepsilon, S^\varepsilon)$$
$$= (S^\tau, S^\tau) - 2(S^\tau, S^\varepsilon) + (S^\varepsilon, S^\varepsilon), \tag{15.17}$$

where S is the position vector of the point of intersection of the subspaces $\mathscr{S}^\tau$ and $\mathscr{S}^\varepsilon$, corresponding to the exact solution of the problem at hand.

The following inequalities follow easily from equation (15.16):

$$(S - S^\tau, S - S^\tau) \le (S^\tau - S^\varepsilon, S^\tau - S^\varepsilon), \tag{15.18}$$

$$(S - S^\varepsilon, S - S^\varepsilon) \le (S^\tau - S^\varepsilon, S^\tau - S^\varepsilon). \tag{15.19}$$

---

† Looked upon as a system of six differential equations for the determination of three displacement components when six strain components are prescribed. They are usually represented (after differentiation) in the form: $e_{ij,kl} + e_{kl,ij} - e_{ik,jl} - e_{jl,ik} = 0$; compare equation (12.4).

Likewise, equation (15.17) yields

$$(S, S) - 2(S, S^\tau) \le (S^\varepsilon, S^\varepsilon) - 2(S^\tau, S^\varepsilon) \tag{15.20}$$

and

$$(S, S) - 2(S, S^\varepsilon) \le (S^\tau, S^\tau) - 2(S^\tau, S^\varepsilon). \tag{15.21}$$

Direct consequences of the foregoing inequalities are the following three *minimum principles*[40]:

(a) The difference $\|S^\tau - S^\varepsilon\|^2$ between two arbitrary states S^τ and S^ε reaches its minimum (actually, zero) for $S^\tau = S$ and $S^\varepsilon = S$.

(b) If S^ε is a generic state in $\mathscr{S}^\varepsilon$ and S^τ is a fixed state in $\mathscr{S}^\tau$, then the difference $\|S^\tau - S^\varepsilon\|^2$ reaches its minimum for $S^\varepsilon = S$. The same conclusion concerns the difference $(S^\varepsilon, S^\varepsilon) - 2(S^\tau, S^\varepsilon)$.

(c) If S^τ is a generic state in $\mathscr{S}^\tau$ and S^ε is a fixed state in $\mathscr{S}^\varepsilon$, then the difference $\|S^\tau - S^\varepsilon\|^2$ reaches its minimum for $S^\tau = S$. The same conclusion concerns the difference $(S^\tau, S^\tau) - 2(S^\tau, S^\varepsilon)$.

By appeal to the definition (12.8), we now have (by symmetry of e_{ij} and τ_{ij})

$$(S^\varepsilon, S^\tau) = \int_V e_{ij}'' \tau_{ij}' \, dV$$

$$= \int_V u_{i,j}'' \tau_{ij}' \, dV$$

$$= \int_V (u_i'' \tau_{ij}')_{,j} \, dV - \int_V u_i'' \tau_{ij,j}' \, dV$$

$$= \int_\Omega u_i'' \tau_{ij}' n_j \, d\Omega - \int_V u_i'' \tau_{ij,j}' \, dV, \tag{15.22}$$

the last step by applying the Gauss–Green theorem.

The definitions (15.12)–(15.15) of the states S^τ and S^ε enable one to cast the preceding equation into the form

$$(S^\varepsilon, S^\tau) = \int_{\Omega_\tau} u_i'' f_i \, d\Omega + \int_{\Omega_u} g_i \tau_{ij}' n_j \, d\Omega + \int_V u_i'' F_i \, dV. \tag{15.23}$$

Accordingly,

$$\tfrac{1}{2}(S^\tau - S^\varepsilon, S^\tau - S^\varepsilon) = \tfrac{1}{2}(S^\tau, S^\tau) + \tfrac{1}{2}(S^\varepsilon, S^\varepsilon) - (S^\varepsilon, S^\tau)$$

$$= W^\tau + W^\varepsilon - \int_{\Omega_\tau} u_i'' f_i \, d\Omega - \int_{\Omega_u} g_i \tau_{ij}' n_j \, d\Omega - \int_V u_i'' F_i \, dV,$$

$$\tag{15.23a}$$

where W^τ and W^ε denote the strain energies associated with the states S^τ and S^ε, respectively [cf. equations (8.25) and (12.8)].

We now refer to principles (b) and (c) above, and carry out the minimization of the difference $(S^\tau - S^\varepsilon, S^\tau - S^\varepsilon)^2$ by simultaneously holding either S^ε or S^τ fixed. We examine two cases.

Case (b_1). S^τ held fixed. In this case, the actual state of equilibrium is associated with the value $S^\varepsilon = S$ and the minimum of the expression

$$\Pi \equiv W(e_{ij}) - \int_{\Omega_\tau} u_i f_i \, d\Omega - \int_V u_i F_i \, dV, \qquad (15.24)$$

provided the displacements satisfy the boundary conditions (15.15) on Ω_u; $W(e_{ij})$ is here the actual strain energy expressed in terms of strains.

A glance at the preceding equation convinces us that Π represents the potential energy of the system. Accordingly, the minimum principle (b_1) turns out to be the classical principle of *minimum potential energy* of an elastic system (Ref. 31, p. 385).

Case (c_1). S^ε held fixed. In this case, the actual state of equilibrium is associated with the value $S^\tau = S$ and with the minimum of the expression

$$\Pi^* \equiv W(\tau_{ij}) - \int_{\Omega_u} g_i t_{(n)i} \, d\Omega, \qquad (15.25)$$

provided the state of stress satisfies the equilibrium equations (15.12) and the boundary conditions (15.13) prescribed on Ω_τ; $W(\tau_{ij})$ is here the actual strain energy expressed in terms of stresses, and $t_{(n)i}$ denotes the surface tractions $\tau_{ij} n_j$ on Ω_u. In this case, Π^* represents the complementary energy of the system, and the minimum principle (c_1) turns out to be the classical principle of *minimum complementary energy* (Ref. 31, p. 389).

These principles can be written briefly in the forms

$$\delta\Pi = 0 \quad \text{and} \quad \delta\Pi^* = 0, \qquad (15.26)$$

respectively, where δ designates the variation.

Alternatively, we can turn to the inequalities (15.20) and (15.21), for simplicity setting the body force equal to zero, and convert the principles (15.26) into the forms

$$(S^\varepsilon, S^\varepsilon) - 2 \int_{\Omega_\tau} u_i^\varepsilon f_i \, d\Omega \geq (S, S) - 2 \int_{\Omega_\tau} u_i f_i \, d\Omega \qquad (15.27)$$

and

$$(S^\tau, S^\tau) - 2 \int_{\Omega_u} t_{(n)i}^\tau g_i \, d\Omega \geq (S, S) - 2 \int_{\Omega_u} t_{(n)i} g_i \, d\Omega. \qquad (15.28)$$

15.3. Laws' Approach

A remarkably ingenious derivation of the work and energy principles of linear elasticity (the latter having already been discussed in the preceding section in the context of orthogonal subspaces) was given by Laws.† Laws is concerned with the development of "energy" (that is, integral) bounds for the solution of the class of problems (11.85), i.e.,

$$L^*Lu = f, \tag{15.29}$$

where L^* is the *formal* adjoint of the linear operator L, defined by

$$(\tau, Lu) = \langle L^*\tau, u \rangle + \{\tau, u\}. \tag{15.30}$$

Here, $\langle u_1, u_2 \rangle$ is the inner product of *vector*-valued functions u_1 and u_2 belonging to one function space, $\mathscr{H}$, and (τ_1, τ_2) is the inner product of *tensor*-valued functions τ_1 and τ_2 belonging (in general) to a function space H;‡ braces, such as $\{\tau, u\}$, stand for a bilinear expression involving functions defined on the boundary S of the volume V occupied by the material body under consideration.

As follows from Section 11.1.1 of Chapter 11, the class of equations of the type (15.29) is rather comprehensive, including the classical equations of Laplace, Sturm–Liouville, and plate and beam theories, to mention only a few.

A well-known realization of equation (15.30) is Green's second identity

$$\int_V v\nabla^2 u \, dV = \int_V u\nabla^2 v \, dV + \int_\Omega \left(v\frac{\partial u}{\partial n} - u\frac{\partial v}{\partial n} \right) d\Omega. \tag{15.31}$$

For our future purpose, and to appreciate the extraordinary elegance of Laws' procedure, it is appropriate to make use of the so-called direct notation,§ which represents the ultimate in simplicity and coincides, in fact, with the notation used in functional analysis. To this end, we shall denote ordinary vectors such as the displacement u_i, $i = 1, 2, 3$, by the letter u, second-order tensors such as the stress tensor τ_{ij} or the strain tensor e_{ij} by the letters τ and e, respectively, and fourth-order tensors such as the stiffnesses C_{ijkl} and the compliances S_{ijkl} by C and S, respectively.

---

† Laws.[94] Compare also Arthurs,[95] as well as Noble and Sewell.[96]

‡ Generally speaking, Latin lowercase letters may also denote, for example, scalar-valued functions (such as the temperature) and Greek lowercase letters denote vector-valued functions (such as the heat flux vector). Of course, from the standpoint of functional analysis, either of these functions is a vector in a vector space.

§ Also known as matrix, or dyadic, notation.

With this in mind, we write the constitutive equation of a general anisotropic elastic body in the form

$$\tau = Ce, \tag{15.32}$$

which, in index notation, translates into[†]

$$\tau_{ij} = C_{ijkl}e_{kl}. \tag{15.32a}$$

Inverting (15.32) gives

$$e = S\tau, \tag{15.33}$$

so that $CS = 1$, where 1 is the unit fourth-order symmetric tensor.[‡] The vector of surface tractions (15.13) is represented by

$$t = \tau n, \tag{15.33a}$$

where n is a unit normal, and the equations of equilibrium (15.12) by

$$\operatorname{div} \tau + f = 0. \tag{15.33b}$$

By appeal to the strain–displacement relations (15.14), we can cast the constitutive equation (15.32) in the form

$$\tau = Lu \tag{15.34}$$

(in index notation, $\tau_{ij} = C_{ijkm}u_{k,m}$). It is now not difficult to show that

$$L^*\tau = -\operatorname{div} \tau, \tag{15.35}$$

so that the equilibrium equations (15.33b) take the desired form (of the Lamé equations)

$$L^*Lu = f. \tag{15.36}$$

In fact, let us accept the following definitions of inner products[§]:

$$\langle u, v \rangle = \int_V uv \, dV \quad \text{in } \mathscr{H} \tag{15.37}$$

and

$$(\tau^{(1)}, \tau^{(2)}) = \int_V \tau^{(1)}(S\tau^{(2)}) \, dV \quad \text{in } H. \tag{15.38}$$

[†] Note that, in the direct notation, the summation is always understood to be performed over the closest indices of the adjacent symbols, e.g., over k and l in (15.32), over i in $uv \equiv u_i v_i$, and so on.

[‡] In index notation: $C_{ijkl}S_{klmn} = \delta_{im}\delta_{jn}$.

[§] In index notation: $\langle u, v \rangle = \int_V u_i v_i \, dV$ and $(\tau^{(1)}, \tau^{(2)}) = \int_V \tau_{ij}^{(1)}S_{ijkl}\tau_{kl}^{(2)} \, dV = \int_V \tau_{ij}^{(1)}e_{ij}^{(2)} \, dV$, by (15.33). Thus, the product $(\tau^{(1)}, \tau^{(2)})$ is a measure of elastic energy.

We then have, using equations (15.33a), (15.34), and the divergence theorem,

$$(\tau, Lu) = \int_V \tau e \, dV$$

$$= -\int_V (\operatorname{div} \tau) u \, dV + \int_\Omega u(\tau n) \, d\Omega. \qquad (15.39)$$

A comparison of the preceding equation with the defining equation (15.30) gives the anticipated result

$$\langle L^*\tau, u \rangle = -\int_V (\operatorname{div} \tau) u \, dV, \qquad (15.40)$$

so that

$$L^*\tau = -\operatorname{div} \tau, \qquad (15.40a)$$

as well as

$$\{\tau, u\} = \int_\Omega u(\tau n) \, d\Omega. \qquad (15.41)$$

With these rather formal preliminaries out of the way, a derivation of the four central theorems of linear elasticity lies, so to speak, on the surface.

The *reciprocal theorem* of Betti and Rayleigh involves two different systems of forces acting on the same body; the boundary conditions may differ in the two cases. Let u_1 and u_2 be solutions of the equations

$$L^*Lu = f_1 \quad \text{and} \quad L^*Lu = f_2, \qquad (15.42)$$

respectively. By appeal to the constitutive equation (15.34), we set

$$\tau_1 = Lu_1 \quad \text{and} \quad \tau_2 = Lu_2. \qquad (15.43)$$

From (15.33) and (15.38), we now have

$$(\tau_1, \tau_2) = \int_V \tau_1 e_2 \, dV,$$

$$(\tau_2, \tau_1) = \int_V \tau_2 e_1 \, dV$$

$$= \int_V \tau_2 S\tau_1 \, dV$$

$$= \int_V \tau_1 e_2 \, dV, \qquad (15.44)$$

which proves the theorem. An alternative form of (15.44) is obtained by making use of the defining equation (15.30), as well as the formulas (15.42) and (15.43). We find

$$\langle f_2, u_1 \rangle + \{ \tau_2, u_1 \} = \langle f_1, u_2 \rangle + \{ \tau_1, u_2 \}. \tag{15.44a}$$

The proof of the *Clapeyron formula*[31] turns out to be just as elementary. It states that if

$$\tau = Lu \tag{15.45}$$

is the solution of equation (15.36), then, by (15.30) in combination with the relations (15.33b), (15.34), and (15.35), there is

$$(\tau, \tau) = \langle f, u \rangle + \{ \tau, u \}. \tag{15.46}$$

By virtue of (15.38), however, $(\tau, \tau) = 2W$, where W is the elastic strain energy stored in the body; similarly, in view of (15.37) and (15.41), the sharp brackets and the braces in the preceding equation denote the work of the body forces and the surface tractions, respectively. Expressed in words, therefore, equation (15.46) declares that the energy of deformation of a body in equilibrium under a given system of forces equals half the work done by these forces acting through the displacements of the load points.

In order to derive the energy theorems, it is necessary to assume that the distance—measured in terms of the elastic strain energy—between a varied state and the actual state is positive. In physical language, this hypothesis translates into the conventional assumption that the increase of strain energy from the actual value (τ, τ) to a varied value, $(\bar\tau, \bar\tau)$ say, is a positive-definite function. In symbols,

$$(\bar\tau - \tau, \bar\tau - \tau) \geq 0$$

or

$$(\tau, \tau) \geq 2(\tau, \bar\tau) - (\bar\tau, \bar\tau). \tag{15.47}$$

Now let not only the true state, but also the varied state be compatible with the constitutive equation (15.34):

$$\tau = Lu \quad \text{and} \quad \bar\tau = L\bar u. \tag{15.48}$$

We insert (15.48) into (15.47) and make use of relation (15.29). This yields

$$(\tau, \tau) \geq 2\langle f, \bar u \rangle + 2\{ \tau, \bar u \} - (L\bar u, L\bar u) \tag{15.48a}$$

or, by appeal to Clapeyron's theorem (15.46), alternatively,

$$(\tau, \tau) - 2\langle f, u \rangle - 2\{ \tau, u \} \leq (\bar\tau, \bar\tau) - 2\langle f, \bar u \rangle - 2\{ \tau, \bar u \}. \tag{15.48b}$$

Since $(\tau, \tau) = 2W$ and $(\bar{\tau}, \bar{\tau}) = 2\bar{W}$, either of the last two inequalities expresses the theorem of *minimum potential energy*.

In order to derive the theorem of *minimum complementary energy*, we assume that the varied state satisfies the equation of equilibrium (15.42). By (15.34), then,

$$L^*\bar{\tau} = f. \tag{15.49}$$

Taking advantage of the inequality (15.47) and relations (15.30) and (15.34) gives easily

$$(\tau, \tau) - 2\langle f, u \rangle \leq 2\{\bar{\tau}, u\} - (\bar{\tau}, \tau) \tag{15.50}$$

or, finally, with the help of Clapeyron's theorem,

$$(\tau, \tau) - 2\{\tau, u\} \leq (\bar{\tau}, \bar{\tau}) - 2\{\bar{\tau}, u\}. \tag{15.50a}$$

This completes the proof because each side of the preceding inequality represents the complementary energy of the body in the appropriate state.

As observed by Laws, a trick to rid oneself of the unwelcome terms $\{\tau, \bar{u}\}$ and $\{\bar{\tau}, u\}$ in the inequalities (15.48b) and (15.50), respectively, consists of a special choice of the trial functions, that is, $\bar{u}$ and $\bar{\tau}$. In the first boundary value problem of elasticity, for instance, in which the displacements on the boundary of the body are prescribed, it is sufficient to select the trial functions so that the displacements $\bar{u}$ obey these conditions. We then find that $\{\tau, u\} = \{\tau, \bar{u}\}$ and that, consequently, the unwanted terms in (15.48b) cancel. In order to arrive at a similar cancellation in the inequality (15.50a), we require that $\{\bar{\tau}, u\} = \{\tau, u\}$ for all $\bar{\tau}$.

The discussion in the case of the second boundary value problem of elasticity proceeds in a similar fashion.

A combination of the inequalities (15.48b) and (15.50) finally gives upper and lower bounds for the expression $(\tau, \tau) - 2\langle f, u \rangle$:

$$2\{\bar{\tau}, \bar{u}\} - (\bar{\tau}, \bar{\tau}) \leq (\tau, \tau) - 2\langle f, u \rangle \leq (\bar{\tau}, \bar{\tau}) - 2\langle f, \bar{u} \rangle, \tag{15.51}$$

where we have used another trial function, $\bar{u}$, which satisfies the displacement boundary conditions, so that $\{\bar{\tau}, u\} = \{\tau, \bar{u}\}$.

15.4. A Plane Tripod

We wish to conclude this predominantly theoretical discussion with a simple, but at the same time delightful, illustration given by Prager[97] of the effectiveness of the geometric approach to variational principles. Figure 15.3a displays a plane tripod with elastic legs connecting the joint D with three fixed points A, B, and C. A force $\vec{F}$ is applied to the joint, generating

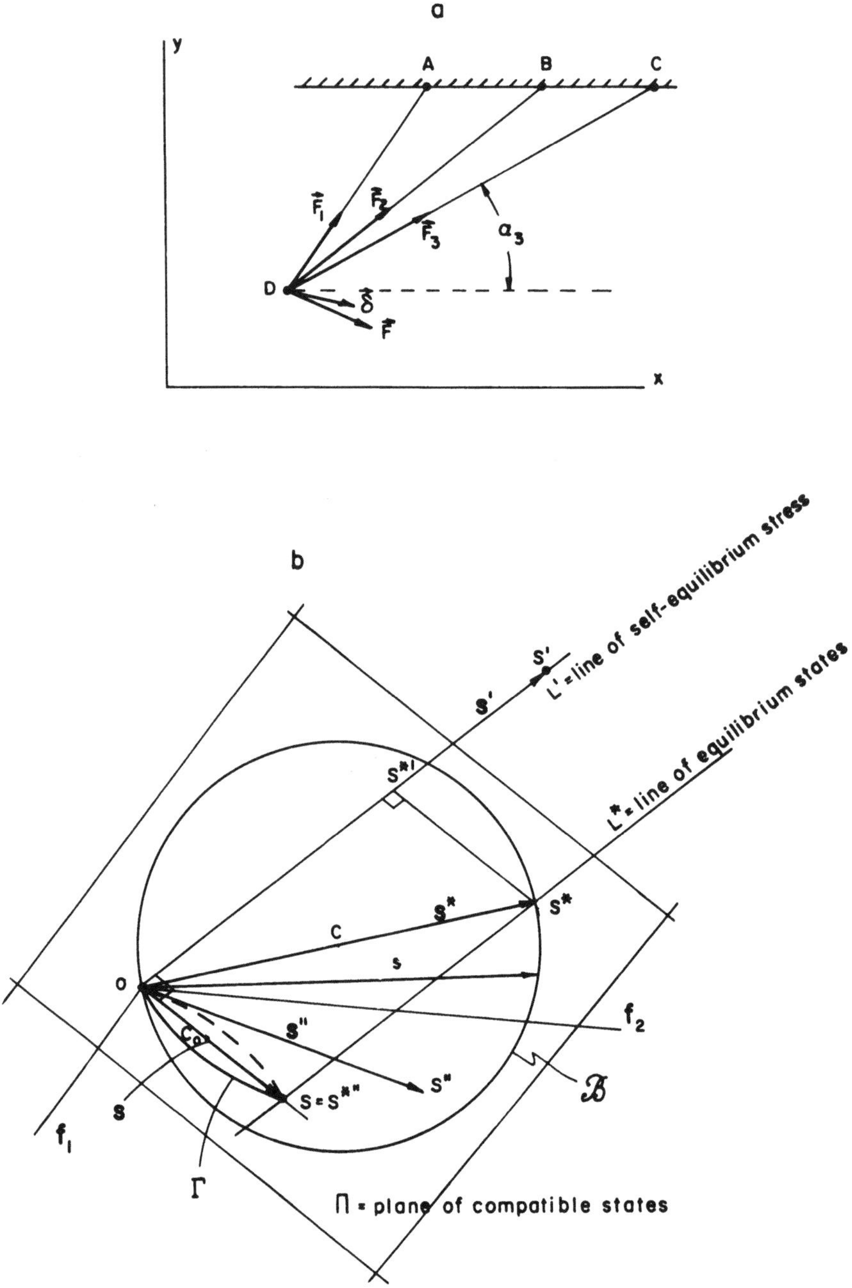

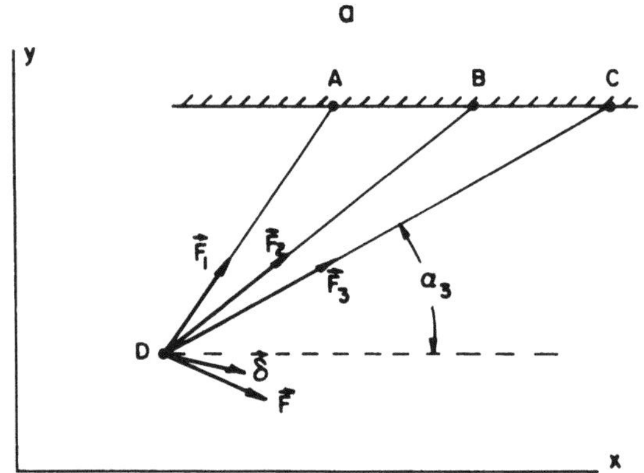

Figure 15.3. A plane tripod.

forces $\vec{F}_i$, $i = 1$, 2, 3, in the legs DA, DB, and DC, respectively. In order for the joint to remain in equilibrium, the resultant of the forces acting on the joint should vanish:[†]

$$\vec{F} + \sum_{i=1}^{3} \vec{F}_i = 0. \tag{15.52}$$

Assume that the elastic elongations of the bars under the action of the axial forces F_i are[‡]

$$\delta_i = c_i F_i, \qquad i = 1, 2, 3, \tag{15.53}$$

where the c_i denote the elastic compliances; the elongations are supposed to be small.

Let us now impose an arbitrary, but small, displacement $\vec{\delta}$ on the joint D and form the dot product of $\vec{\delta}$ with each member of equation (15.52). The elongations δ_i, $i = 1$, 2, 3, that the imposed displacement produces in the bars are equal to the respective orthogonal projections of $\vec{\delta}$ on the axes of the bars. Accordingly,

$$F \, \delta_F + \sum_{i=1}^{3} F_i \, \delta_i = 0, \tag{15.54}$$

where δ_F is the component of $\vec{\delta}$ along the line of action of the force $\vec{F}$. Since the displacement $\vec{\delta}$ is arbitrary (except that it is small and compatible with the geometric conditions), the equation above expresses the *principle of virtual work* applied to the given structure. It is important to realize that there is *no interdependence* here between the forces F_i and the displacements δ_i produced by the virtual displacement of the joint.

It is not difficult to convince oneself that the tripod shown in Figure 15.3a is a statically indeterminate structure. This means, among other things, that by a suitable operation (e.g., by turning a turnbuckle in one of the bars), it is possible to produce forces in the bars, even in the complete absence of an external load (i.e., for $\vec{F} = \vec{0}$). Assuming that the forces so generated satisfy the condition of equilibrium of the joint D, we turn the corresponding state of the structure a *state of self-equilibrated stress*.

A second state of interest is that in which the external force $\vec{F}$, in combination with a certain system of internal forces F_i, satisfies[*] the equation of equilibrium (15.52).[§]

We shall call this state a *state of equilibrium* $(\vec{F} \neq \vec{0})$.

[†] We find it convenient to denote vectors in physical space by an overhead arrow, function vectors by boldface type, and points in function space by lightface type.

[‡] Do not sum here or hereafter.

[§] For the statically indeterminate system under discussion, there exists an infinite number of such systems $\{F_i\}$. Only one of these is the system actually generated by the given force F.

Finally, it is convenient to consider a state in which, in the presence of the force $\vec{F}$, the joint D becomes shifted by an amount $\vec{\delta}$. In this case, we determine the associated elongations of the bars, δ_i, $i = 1, 2, 3$, and evaluate the forces in the bars from the formula $F_i = \delta_i/c_i$. The last-named forces aare not bound, in general, to be in equilibrium with the force $\vec{F}$ (unless they accidentally turn out to be the actual forces produced by the force $\vec{F}$). We shall call the state of stress just described a *compatible state of stress*.

It is now convenient to introduce the reduced forces

$$f_i = F_i(c_i/2)^{1/2}, \qquad i = 1, 2, 3, \tag{15.55}$$

and to treat these forces as coordinates relative to a Cartesian rectangular system (Figure 15.3b) in a function space of elastic states. In such a space, the components of a state vector **S** are $\mathbf{f}_1, \mathbf{f}_2, \mathbf{f}_3$, and the square of the distance of a point S (defined by the position vector **S**) from the space origin O is

$$\begin{aligned}
\sum_{i=1}^{3} f_i^2 &= \frac{1}{2} \sum_{i=1}^{3} c_i F_i^2 \\
&= \frac{1}{2} \sum_{i=1}^{3} F_i \delta_i,
\end{aligned} \tag{15.56}$$

the last expression following in view of the relation (15.53). This result implies that the square of the distance OS is equal to the *strain energy* associated with the state of stress represented by the point S.

In our new notation, a state of *self-equilibrated stress* is defined by the equations of equilibrium [see equation (15.52) and Figure 15.3a],

$$\begin{aligned}
f_1 \frac{\cos \alpha_1}{(c_1)^{1/2}} + f_2 \frac{\cos \alpha_2}{(c_2)^{1/2}} + f_3 \frac{\cos \alpha_3}{(c_3)^{1/2}} &= 0, \\
f_1 \frac{\sin \alpha_1}{(c_1)^{1/2}} + f_2 \frac{\sin \alpha_2}{(c_2)^{1/2}} + f_3 \frac{\sin \alpha_3}{(c_3)^{1/2}} &= 0,
\end{aligned} \tag{15.57}$$

each of which represents a plane passing through the origin of the coordinates $\mathbf{f}_1, \mathbf{f}_2, \mathbf{f}_3$. The line of intersection of these planes OS' includes all possible states of self-equilibrated stress, and will be called the *line of self-equilibrated stress*.

In those states called above the *states of equilibrium*, the left-hand members of equations (15.57) remain unchanged, but the zeros on the right-hand sides must be replaced by the components $-F_x/(2)^{1/2}$ and $-F_y/(2)^{1/2}$ of the force $\vec{F}$, respectively. The resulting system of equations determines a line parallel† to the line OS'; we shall call this line, which passes through a point S^*, say, a *line of equilibrium states* (Figure 15.3b).

---

† This obtains in view of the fact that the coefficients of like variables in the equations of the corresponding planes are proportional to each other (here, even equal).

With regard to the *compatible states of stress*, let S'' denote a point corresponding to such a state, and let S_h'' and S_v'' be the vectors of compatible states corresponding to unit horizontal and vertical displacements of the joint D, respectively. In this case, the general joint displacement for horizontal and vertical displacements δ_h and δ_v, respectively, is associated with the state

$$S'' = \delta_h S_h'' + \delta_v S_v''. \tag{15.58}$$

As δ_h and δ_v vary, the point S'' moves over a plane Π, passing through the origin and determined by the vectors S_h'' and S_v''. We shall call this plane the *plane of compatible states* (Figure 15.3b).

It is easy to demonstrate that the line of self-equilibrated stress (respectively, of equilibrium states) is perpendicular to the plane Π so that

$$S' \cdot S'' = 0, \tag{15.59}$$

where S' is a state of self-equilibrated stress and S'' is some compatible state. In fact,

$$S' \cdot S'' = \sum_{i=1}^{3} f_i' f_i''$$

$$= \sum_{i=1}^{3} F_i' \delta_i'', \tag{15.60}$$

the last expression produced by virtue of the relations (15.53) and (15.54). The right-hand member of the foregoing equation represents the work of the residual stress done on the displacements of a compatible stress. According to the principle of virtual work (15.54), this work vanishes because $\vec{F} = \vec{0}$ in this case; this proves our assertion.

Returning to Figure 15.3b, one must keep in mind that in the *actual* state of stress, the forces acting on the joint D are in equilibrium and the deformations of the bars produced by these forces are compatible. It follows that the *actual state of stress*[†] is represented by the point S of intersection of the line L^* and the plane Π. Since the segment OS also represents a compatible component of an arbitrary equilibrium state S^*, then the point S coincides with the orthogonal projection, $S^{*''}$, of S^* on Π. The self-equilibrated component, $S^{*'}$, of S^* is $OS^{*'}$, where $S^{*'}$ is the foot of the perpendicular from S^* to the line L'.

Inasmuch as the point S is that point on the line of equilibrium states which is closest to the origin, and since the distance in the state space is measured in terms of the elastic strain energy, our geometric analysis suggests the following:

[†] It is interesting to note that this fact, expressed here in geometric terms, proves that the problem under discussion has a unique solution.

Minimum Principle. The strain energy of a given structure remaining in equilibrium under the action of external load becomes minimum for the actual state of stress.

It is not difficult to recognize that the just-formulated principle is the well-known *principle of minimum strain energy*, commonly associated with the name of Castigliano.

A companion maximum principle was established by Prager in the following way. Imagine that we are given a point S^* corresponding to an equilibrium state, but that we do not know the locations of the line L^* and the plane Π. The only conclusion that is justified in this case is that the angle OSS^*, where S denotes the actual state of stress, is a right angle. If this is so, we are permitted to identify the point S with a point on a sphere, say $\mathscr{B}$, a diameter of which is OS^*. After Prager, we call the sphere a sphere of states *standardized* with respect to S^*. If $\mathbf{s}$ is a state vector in the state space, then the tip of this vector lies on the sphere of standardized states if

$$\mathbf{s} \cdot (\mathbf{S^*} - \mathbf{s}) = 0. \tag{15.61}$$

It is not difficult to convince oneself that the standardized compatible states are represented by the points on the circle Γ along which the plane of compatible states Π intersects the sphere $\mathscr{B}$. One of the diameters of this circle (the circle of standardized compatible states) is the segment OS. In terms of strain energy, this result can be stated as the following:

Maximum Principle. The strain energy of a given structure subject to a compatible deformation reaches its maximum for the actual state of stress.

We note that if $\mathbf{s}$ in equation (15.61) denotes a compatible state, then by the definition (15.55) of the coordinates of the state space,

$$\mathbf{S^*} \cdot \mathbf{s} = \sum_{i=1}^{3} S_i^* s_i = \sum_{i=1}^{3} F_i^* \left(\frac{c_i}{2}\right)^{1/2} F_i \left(\frac{c_i}{2}\right)^{1/2},$$

where F_i^* and F_i are the forces in the bars in a state of equilibrium due to a compatible displacement of the joint D. By virtue of equations (15.53) and (15.54), therefore,

$$\mathbf{S^*} \cdot \mathbf{s} = \frac{1}{2} \sum_{i=1}^{3} F_i^* \, \delta_i. \tag{15.62}$$

However, the product $\mathbf{s} \cdot \mathbf{s}$ represents the strain energy associated with the state $\mathbf{s}$; thus, rephrased in terms of energy, the equality $\mathbf{S^*} \cdot \mathbf{s} = \mathbf{s} \cdot \mathbf{s}$ states that the strain energy of a standardized compatible state is equal to half the work of the given load done on the displacement of the joint associated with a compatible state.

Problems

1. The presence of a derivative in the functional (15.10) may be occasionally undesirable. Show that this can be avoided by selecting in the problem considered the form $\langle u, v \rangle = \int_0^1 u(t)[\int_0^{1-t} v(\tau)\, d\tau]\, dt$ instead of (15.2). Is the latter form symmetric?

2. Show that the definition $\langle u, v \rangle = (u, Lv)$, where $(,)$ denotes an inner product, implies (15.5). Find $F[u]$ from (15.6) for the case considered in the preceding problem, selecting the inner product in the Hilbert form (8.2): $(u, v) = \int_\Omega u(P)v(P)\, d\Omega$.

3. Prove that two subspaces, $\mathscr{S}_1$ and $\mathscr{S}_2$, comprising vectors S_1 and S_2 of two equilibrium states of an arbitrary elastic body, respectively, are orthogonal to each other provided that the body force is absent and the surface tractions (surface displacements) in one state are associated with zero displacements (zero tractions) in the second state.

4. A unit orthotropic cube with elastic constants E_i, v_{ij}, $i = 1, 2, 3$ is subject to consecutive actions of oppositely directed tractions $\tau_{11}^{(1)}$ and $\tau_{22}^{(2)}$ applied to two pairs of parallel faces with normals n_1 and n_2, respectively. Find the condition for the orthogonality of the corresponding states.

5. Any number λ for which there exists a nonzero vector u satisfying $(L - \lambda)u = 0$, L being a linear operator,[†] is an *eigenvector* of L corresponding to the *eigenvalue* λ.[‡] Assume that L is symmetric, i.e., $(Lu, v) = (u, Lv)$, and denote Rayleigh's quotient by $\rho(u) \equiv (u, Lu)/(u, u)$. If $u = u_n$, an eigenvector of L, then $\rho(u_n) = \lambda_n$, the associated eigenvalue. Show that $\rho(u)$ has a stationary value whenever u is an eigenvector u_n.

6. The best approximation to a vector y by a linear combination of n vectors $\{x_i\}$ is the vector $\sum_{i=1}^n a_i x_i$ such that the (error) norm $\|y - \sum_{i=1}^n a_i x_i\|$ is minimized. Find the best approximations to the function $e^{\alpha t}$, $\alpha = \text{const}$, in the space of functions $C_{0 \leq t \leq 1}$ selecting as the norm: (1) the $p = \infty$-norm [equation (5.34)], i.e., $\|y\| = \max_{0 \leq t \leq 1} |y(t)|$; (2) $\|y\| = \{\int_0^1 [y(t)]^2\, dt\}^{1/2}$, taking $n = 1$ and $x_1 = c$, a constant.

7. The deflection $u(x)$ of an elastic string, acted upon by a transverse continuous load $p(x)$ and subject to tension S, is governed by the equation $-Su'' = p(x)$, $0 < x < l$, (1). The string has length l and is fixed at the ends: $u(0) = u(l) = 0$, (2). Show that the operator L, $Lu = -Su''$ on the domain D of all functions $u(x)$ in $C^2(0, 1)$ satisfying conditions (2) is symmetric. Show that if u is a solution of (1) and (2), then u is also a solution of the problem $\Pi(u) = \min$, where

$$\Pi(u) = \frac{S}{2} \int_0^l (u')^2\, dx - \int_0^l pu\, dx$$

is the total potential energy of deformation.

---

† Compare the text concerning equation (8.16).
‡ Compare the text concerning equation (8.52).

8. Let $F[u] = b(u, u) - 2l(u)$ be a functional defined on a subspace S of a Hilbert space, where $b(u, v)$ is a symmetric bilinear form $[b(u, v) = b(v, u), b(\alpha u_1 + \beta u_2, v) = \alpha b(u_1, v) + \beta b(u_2, v)]$ and $l(u)$ is a linear functional $[l(\alpha u_1 + u_2) = \alpha l(u_1) + l(u_2)]$. Show that: (a) if $F[v]$ is stationary for $u \in S$, then $b(u, f) = l(f)$ for each $f \in S$; (b) if u is such a function, then $F[u] = -b(u, u) = -l(u)$.

9. During the last decade, there has been interest in studying boundary value problems in which *boundary conditions* are expressed by *inequalities*. This happens, for example, in the case of *unilateral constraints* (as when a beam, after its deflections reach certain values, comes in contact with an obstacle), or when friction is present on the boundary of a body[24, 98]. Now let $\{v(x)\}$ be a set S of sufficiently smooth functions defined on the interval $[0, 1]$ and satisfying the condition $v(1) \geq 0$. Show that S is a convex set, and that if u satisfies the inequality $b(u, v - u) - l(v - u) \geq 0$, (1), for all $v \in S$, then u also makes $F[v] = b(v, v) - 2l(v)$ minimum (b and l are as in Problem 8).

10. Show that if a bilinear form $b(v, w)$, with v and w in a Hilbert space, has the properties of an inner product, then $[b(v, w)]^2/b(w, w) \leq b(v, v) \leq b(w, w)$ provided that $b(w, w) \neq 0$ in the case of the left-hand bound and $b(w - v, v) = 0$ in the case of the right-hand bound.

16

Distributions. Sobolev Spaces

We conclude this book with a glimpse at the means available for treating some unorthodox situations encountered in various areas of physics and engineering, in particular, in dynamics, electromagnetism, quantum field theory, and optimization. We here have in mind problems in which such classical requirements as the regularity and differentiability of functions, convergence of series, and smoothness of boundary conditions cannot be maintained, but must be relaxed, either in part or completely. This happens most frequently when we are concerned with temporal impulses (i.e., high intensity actions of very short duration), localized (concentrated) forces and couples acting on structural systems, point sources of various kinds, point masses, point charges, and finally, supports along lines or at isolated points.

16.1. Distributions

To analyze such singular, or outright "improper" situations, it has been found imperative to extend the conventional concept of an "ordinary" function to the new, partially revolutionary, concept of *generalized* function. This extension was accomplished with the construction of the theory of *distributions* (or *generalized functions*, as suggested in a somewhat different form by Sobolev[99]) developed comprehensively by Schwartz.[100 103] Including "improper" functions along with "ordinary" functions, the class of generalized functions is enriched beyond the family of point functions in a manner similar to that in which the transcendental functions enlarge the class of algebraic functions, or the infinitely distant points enlarge the class of ordinary points.† As is well known, the first "improper" ("symbolic," "patholo-

---

† It should be emphasized that the extension of the classical ideas was prompted less by a purely academic interest than by the desire to solve practical problems.

gic") function was that introduced by Dirac (in 1927) as the *delta function* $\delta(x)$, considered nowadays to be a type of limit of any one of a family of "delta sequences," such as

$$\delta_k(x) = \begin{cases} \dfrac{1}{k} & \text{for } |x| < k, \\[2mm] 0 & \text{for } |x| > k, \end{cases} \tag{16.1}$$

for $k > 0$. It is easily seen that, for each k,

$$\int_{-\infty}^{\infty} \delta_k(x)\, dx = 1. \tag{16.2}$$

Clearly, for any function $\phi(x)$ that is continuous at $x = 0$, we have

$$\lim_{k \to 0} \int_{-\infty}^{\infty} \delta_k(x)\phi(x)\, dx = \lim_{k \to 0} \frac{1}{2k} \int_{-k}^{k} \phi(x)\, dx$$

$$= \lim_{k \to 0} \frac{1}{2k} \phi(\xi_k) 2k = \phi(0), \tag{16.3}$$

where $-k \leq \xi_k \leq k$, the next-to-last step following from the first mean value theorem of the integral calculus (involving definite integrals of continuous functions). It is, therefore, tempting (but, from the classical viewpoint, incorrect) to conclude that, if we define the point function δ by $\delta(x) = \lim_{k \to 0+} \delta_k(x)$, we obtain a function with the curious properties

$$\delta(x) = 0, \quad x \neq 0, \qquad \int_{-\infty}^{\infty} \delta(x)\, dx = 1,$$

$$\tag{16.4}$$

$$\int_{-\infty}^{\infty} \delta(x)\phi(x)\, dx = \phi(0)$$

for any continuous function $\phi(x)$. It is customary to call the latter a *test† function* and consider it instead to be a member of the relatively narrow class of test functions belonging to the space C_0^{∞}. This space consists of all *infinitely differentiable functions* (1) defined for $-\infty < x < \infty$ (or, more generally, $\mathscr{R}_n$) and (2) vanishing outside of some finite interval (or a finite

---

† Some authors call the functions of the class C_0^{∞} "finite" functions; see, e.g., Ref. 26. The notation C_0^{∞} (where ∞ stands for "infinite differentiability" and 0 for "vanishing outside some finite interval") is often replaced by K.

portion Ω of $\mathscr{R}_n$).† The most decisive of the equations (16.4), namely, the last, is now accepted as the *definition* of the improper function $\delta(x)$. A similar device is employed to define *other symbolic* functions so that, figuratively speaking, these functions are determined by the values they produce [here $\phi(0)$] from the test functions ϕ. Since the just-mentioned equation of the group (16.4) assigns a number $(\phi(0))$ to a function (ϕ), it constitutes a *functional*,‡ say, f. The result of the action of f on ϕ is often written $f\phi$ rather than $f(\phi)$, but we shall denote this number more conveniently by $\langle f, \phi \rangle$. As we shall see shortly, it is often possible (and, for the most part, convenient) to represent the functional $\langle f, \phi \rangle$ in the form

$$\langle f, \phi \rangle = \int_{-\infty}^{\infty} f(x)\phi(x)\, dx, \tag{16.4a}$$

similar to that appearing in (16.4).

What emerges here is the idea of extending the notion of a function via its identification with a *functional* associated with (or generated by) the function. Such an approach is not new, but has long been utilized in the theory of the well-known Fourier series. In the latter, the classical definition of a (real) function, f, as a rule associating a unique real number $y = f(x)$ with each element x of a set of real numbers, is actually replaced by a characterization of f by means of a set of Fourier coefficients, the latter constituting nothing else but functionals (integrals) involving certain auxiliary functions. And so, for example, a cosine expansion of a function f defined in $[0, \pi]$ includes coefficients in the form

$$b_n = \frac{2}{\pi} \int_0^{\pi} f(x)\cos nx\, dx, \tag{16.4b}$$

in which the presence of the auxiliary functions $\{\cos nx\}$ is clearly visible.

The analogy between the formulas (16.4a) and (16.4b) is thus apparent, with the auxiliary functions playing the role of test functions.

With regard to the space of test functions C_0^{∞}, the latter is clearly a linear space, with the usual pointwise definitions of addition and scalar

---

† Of greatest interest here is the so-called *support* of a function on $\mathscr{R}_n$. Precisely, the support of a function $f(x)$ in Ω is the *closure* of the set of elements x in Ω such that $f(x) \neq 0$. For the definition of "closure" of a set, as the set in addition to all its accumulation points, compare the text preceding equations (6.3b). The reader is reminded that the symbol $\mathscr{R}$ denotes the set of real numbers and the symbol $\mathscr{R}_n$ denotes the set of ordered n-tuples of real numbers [cf. the text following equation (5.9)]. We note that the space of test functions can be selected in various ways (e.g., as a space of continuous functions), but it turns out most useful to impose on this space the severe smoothness requirements accepted above.

‡ See the remarks about functionals and their linearity in the second footnote following equation (8.20).

multiplication. The introduction of the standard topology on this space is slightly complicated, requiring acquaintance with concepts whose clarification lies beyond the scope of this book. Thus, rather than elaborate upon the topological structure postulated for C_0^∞ in more complete treatments (e.g., Ref. 10), we merely note that there is no norm compatible with this structure; the space C_0^∞ with this standard topology does, however, admit a rather simple characterization, of the concept of *convergence*, namely, a sequence of functions $\phi_1, \ldots, \phi_k, \ldots$ in $C_0^\infty(\mathscr{R}_n)$ *converges* to a function ϕ in $C_0^\infty(\mathscr{R}_n)$ if and only if†

(a) there exists a common bounded region Ω in $\mathscr{R}_n$ outside of which all the functions ϕ_k vanish (this is, of course, equivalent to saying that the supports of all these functions are included within a sufficiently large ball), and

(b) the sequence of partial derivatives

$$\{D_n^r \phi_k\} \equiv \left| \frac{\partial^{|r|} \phi_k}{\partial x_1^{r_1} \cdots \partial x_n^{r_n}} \right|, \qquad |r| = \sum_{i=1}^n r_i, \tag{16.5}$$

converges uniformly in Ω to the partial derivative

$$D_n^r \phi \equiv \frac{\partial^{|r|} \phi}{\partial x_1^{r_1} \cdots \partial x_n^{r_n}} \tag{16.5a}$$

for any multi-index $r = (r_1, \ldots, r_n)$ of "length" $|r| \geq 0$. Returning now to the idea of a functional, it is said that a *linear functional* f on C_0^∞ is *continuous* if, whenever a sequence of test functions $\{\phi_k\}$ converges to ϕ in C_0^∞, i.e., satisfies the just-listed conditions (a) and (b), then the numerical sequences $\langle f, \phi_k \rangle$ tend to $\langle f, \phi \rangle$.

Following the nomenclature suggested by Schwartz, each *continuous linear functional* on the space $C_0^\infty(\mathscr{R}_n)$ is *said* to be a *distribution* (or, more exactly, an *n*-dimensional distribution).

Distributions can be "generated" by functions of various degrees of smoothness; this construction can be effected (Stakgold,[24] p. 92) even for those of such a *broad* class as the *locally integrable* functions in $\mathscr{R}_n$. For such a function $f(x)$, the integral $\int_\Omega |f(x)|\, dx$ exists for every bounded domain Ω in $\mathscr{R}_n$‡ and the associated "*n*-dimensional" distribution takes the form

$$\langle f, \phi \rangle = \int_{\mathscr{R}_n} f(x)\phi(x)\, dx, \tag{16.6}$$

where $x = (x_1, x_2, \ldots, x_n)$ and $\int_{\mathscr{R}_n}$ denotes the *n*-tuple integral $\int_{-\infty}^\infty \cdots \int_{-\infty}^\infty$.

---

† Many authors give a slightly different formulation of the condition (b) by introducing the notion of the so-called null sequences (tending to zero). See, e.g., Stakgold (Ref. 24, p. 90).

‡ More generally, a function is *locally integrable* in a domain if it is integrable over every closed and bounded portion of the given domain (here $\mathscr{R}_n$). The condition of local integrability is less stringent than, e.g., either continuity or piecewise continuity.

A distribution which can be represented in the form (16.6), via some locally integrable function $f(x)$, is called *regular*.[†] All other distributions are termed *singular*, but are often symbolically represented in the same form (and we shall follow this custom) by considering $f(x)$ as some sort of *generalized function*. In this case, of course, there is no function f on $\mathscr{R}_n$—in the classical sense—which, when multiplied by ϕ and integrated over $\mathscr{R}_n$, produces $\langle f, \phi \rangle$. Rather, the right-hand side of (16.6) is then to be regarded as merely a collection of symbols, another expression for $\langle f, \phi \rangle$, and not as an integral. We have already encountered an example of this usage in the last of equations (16.4), which can now be given the more general form

$$\langle \delta, \phi \rangle = \int_{R_n} \delta(x)\phi(x)\, dx = \phi(0), \tag{16.6a}$$

where $\phi(x) \in C_0^\infty$. It can be shown that there is no point function δ on $\mathscr{R}_n$ which can give such a result upon "acting" on each test function ϕ. It is worth emphasizing that, in practice, the identification of a distribution-generating function with the associated distribution itself is so complete that it is quite normal to speak of distributions as if they were not merely functionals, but (whether ordinary or improper) functions themselves. This is, in particular, true of the Dirac function, by which is usually meant the δ-function itself and the functional (16.6a). Likewise, for the moment assuming that ordinary manipulations are valid for the delta function, we can write formally for its "derivative" δ' and $\phi \in C_0^\infty$,

$$\langle \delta', \phi \rangle = \int_{-\infty}^{\infty} \delta'(x)\phi(x)\, dx = \delta(x)\phi(x)\Big|_{-\infty}^{\infty} - \int_{-\infty}^{\infty} \delta(x)\phi'(x)\, dx = -\phi'(0), \tag{16.7}$$

where we have considered that, by definition, the test functions vanish at infinity.

At this stage, we also remark that, while the *ordinary* derivative of the Heaviside unit step function,

$$H(x) = \begin{cases} 0 & \text{for } x < 0, \\ 1 & \text{for } x > 0, \end{cases} \tag{16.8}$$

is zero for both $x < 0$ and $x > 0$, we should have, for its "distributional derivative,"

$$\langle H', \phi \rangle = \int_{-\infty}^{\infty} H'(x)\phi(x)\, dx = -\int_{-\infty}^{\infty} H(x)\phi'(x)\, dx$$

$$= -\int_{0}^{\infty} \phi'(x)\, dx = \phi(0). \tag{16.9}$$

[†] For regular distributions alone, strictly speaking, the sign $\langle , \rangle$ becomes identical to that of the inner product $(,)$ of the Hilbert type [cf. equation (8.2)].

Comparison of (16.9) with the last of equations (16.4) implies that $H'(x) = \delta(x)$.† These (of course, purely formal) properties of the delta function and its "derivative" of generating linear functionals prompted Schwartz to replace the questionable "improper" functions by certain well-defined linear functionals.

It should be clear that the relation (16.7) actually suggests that the "derivative" of δ should be the distribution δ' given by $\langle \delta', \phi \rangle = -\phi'(0)$ for $\phi \in C_0^\infty$. In (16.13), we shall set down a definition of (partial) derivative of a distribution which is consistent with this observation (this is not surprising—the basis for the definition will be nothing more than the formula for integration by parts!).

Now let h denote an infinitely differentiable function and $\phi \in C_0^\infty$; then also $h\phi \in C_0^\infty$. This observation allows us to *define* multiplication of a distribution f by an infinitely differentiable function h: we set (by analogy to the familiar $\int_{\mathcal{R}_n} h(x)f(x)\phi(x)\, dx = \int_{\mathcal{R}_n} f(x)h(x)\phi(x)\, dx$ for three well-behaved functions)

$$\langle hf, \phi \rangle = \langle f, h\phi \rangle.$$

As an illustration, take $f = \delta$ and h to be given simply by $h(x) = x$. We claim that then $h\delta = 0$, that is,

$$\langle h\delta, \phi \rangle = 0,$$

0 also denoting here the null distribution, $\langle 0, \phi \rangle = 0$ for each $\phi \in C_0^\infty$. Indeed,

$$\langle h\delta, \phi \rangle = \langle \delta, h\phi \rangle = h(0)\phi(0) = 0 \cdot \phi(0) = 0 \qquad (16.10)$$

for each $\phi \in C_0^\infty$, proving our claim.‡

The Dirac function is not the only improper function. Examples of others are dipoles, quadrupoles, and similar derivatives of the δ-function or combinations of the latter with its derivatives.[104]

If $f(x)$ is a locally integrable function on a domain Ω, and D^r denotes a partial differentiation of order r with respect to certain of the n variables, we can introduce a distribution f by the equation

$$\langle f, \phi \rangle = \int_{\mathcal{R}_n} f(x)D^r\phi\, dx, \qquad \phi \in C_0^\infty. \qquad (16.11)$$

---

† The two distinct derivatives coincide in a sense if, in the ordinary calculus, one accepts the traditionally rejected idea that at the point $x = 0$ of the discontinuity of $H(x)$, a derivative exists and is infinite. A "graph" of $\delta'(t)$ displays two spikes along the axis of ordinates pointing in opposite directions. Compare, e.g., Lighthill (Ref. 103, p. 12).

‡ Note that two distributions are *equal* if they produce the same action on each test function.

As an example, for $n > 1$, we can take

$$\langle f, \phi \rangle = \int_{\mathscr{R}_n} \frac{1}{\|x - x_0\|} D_n{}^r \phi(x)\, dx \tag{16.12}$$

for some x_0 in $\mathscr{R}_n$. Below we use the equation (16.11) to define a central object in (16.17). It is quite natural to consider *differentiation of distributions.* To this end, letting $\partial_j f$ denote the partial derivative with respect to the variable x_j of a distribution f on $\mathscr{R}_n$, we define

$$\langle \partial_j f, \phi \rangle \equiv \int_{\mathscr{R}_n} \frac{\partial f}{\partial x_j} \phi(x)\, dx$$

$$= -\langle f, \partial_j \phi \rangle \tag{16.13}$$

for each test function ϕ. For a derivative of *order r*, we shall then have, after repeated applications of (16.13),

$$\langle \partial^r f, \phi \rangle = (-1)^{|r|} \langle f, \partial^r \phi \rangle, \tag{16.14}$$

where ∂^r denotes the derivative $\partial_1^{r_1}, \partial_2^{r_2}, \ldots, \partial_n^{r_n}$ with respect to the n variables, and the multi-index $r = (r_1, r_2, \ldots, r_n)$ has "length"

$$|r| = r_1 + r_2 + \cdots + r_n. \tag{16.15}$$

Suppose now that the distributions appearing in (16.14) are induced by locally integrable functions u and v, respectively. Then (16.14) becomes, explicitly, upon constructing each member of (16.14),

$$\int_{\mathscr{R}_n} u\phi\, dx = (-1)^{|r|} \int_{\mathscr{R}_n} v \frac{\partial^r \phi}{\partial x_1^{r_1} \cdots \partial x_n^{r_n}}\, dx. \tag{16.16}$$

In this case, many authors (e.g., Refs. 8 and 99) say that u is the *r*th *generalized derivative* of v, and denote u by the collection of symbols $\partial^r v / \partial x_1^{r_1} \cdots \partial x_n^{r_n}$, i.e., by using the same notation as is used in the case in which v has an *r*th derivative in the usual (classical) sense. The preceding equality is then written

$$\int_{\mathscr{R}_n} v \frac{\partial^r \phi}{\partial x_1^{r_1} \cdots \partial x_n^{r_n}}\, dx = (-1)^{|r|} \int_{\mathscr{R}_n} \frac{\partial^r v}{\partial x_1^{r_1} \cdots \partial x_n^{r_n}} \phi\, dx, \tag{16.17}$$

which is formally the rule for integration by parts (since $\phi \in C_0^\infty$); if v has continuous partial derivatives of all orders $\leq |r|$, then (16.17) follows in the classical manner, of course.

As another example[22, 24, 105], consider the discontinuous sawtooth-like function $f(x)$ on $\mathscr{R}$ of period 2π [i.e., $f(x + 2\pi) = f(x)$] and given in the interval $[-\pi, \pi]$ by the equations (Figure 16.1a)

$$f(x) = \begin{cases} \dfrac{\pi - x}{2} & \text{for} \quad 0 < x \le \pi, \\[2mm] 0 & \text{for} \qquad x = 0, \\[2mm] -\dfrac{\pi + x}{2} & \text{for} \;\; -\pi \le x < 0. \end{cases} \qquad (16.18)$$

Clearly, the function is continuous except at the points $x_k = 2k\pi$, $k = 0, \pm 1$, $\pm 2, \ldots$, where it suffers jumps of magnitude π. The function satisfies the Dirichlet conditions (Ref. 27, p. 181) and has the Fourier series

$$f(x) = \sum_{k=1}^{\infty} \frac{\sin kx}{k}. \qquad (16.19)$$

As seen from (16.18), $f(x)$ is differentiable except at the points $x = 2k\pi$, where, however, the left- and right-hand derivatives exist.

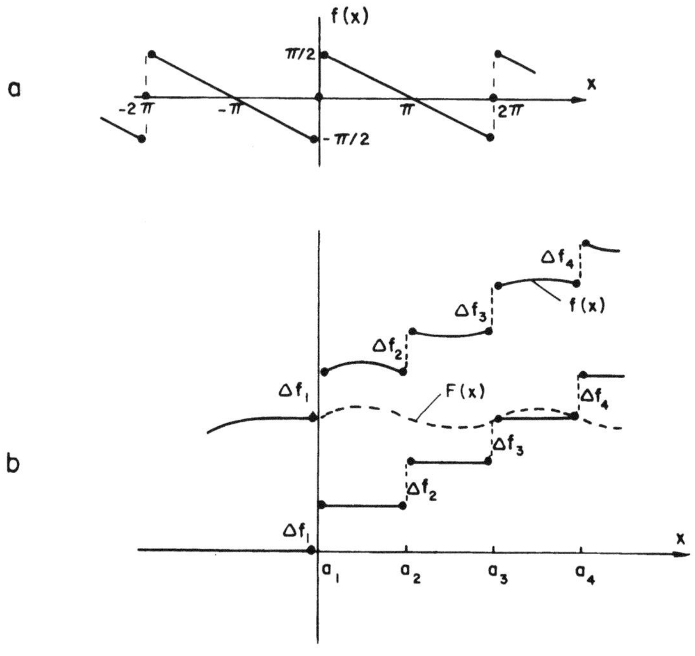

Figure 16.1. Illustration for equations (16.18) and (16.20).

Before we proceed further, let us note that, while conventionally the Heaviside function (16.8) is not differentiable at the point $x = 0$, from the standpoint of the distributional calculus, its derivative is found to be equal to the delta function [cf. equations (16.4) and (16.9)]. This result is easily extended to functions with *multiple simple jumps* (Figure 16.1b).

Then let $f(x)$ be a function infinitely differentiable for all values of x except $a_1, a_2, \ldots, a_n$, where it has left- and right-hand derivatives, but suffers jumps of amounts $\Delta f_1, \ldots, \Delta f_n$, respectively. Denote by $\bar{f}'$ the purely formal derivative of f obtained by disregarding the jumps; thus, for instance, $\bar{H}' = 0$. Clearly, $\bar{f}'$ is a piecewise continuous function. Next, construct the function (Figure 16.1b)

$$F(x) = f(x) - \sum_{k=1}^{n} \Delta f_k H(x - a_k), \tag{16.20}$$

evidently continuous and having a piecewise continuous derivative, F', in the ordinary sense. In view of this, F' represents the distributional derivative of F,[†] *as well.* Thus, differentiating (16.20) in the distributional sense, we have

$$F' = f' - \sum_{k=1}^{n} \Delta f_k \delta(x - a_k). \tag{16.20a}$$

But, as illustrated by Figure 16.1b, the derivative of the continuous function F is indistinguishable from the purely formal derivative, $\bar{f}'$, of the discontinuous function f. Therefore, considering that $F' \equiv \bar{f}'$, one obtains the distributional equation

$$f' = \bar{f}' + \sum_{k=1}^{n} \Delta f_k \delta(x - a_k). \tag{16.21}$$

For $f = H(x)$, $\bar{f}' = \bar{H}' = 0$, $n = 1$, and $a_1 = 0$, this yields the already-noted relation $H'(x) = \delta(x)$. In words, equation (16.21) states that in order to obtain the distributional derivative of a function with simple jumps, one must augment the purely formal derivative of this function by the contributions of the jumps of amount $\Delta f_k \delta(x - a_k)$ for the jump at a_k, $k = 1, 2, \ldots, n$.

Now, returning to the main problem under scrutiny [equation (16.18)], observe that in the present case we must take $a_k = 2k\pi$, $\Delta f_k = \pi$, and $\bar{f}' = -\frac{1}{2}$. With these in mind, equation (16.21) becomes

$$f'(x) = -\tfrac{1}{2} + \pi \sum_{k=-\infty}^{\infty} \delta(x - 2k\pi). \tag{16.21a}$$

With regard to the Fourier representation (16.19), we note that, while the

---

† Compare the remark following equation (16.29), infra.

series does not converge uniformly to $f(x)$, it does converge in the distributional sense. In fact, by integrating the series (16.19), we arrive at the series $-\sum_{k=1}^{\infty} (\cos kx/k^2)$, converging uniformly (and, therefore, distributionally†) to a function whose derivative is $f(x)$. This being so, it is legitimate to differentiate the integrated series term-by-term in order to regain (16.19). A repeated differentiation in the distributional sense now gives

$$f'(x) = \sum_{k=1}^{\infty} \cos kx, \tag{16.22}$$

so that, comparing (16.22) with (16.21a), we arrive at the classically awkward result

$$\sum_{k=-\infty}^{\infty} \delta(x - 2k\pi) = \frac{1}{2\pi} + \frac{1}{\pi} \sum_{k=1}^{\infty} \cos kx. \tag{16.23}$$

From a conventional viewpoint, the left-hand member of the preceding equation is absurd, while the right-hand member stands for a series that does not converge. In the distributional sense, however, equation (16.23) is meaningful, insofar as it expresses the fact that the operation of both members on a test function $\phi(x) \in C_0^{\infty}$ is the same. Indeed, multiplying by $\phi(x)$ and integrating from $-\infty$ to ∞, we find‡ that, for every test function $\phi(x)$,

$$\sum_{k=-\infty}^{\infty} \phi(2k\pi) = \frac{1}{2\pi} \int_{-\infty}^{\infty} \phi(x)\, dx + \frac{1}{\pi} \sum_{k=1}^{\infty} \int_{-\infty}^{\infty} \cos(kx)\phi(x)\, dx. \tag{16.24}$$

In applications, one often meets linear partial differential equations with constant coefficients. Denoting the corresponding operators concisely by

$$P(\partial) \equiv \sum_{\tau} \alpha_{\tau} \partial^{\tau}, \tag{16.25}$$

we find, for two functions $\phi, \psi \in C_0^{\infty}$, after using Green's theorem,

$$\int_{\mathscr{R}_n} \psi P\phi\, dx = \int_{\mathscr{R}_n} \phi P^*\psi\, dx, \tag{16.26}$$

where

$$P^*(\partial) = \sum_{\tau} (-1)^{|\tau|}\alpha_{\tau} \partial^{\tau} \tag{16.26a}$$

† Note our earlier definition of convergence.
‡ Compare additional remarks in Stakgold (Ref. 24, p. 139).

is the adjoint (Ref. 72, p. 8) of $P(\partial)$. This suggests introduction of the following distribution, Pf, corresponding to the distribution f:

$$\langle Pf, \phi \rangle = \int_{\mathscr{R}_n} Pf\phi \, dx$$

$$= \int_{\mathscr{R}_n} fP^*\phi \, dx = \langle f, P^*\phi \rangle. \tag{16.27}$$

In particular, for the (self-adjoint) Laplace operator,

$$\nabla^2 \equiv \sum_{i=1}^{n} \frac{\partial^2}{\partial x_i^2}, \tag{16.28}$$

we have $P^* = P$, so that the distribution $\nabla^2 f$ is determined by the equation

$$\langle \nabla^2 f, \phi \rangle = \int_{\mathscr{R}_n} \nabla^2 f \phi \, dx$$

$$= \int_{\mathscr{R}_n} f \nabla^2 \phi \, dx = \langle f, \nabla^2 \phi \rangle. \tag{16.29}$$

Elaborating upon our earlier example, it is not difficult to show that if a function f is continuously or piecewise continuously differentiable, then its distributional derivative coincides with that taken in the ordinary sense. On the other hand, existence of a distributional derivative does not imply existence of a derivative in the ordinary sense (Ref. 99, sect. 5). Naturally, ordinary and generalized derivatives of a function (if they exist) may appear as quite different objects. As an illustration, we need only refer to the ordinary and distributional derivatives of the Heaviside function examined above [cf. equation (16.9)].

It is of interest to note a rather classically unexpected fact that existence of distributional derivatives of some order $|\tau| > 1$ does not automatically imply existence of distributional derivatives of order less than $|\tau|$.

16.2. Sobolev Spaces

As an example of application of the calculus of distributions, consider a point source of heat of constant intensity $Q = q\rho c$, equal, say, to k, for simplicity, where k is the conductivity of a medium considered to be infinite in extent, q is the strength of the source, and ρ and c are the mass density and specific heat of the medium, respectively. The equation governing the temperature field $\theta = \theta(r)$ outside the source is[†]

$$\nabla^2\theta \equiv \frac{1}{r^2}\frac{d}{dr}\left(r^2\frac{d\theta}{dr}\right) = 0, \qquad r = \|x\|. \tag{16.30}$$

Considering the presence of the singularity at $r = 0$, we select as a function space the *Sobolev space* $W^{2,1}$, including all functions which, together with their first and second generalized partial derivatives, are integrable over the entire space $\mathscr{R}_3$. The space is discussed briefly below.

It is immediately seen that for $r \neq 0$, $\theta = 1/r$ is a solution of equation (16.30), and one can introduce the distribution induced by $1/r$.

That is, the function given by $1/r$, locally integrable,

$$\int_K \frac{1}{r}\, dx,$$

exists for any closed and bounded set K in $\mathscr{R}_3$. Consequently, the integral

$$\langle r^{-1}, \phi \rangle = \int_{\mathscr{R}_3} \frac{1}{r}\phi\, dx$$

exists for any (test) function $\phi \in C_0^\infty(\mathscr{R}_3)$. In this manner, we see that the function r^{-1} can be considered as a continuous linear functional on $C_0^\infty(\mathscr{R}_3)$, i.e., as a distribution.

We thus have [cf. the symbolism in (16.29)]

$$\langle \nabla^2 r^{-1}, \phi \rangle = \int_{\mathscr{R}_3} r^{-1}\nabla^2\phi\, dx, \qquad \phi \in C_0^\infty(\mathscr{R}_3). \tag{16.31}$$

Applying the third Green's identity to the region $\|x\| > \varepsilon > 0$, and bearing in mind the properties of a test function, we transform the preceding integral into

$$-\int_{S_\varepsilon} \left[\frac{1}{r}\frac{d\phi}{dr} + \phi\frac{1}{r^2}\right] dS,$$

where S_ε denotes the surface of the ball of radius ε, centered at the origin. Note that $dS = \varepsilon^2 d\omega$, where $d\omega$ is the element of surface area of the unit sphere. In the limit, as $\varepsilon \to 0$, we find[‡]

$$\langle \nabla^2 r^{-1}, \phi(r) \rangle = -4\pi\phi(0), \tag{16.32}$$

[†] See, e.g., Nowinski (Ref. 83, equation (9.66)]. It is assumed that the source is located at the origin of a spherical coordinate system r, α, β.
[‡] A more general discussion can be found in Sneddon.[10]

so[†]

$$\nabla^2\left(\frac{1}{4\pi r}\right) = -\delta(r).\tag{16.33}$$

This gives the temperature field in the form $\theta(r) = \frac{1}{4}\pi r$.[‡]

In applications of the distributional calculus to differential equations of mathematical physics and engineering, it is often appropriate to work within one of the Banach spaces (that is, normed complete linear spaces, as noted at the beginning of this chapter) known as Sobolev spaces[§] and denoted by $W^{k,\,p}(\Omega)$. $W^{k,\,p}(\Omega)$ is the space of real-valued functions $\phi \in \mathscr{L}_p(\Omega)$ whose generalized derivatives of orders $\le k$ exist and belong to $\mathscr{L}_p(\Omega)$.[¶] Functions equal almost everywhere (that is, everywhere except perhaps on a set of zero "measure" or, imprecisely, zero volume) are considered to be identical. In the space $W^{k,\,p}(\Omega)$, the norm is defined by

$$\|\phi\|_{W^{k,\,p}(\Omega)} = \left\{\sum_{|s|\le k}\int_\Omega |\partial^s\phi|^p\,dx\right\}^{1/p}.\tag{16.34}$$

The main reason for introducing Sobolev spaces is to be found in the search for so-called *weak solutions* of boundary value problems.[††] Suppose, for instance, that one considers the torsion problem for a shaft of square cross section Ω, of side a, in terms of Prandtl's stress function F:

$$\frac{\partial^2 F}{\partial x^2} + \frac{\partial^2 F}{\partial y^2} = g(x,\,y)\qquad \text{for } -\frac{a}{2}\le x,\,y\le\frac{a}{2},\tag{16.35}$$

$$F = 0\qquad\qquad \text{for } x,\,y = \pm\frac{a}{2},\tag{16.35a}$$

where $g(x,\,y) = -2G\alpha$, G is the shear modulus, and α is the angle of twist per unit length. Let F be a solution in the usual sense; we multiply both sides of (16.35) by a test function $\phi \in C_0^\infty(\Omega)$ and integrate over Ω. We have, applying the divergence theorem in Ω,

$$\int_\Omega \nabla^2 F\phi\,dx\,dy = \int_\Omega F\nabla^2\phi\,dx\,dy,\tag{16.36}$$

[†] Recall (16.3) and what was said earlier about equality of improper functions [see footnote following equation (16.10)].

[‡] Compare Carslaw and Jaeger (Ref. 106, p. 261). Note that the general form of the heat conduction equation in the stationary case is $\nabla^2\theta = -Q(x,\,y)/k$. Clearly, $-\frac{1}{4}\pi r$ is the fundamental solution of the Laplace equation in three dimensions. See Greenberg (Ref. 72, p. 114).

[§] Compare Sobolev (Ref. 99, Sec. 7) and Kufner, Oldrich, and Fucik (Ref. 107, Chap. 5).

[¶] We recall that $\mathscr{L}_p$ denotes the set of all functions whose pth powers are integrable.

[††] See, e.g., Stakgold (Ref. 24, p. 171), as well as the elegant paper by Weinacht.[108]

so that

$$\int_{\Omega} F \nabla^2 \phi \, dx \, dy = \int_{\Omega} g(x, y) \phi(x, y) \, dx \, dy. \tag{16.36a}$$

A function $F(x, y)$ satisfying the latter equation is known as a solution in the *weak sense* of the original problem, provided that the equality holds for each $\phi \in C_0^{\infty}(\Omega)$. More generally, if g is a locally integrable function, then a locally integrable function† F is a *weak solution* of (16.35) if it satisfies (16.36a) for each $\phi \in C_0^{\infty}(\Omega)$.

Clearly, equation (16.35) can also be treated *distributionally*. That is, if g is a distribution, it is said that a distribution F is a solution of (16.35) if, for each $\phi \in C_0^{\infty}(\Omega)$,

$$\langle F, \nabla^2 \phi \rangle = \langle g, \phi \rangle. \tag{16.37}$$

In this way, we recognize *three classes* of solutions of differential equations: the *classical* solutions, the *weak* solutions, and the *distributional* solutions.

Evidently, a solution in the classical sense is a solution in the weak sense. Furthermore, by combining the left-hand side of equation (16.36) with the right-hand side of (16.36a), and selecting the test function so as to coincide with an approximate solution to (16.35) and satisfy, say, the boundary condition (16.35a), we find a close connection between the definition of a *weak* solution and the *Galerkin* method. In any case, the function $F(x, y)$ in (16.36a) is subjected to no other restrictions, a requirement considerably more liberal than the original, requiring that the function obey condition (16.35a) and belong to the class C^2. Indeed, in the present case, the demand that $F \in W^{0, 1}$ would do.

We could also proceed slightly differently, applying the divergence theorem to the left-hand side of equation (16.36) and, bearing in mind condition (16.35a) and the properties of a test function, arrive at

$$\int_{\Omega} \left(\frac{\partial F}{\partial x} \frac{\partial \phi}{\partial x} + \frac{\partial F}{\partial y} \frac{\partial \phi}{\partial y} \right) dx \, dy = - \int_{\Omega} g(x, y) \phi(x, y) \, dx \, dy. \tag{16.38}$$

Also here we note requirements on the function F weaker than membership in the class C^2. It suffices, namely, that $F \in W^{1, 1}$. In general, if one assumes that a weak solution is an element of a Sobolev space,‡ one understands the partial derivatives in the distributional sense.

---

† The restriction of local integrability is understandable if one recalls the remark concerning the representation (16.6).

‡ An interesting application of Sobolev spaces is due to Szefer and Demkowicz.[109]

Answers to Problems

Chapter 2

1. No, because of the absence of the concepts of length and angle, there is no means of distinguishing, say, the major axis of the ellipse from its minor axis, or a right angle in the square from an acute or obtuse one in the rhombus.

2. By (2.10), $\mathbf{0} + \mathbf{x} = \mathbf{x}$. If also $\mathbf{x} + \mathbf{x} = \mathbf{x}$, then $\mathbf{0} + \mathbf{x} = \mathbf{x} + \mathbf{x}$, so that

$$\mathbf{0} + [\mathbf{x} + (-\mathbf{x})] = \mathbf{x} + [\mathbf{x} + (-\mathbf{x})].$$

Hence, by (2.11), $\mathbf{0} + \mathbf{0} = \mathbf{x} + \mathbf{0}$ or, by (2.10), $\mathbf{0} = \mathbf{x} + \mathbf{0}$. A comparison with (2.10) proves the assertion.

3. We have $\mathbf{x} + \overrightarrow{CM} = \mathbf{z}$, $\mathbf{y} - \overrightarrow{MB} = \mathbf{z}$ or, upon adding, $\mathbf{z} = \frac{1}{2}(\mathbf{x} + \mathbf{y})$.

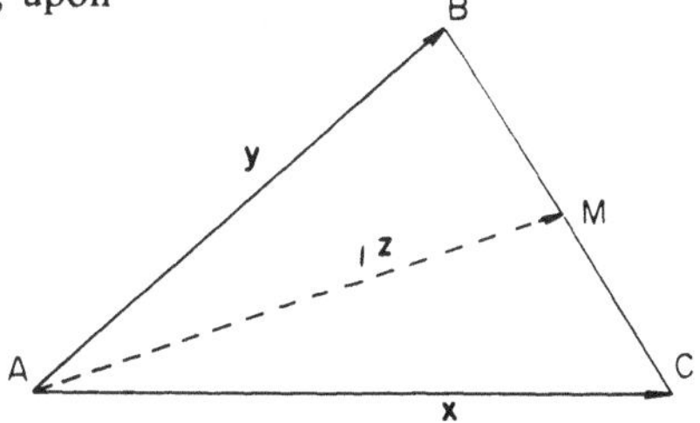

4. We first note that by a midpoint of a side, say AB, is meant the point M_1 such that $\overrightarrow{AM_1} = \overrightarrow{M_1B}$ in the sense of Axiom 2.2. We have: $\overrightarrow{M_1M_2} = (\mathbf{x} + \mathbf{y})/2$, $\overrightarrow{M_2M_3} = (\mathbf{y} + \mathbf{z})/2$, $\overrightarrow{M_3M_4} = (\mathbf{z} + \mathbf{w})/2$, $\overrightarrow{M_4M_1} = (\mathbf{w} + \mathbf{x})/2$. But $\mathbf{x} + \mathbf{y} + \mathbf{z} + \mathbf{w} = \mathbf{0}$. Thus,

$$\overrightarrow{M_1M_2} = (\mathbf{x} + \mathbf{y})/2 = -(\mathbf{z} + \mathbf{w})/2 = -\overrightarrow{M_3M_4} = \overrightarrow{M_4M_3};$$

similarly, $\overrightarrow{M_4M_1} = \overrightarrow{M_3M_2}$. By Axiom 2.3, the assertion is verified.

5. We have, from the triangle ABC, $2\overrightarrow{AM_1} + 2\overrightarrow{BM_2} = 2\overrightarrow{AM_3}$. From the triangle AM_1M_3, $\overrightarrow{AM_1} + \overrightarrow{M_1M_3} = \overrightarrow{AM_3}$. Comparing the preceding equations gives $\overrightarrow{M_1M_3} = \overrightarrow{BM_2} = \overrightarrow{M_2C}$, as asserted.

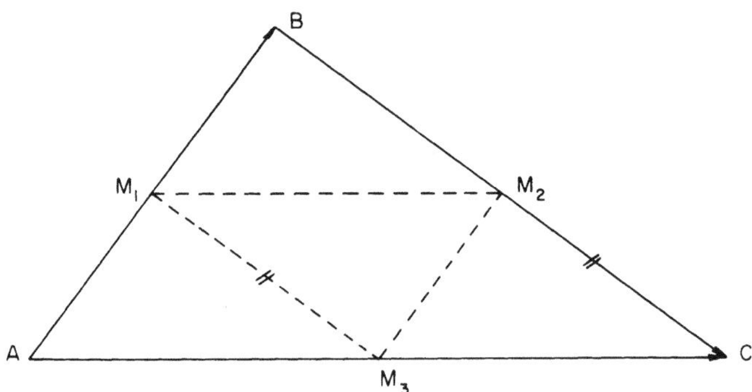

6. We have

$$\overrightarrow{AS} = \alpha\overrightarrow{AM}$$

$$= \alpha(\overrightarrow{AB} + \overrightarrow{BM}) = \alpha(\overrightarrow{AB} + \tfrac{1}{2}\overrightarrow{AD}). \tag{a}$$

Now if $\overrightarrow{DS} = \beta\overrightarrow{DB}$, then

$$\overrightarrow{AS} = \overrightarrow{AD} + \overrightarrow{DS} = \overrightarrow{AD} + \beta\overrightarrow{DB} = \overrightarrow{AD} + \beta(\overrightarrow{AB} - \overrightarrow{AD})$$

$$= (1 - \beta)\overrightarrow{AD} + \beta\overrightarrow{AB} = \beta\overrightarrow{AB} + (1 - \beta)\overrightarrow{AD}. \tag{b}$$

Comparison of (a) and (b) gives $\alpha = \tfrac{2}{3}$, so that $\overrightarrow{AS} = \tfrac{2}{3}\overrightarrow{AB} + \tfrac{1}{3}\overrightarrow{AD}$. But $\overrightarrow{BS} = \overrightarrow{AB} - \overrightarrow{AS} = \tfrac{1}{3}(\overrightarrow{AB} - \overrightarrow{AD})$ and $\overrightarrow{DS} = \overrightarrow{AS} - \overrightarrow{AD} = \tfrac{2}{3}(\overrightarrow{AB} - \overrightarrow{AD})$, which proves the assertion.

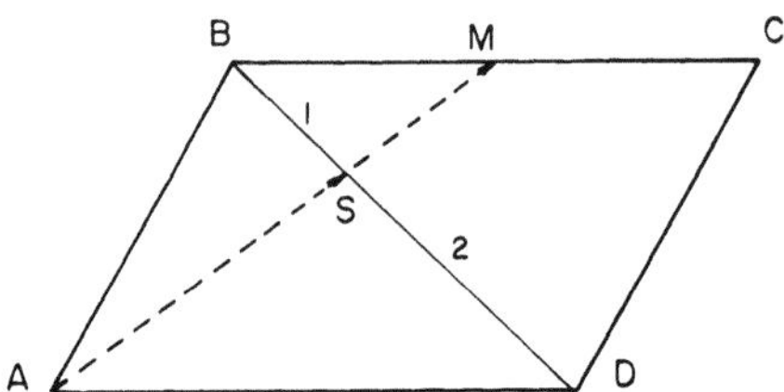

7. Since $\overrightarrow{A_2B_2} = \overrightarrow{A_1B_1}$, the lines l_1 and l_2 are "parallel"; by Axiom 2.2, then the line l_1 is the unique line through A_1 "parallel" to l_2. If l did not meet l_2, it would be a second "parallel" to l_2 through A_1, contrary to Axiom 2.2.

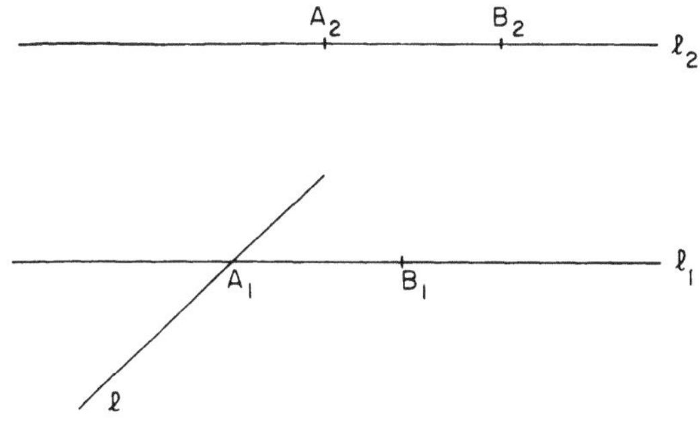

8. Verification of Axiom 2.5, for example:

$$1\mathbf{x} = (1x_1,\ 1x_2,\ 1x_3)$$
$$= (x_1,\ x_2,\ x_3)$$
$$= \mathbf{x}.$$

9. $(1, 0, 0) = (0, \alpha, 0)$. By the definition of equality in Problem 8, there would follow $1 = 0$, which is absurd.

10. Denote the position vectors of the vertices of the triangle by $\mathbf{x}_i$, $i = 1, 2, 3$, respectively, and that of the hypothetical point of intersection of the medians S by $\mathbf{s}$. For the median $A_1 B_1$, we have, by Problem 3, $\overrightarrow{A_1 B_1} = \tfrac{1}{2}(\overrightarrow{A_1 A_2} + \overrightarrow{A_1 A_3})$. Thus,

$$\mathbf{t}_1 = \mathbf{x}_1 + \overrightarrow{A_1 B_1} = \mathbf{x}_1 + \tfrac{1}{2}(\overrightarrow{A_1 A_2} + \overrightarrow{A_1 A_3})$$
$$= \mathbf{x}_1 + \tfrac{1}{2}[(\mathbf{x}_2 - \mathbf{x}_1) + (\mathbf{x}_3 + \mathbf{x}_1)] = \tfrac{1}{2}(\mathbf{x}_3 + \mathbf{x}_2).$$

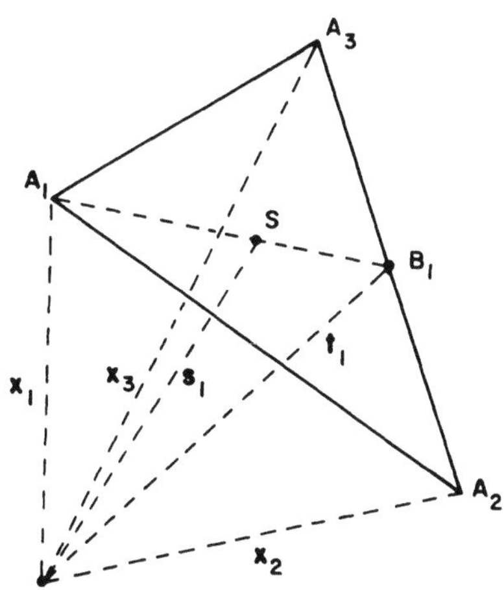

Let

$$\mathbf{s}_1 = \mathbf{x}_1 + \alpha \overrightarrow{A_1 B_1} = \mathbf{x}_1 + \alpha(\mathbf{t}_1 - \mathbf{x}_1)$$

$$= \mathbf{x}_1 + \alpha[\tfrac{1}{2}(\mathbf{x}_3 + \mathbf{x}_2) - \mathbf{x}_1] = (1 - \alpha)\mathbf{x}_1 + (\alpha/2)(\mathbf{x}_2 + \mathbf{x}_3).$$

For the remaining two medians, a cyclic permutation of indices yields, e.g., $\mathbf{s}_2 = (1 - \beta)\mathbf{x}_2 + (\beta/2)(\mathbf{x}_3 + \mathbf{x}_1)$, and there is a unique point S of intersection of the medians if $(1 - \alpha) = \beta/2$, $(1 - \beta) = \alpha/2$, or, finally, if $\alpha = \beta = \tfrac{2}{3}$. In this case, $\mathbf{s} = \mathbf{s}_1 = \mathbf{s}_2 = \mathbf{s}_3 = \tfrac{1}{3}(\mathbf{x}_1 + \mathbf{x}_2 + \mathbf{x}_3)$, and $\overrightarrow{A_1 S} = \tfrac{2}{3}\overrightarrow{A_1 B_1}$, for example.

Chapter 3

1. If, in a triangle ABC, there hold $\overrightarrow{AB} \perp \overrightarrow{AC}$ and $\|\overrightarrow{AB}\| = 3$, then, by the Pythagorean theorem (3.16), there follows $\|\overrightarrow{AC}\| = 4$. Consequently, $\tan \measuredangle ACB = \tfrac{3}{4}$.

2. Denote $\overrightarrow{AB} = \mathbf{x}$, $\overrightarrow{AC} = \mathbf{y}$, $\overrightarrow{BC} = \mathbf{z}$. By hypothesis, $(\mathbf{v}, \mathbf{x}) = 0$, $(\mathbf{v}, \mathbf{y}) = 0$. By (3.2), upon subtracting, $(\mathbf{v}, \mathbf{y} - \mathbf{x}) = 0$. But $\mathbf{y} - \mathbf{x} = \mathbf{z}$, and the assertion is proved.

3. Denote the center of the circle by O, the diameter of interest by AOB, and a point on the circumference by C. Also let $\overrightarrow{OB} = \mathbf{x}$ (so $\overrightarrow{OA} = -\mathbf{x}$), $\overrightarrow{OC} = \mathbf{y}$, $\overrightarrow{AC} = \mathbf{z}$, and $\overrightarrow{BC} = \mathbf{w}$. Then $\overrightarrow{AC} = \overrightarrow{OC} - \overrightarrow{OA}$ or $\mathbf{z} = \mathbf{y} + \mathbf{x}$, while $\overrightarrow{BC} = \overrightarrow{OC} - \overrightarrow{OB}$ or $\mathbf{w} = \mathbf{y} - \mathbf{x}$. But $(\mathbf{z}, \mathbf{w}) = (\mathbf{y} + \mathbf{x}, \mathbf{y} - \mathbf{x}) = \|\mathbf{y}\|^2 - \|\mathbf{x}\|^2 = \mathbf{0}$, since $\|\mathbf{x}\| = \|\mathbf{y}\|$.

4. See Figure 9.1a and the pertinent argument in Chapter 9. Note that a straight line is a carrier of all vectors of the form $\alpha \mathbf{x}$, $-\infty < \alpha < \infty$ if it carries the vector $\mathbf{x}$.

5. Suppose that both inner products obey the distributive law. This implies that

$$\|\mathbf{x}\|\,\|\mathbf{y} + \mathbf{z}\|\cos^2(\mathbf{x}, \mathbf{y} + \mathbf{z}) = \|\mathbf{x}\|\,\|\mathbf{y}\|\cos^2(\mathbf{x}, \mathbf{y}) + \|\mathbf{x}\|\,\|\mathbf{z}\|\cos^2(\mathbf{x}, \mathbf{z}), \qquad (\alpha)$$

and that

$$\|\mathbf{x}\|\,\|\mathbf{y} + \mathbf{z}\|\sin(\mathbf{x}, \mathbf{y} + \mathbf{z}) = \|\mathbf{x}\|\,\|\mathbf{y}\|\sin(\mathbf{x}, \mathbf{y}) + \|\mathbf{x}\|\,\|\mathbf{z}\|\sin(\mathbf{x}, \mathbf{z}), \qquad (\beta)$$

respectively. By a general theorem on orthogonal projections, however,

$$\|\mathbf{x}\|\,\|\mathbf{y} + \mathbf{z}\|\cos(\mathbf{x}, \mathbf{y} + \mathbf{z}) = \|\mathbf{x}\|\,\|\mathbf{y}\|\cos(\mathbf{x}, \mathbf{y}) + \|\mathbf{x}\|\,\|\mathbf{z}\|\cos(\mathbf{x}, \mathbf{z}). \qquad (\gamma)$$

We multiply the preceding equation through by $\cos(\mathbf{x}, \mathbf{y} + \mathbf{z})$ and $\cotan(\mathbf{x}, \mathbf{y} + \mathbf{z})$, respectively, and observe that equations (α) and (β) contradict equation (γ), except in particular cases in which $\cos(\mathbf{x}, \mathbf{y}) = \cos(\mathbf{x}, \mathbf{z}) = \cos(\mathbf{x}, \mathbf{y} + \mathbf{z})$ and $\sin(\mathbf{x}, \mathbf{y}) = \sin(\mathbf{x}, \mathbf{z}) = \cotan(\mathbf{x}, \mathbf{y} + \mathbf{z})$, respectively.

6. Consider the distributivity of the inner product (3.2).

7. Consider the triangle inequality (3.20).

8. Use an appropriate notation.

Chapter 4

1. Linear independence of x, y, and z imples that $\alpha_1 x + \alpha_2 y + \alpha_3 z = 0$ only if $\alpha_1 = \alpha_2 = \alpha_3 = 0$. But the e_i- vectors are linearly independent, so that the coefficients in

$$(2\alpha_1 + \alpha_2 - 2\alpha_3)e_1 + (\alpha_1 - 3\alpha_2 - \alpha_3)e_2 + (-\alpha_1 + 2\alpha_2 + 3\alpha_3)e_3 = 0$$

must vanish. The trivial result $\alpha_1 = \alpha_2 = \alpha_3 = 0$ follows, since the determinant

$$\Delta = \begin{vmatrix} 2 & 1 & -2 \\ 1 & -3 & -1 \\ -1 & 2 & 3 \end{vmatrix}$$

differs from zero (actually, $\Delta = -14$). Therefore, x, y, and z are linearly independent.

2. v depends linearly on x, y, z if, and only if, we can find β_1, β_2, and β_3 such that $3e_1 - 2e_2 - 5e_3 = \beta_1 x + \beta_2 y + \beta_3 z$. Inserting from Problem 1 and considering that the e_i-vectors are linearly independent gives $\beta_1 = -2$, $\beta_2 = 1$, $\beta_3 = -3$.

3. x and $x + 10y$ are independent if, and only if, $\alpha_1 x + \alpha_2(x + 10y) = 0$ implies $\alpha_1 = \alpha_2 = 0$. But this is $(\alpha_1 + \alpha_2)x + 10\alpha_2 y = 0$ and, since x and y are independent, there follows $\alpha_1 = \alpha_2 = 0$. Thus, also x and $x + 10y$ are independent.

4. There is $v = 8e_1 + 5e_2 + 7e_3$ and, by (3.9),

$$\|v\| = [(8e_1 + 5e_2 + 7e_3, 8e_1 + 5e_2 + 7e_3)]^{1/2}.$$

Since the e_i-vectors are orthonormal, $\|v\| = (138)^{1/2}$.

5. A unit vector along the y-direction is $y/\|y\|$. Thus, the projection sought is $x_y = (x, y)/\|y\|$ [see equation 3.5)]. But $(x, y) = -36 - 18 - 6 = -60$, and $\|y\| = (81 + 9 + 9)^{1/2} = (99)^{1/2}$. Consequently, $x_y = -60/(99)^{1/2}$.

6. Let the position vector of a generic point P on the plane be $r = x_1 e_1 + x_2 e_2 + x_3 e_3$. The vector n must be perpendicular to $\overrightarrow{AP} = r - \overrightarrow{OA}$, lying on the plane. Thus,

$$(xe_1 + x_2 e_2 + x_3 e_3 - [2e_1 + 10e_2 + 6e_3], 4e_1 + 6e_2 + 12e_3) = 0$$

or $2x_1 + 3x_2 + 6x_3 = 70$, and is the desired equation of the plane.

7. Use equations (4.12) and (4.13).

8. Inspection shows that $x = y/2 + z$, so that x depends linearly on y and z. The remaining vectors are linearly dependent if $\alpha y + \beta z + \gamma v = 0$ for some α, β, γ, not all zero. But the only solution of the system $8\alpha + 2\beta + 3\gamma = 0$, $10\alpha - 6\beta + 4\gamma = 0$, $4\alpha + 4\beta + 5\gamma = 0$ is $\alpha = \beta = \gamma = 0$ (the determinant of the system is nonzero).

9. We have $z^2 = \|z\|^2 = (x_1, x_1)x^2 + 2xy(x_1, x_2) + y^2(x_2, x_2)$. In case (a): $(x_1, x_2) = 0$; in case (b): $(x_1, x_1) = (x_2, x_2) = 1$, in addition to $(x_1, x_2) = 0$.

10. We first observe that $y = 2x$, so that y can be eliminated. The remaining vectors are linearly dependent only if $\alpha x + \beta y + \gamma v = 0$ for some α, β, and γ, not all of which are zero. This requires

$$2\alpha + 2\beta + 8\gamma = 0, \qquad 4\alpha + \beta + 13\gamma = 0, \qquad -3\alpha + 3\beta - 6\gamma = 0.$$

But the determinant of this homogeneous system vanishes, so that one of the equations of the system is a consequence of the remaining two ($v = -2z + \frac{5}{2}x$). The independent subset, therefore, includes only two vectors, x and z, say.

Chapter 5

1. Using (5.27), we normalize g_1 to obtain $f_1 = (1/2^{1/2})\,(1, 1, 0)$. Equation (5.31) then gives $c_{22} = \left(\frac{3}{2}\right)^{-1/2}$, whence, from (5.30), the second orthonormal vector becomes $f_2 = \left(\frac{2}{3}\right)^{1/2}(-\frac{1}{2}, \frac{1}{2}, 1)$. For $k = 3$, equations (5.32b) yield $c_{33} = 3^{1/2}$ and $f_3 = 3^{1/2}(\frac{1}{3}, -\frac{1}{3}, \frac{1}{3})$.

2. We construct an orthonormal basis for the n-space,

$$f_1, f_2, \ldots, f_m, f_{m+1}, \ldots, f_n,$$

which includes the m given n-vectors. In terms of this basis,

$$x = \sum_{i=1}^{n} \frac{(x, f_i)f_i}{(f_i, f_i)}.$$

An inner multiplication of both sides with x gives the desired inequality for $m \leq n$.

3. Assume that, for some scalars, $\alpha_1, \alpha_2, \ldots, \alpha_n$, we have $\alpha_1 f_1 + \cdots + \alpha_k f_k + \cdots + \alpha_n f_n = 0$, and take the inner product with f_k. This gives $\alpha_k \|f_k\| = 0$ for any f_k. Since $f_k \neq \theta$, each $\alpha_k = 0$, and the set $\{f_k\}$ is linearly independent.

4. We represent $f_k = \sum_{i=1}^{n} \alpha_i^{(k)} e_i$ and, taking a self-inner product, find that $\sum_{i=1}^{n} \alpha_i^{(k)} \alpha_i^{(l)} = \delta_{kl}$, where δ_{kl} is the Kronecker delta.

5. Suppose that the x_i are linearly dependent:

$$\alpha(\tfrac{1}{2}, -1, 0, \tfrac{3}{2}) + \beta(1, \tfrac{3}{2}, 0, -\tfrac{1}{2}) + \gamma(1, -\tfrac{1}{2}, 1, \tfrac{1}{2}) = (6, 18, -8, -4).$$

It follows that

$$\tfrac{1}{2}\alpha + \beta + \gamma = 6, \quad -\alpha + \tfrac{3}{2}\beta - \tfrac{1}{2}\gamma = 18, \quad \gamma = -8, \quad \tfrac{3}{2}\alpha - \tfrac{1}{2}\beta + \tfrac{1}{2}\gamma = -4.$$

The preceding system is consistent, its solution being: $\alpha = 4$, $\beta = 12$, $\gamma = -8$; e.g., $x_4 = 4x_1 + 12x_2 - 8x_3$. Since the remaining vectors x_1, x_2, and x_3 are linearly independent (verify!), the space is three-dimensional.

6. Linear independence is trivially verified. The vectors span $\mathscr{E}_4$, since for any vector $x = (x_1, x_2, x_3, x_4)$, there is

$$(x_1, x_2, x_3, x_4)$$

$$= \frac{x_1}{2}\,(2, 0, 0, 0) + \frac{x_2}{3}\,(0, 3, 0, 0) + \frac{x_3}{4}\,(0, 0, 4, 0) + \frac{x_4}{5}\,(0, 0, 0, 5).$$

7. Represent $x = x_1(1, 0, 0) + x_2(0, 1, 0) + x_3(\alpha, \beta, \gamma)$, where $h = (\alpha, \beta, \gamma)$. Orthogonality of h to e_1 and e_2 demands that $\alpha = \beta = 0$, and the foregoing relation yields $x_1 = 2$, $x_2 = 3$, $x_3 = 4$ if we set $\gamma = 1$ or $h = (0, 0, 1)$. From the Pythagorean theorem, $x = (29)^{1/2}$; the orthogonal projections of x on h and $e_1 + e_2$ are 4 and $(13)^{1/2}$, respectively.

8. The vector g in $\mathscr{E}_n$ can be represented in the form $g = \alpha_1 e_1 + \cdots + \alpha_n e_n$, where $\{e_k\}$ is an orthonormal basis in $\mathscr{E}_n$. It is required that $(h, e_i) = (f - g, e_i) = 0$, $i = 1, 2, \ldots, n$. This gives $\alpha_i = (f, e_i)$, $i = 1, 2, \ldots, n$, and the representation is $f = \sum_{i=1}^{n} (f, e_i)e_i + h$.

Chapter 6

1. Let the sequence $\{x^k\}$ tend to x, i.e., by (6.3), $\lim\|x - x^k\| = 0$ for $k \to \infty$. By the triangle inequality (5.25),

$$\|x^m - x^n\| = \|(x^m - x) + (x - x^n)\| \leq \|x^m - x\| + \|x - x^n\|.$$

Since the right-hand side $\to 0 + 0$ as $m, n \to \infty$, the assertion is proved.

2. We have $\|(3 - 1/m) - (3 - 1/n)\| \leq \|1/m\| + \|1/n\| \to 0 + 0$, as $m, n \to \infty$, by triangle inequality. Thus, the sequence is Cauchy, but the only possible limit, 3, lies outside the open interval $(1, 3)$, so that the sequence does not converge in the sense of (6.3).

3. The following sets constitute subspaces of $\mathscr{E}_3$: $\{(\alpha_1, \alpha_2, 0)\}$, $\{(\alpha_1, 0, \alpha_3)\}$, $\{(0, \alpha_2, \alpha_3)\}$, $\{(\alpha_1, 0, 0)\}$, $\{(0, \alpha_2, 0)\}$, $\{(0, 0, \alpha_3)\}$, as well as $\{(\alpha_1, \alpha_2, \alpha_3)\}$, $\{(0, 0, 0)\}$. Note that $\mathscr{E}_3$ and $\theta = (0, 0, 0)$ satisfy all conditions imposed on a subspace.

4. By hypothesis, there are scalars $\gamma_k^{\,j}$, $j = 1, 2, \ldots, n$, $k = 1, 2, \ldots, m$, such that $y^j = \sum_{k=1}^{m} \gamma_k^{\,j} x^k$ for $j = 1, 2, \ldots, n$. It is clear that $\mathscr{S}_1$ is contained in $\mathscr{S}_2$, so we need only show the reverse inclusion. Then let $x \in \mathscr{S}_2$, whence there are scalars α_k, $k = 1, 2, \ldots, m$, and β_j, $j = 1, 2, \ldots, n$, for which $x = \sum_{k=1}^{m} \alpha_k x^k + \sum_{j=1}^{n} \beta_j y^j$. Using the linear dependence of each y^j on the set $\{x^k\}_{k=1}^{m}$, $x = \sum_{k=1}^{m} \alpha_k x^k + \sum_{j=1}^{n} \beta_j \sum_{k=1}^{m} \gamma_k^{\,j} x^k = \sum_{k=1}^{m} (\alpha_k + \sum_{j=1}^{n} \beta_j \gamma_k^{\,j})x^k$. Thus, $x \in \mathscr{S}_1$, showing that $\mathscr{S}_2$ is contained in $\mathscr{S}_1$. Therefore, $\mathscr{S}_1 = \mathscr{S}_2$.

5. We have $\|y^{n+p} - y^n\|^2 = \|\sum_{k=n+1}^{n+p} \alpha_k x^k\|^2 = \sum_{k=n+1}^{n+p} \|\alpha_k x^k\|^2 = \sum_{k=n+1}^{n+p} (\alpha_k^2)$, the last two steps by the orthonormality of $\{x^k\}$. Since, by hypothesis, the series $\sum_{k=1}^{\infty} \alpha_k^2$ converges, the Bolzano–Cauchy condition of convergence implies that $\sum_{k=n+1}^{n+p} \alpha_k^2 \to 0$ as $n, p \to \infty$ (see Ref. 120, p. 383). This proves the assertion.

6. Let $\alpha \le x^n \le \beta$ and $\|x - x^n\| \equiv |x - x^n| \to 0$ as $n \to \infty$. Suppose that x is not in $[\alpha, \beta]$, for instance, $x < \alpha$, so that $\alpha - x = \varepsilon > 0$. Now for sufficiently large n, $x^n - x \le |x^n - x| < \varepsilon = \alpha - x$. Thus, $x^n < \alpha$, which contradicts our assumption. Similarly, we cannot have $\beta < x$.

7. We form a sequence $\{y^n\}$ in $\mathscr{S}^\perp$, convergent to some vector y. By assumption, for any x in $\mathscr{S}$ and for each n, there is $(x, y^n) = 0$. By the Cauchy–Schwarz inequality (5.24), $|(x, y)| = |(x, y) - (x, y^n)| = |(x, y - y^n)| \le \|x\| \|y - y^n\| \to 0$. Thus, $(x, y) = 0$ and y is orthogonal to any x in $\mathscr{S}$ or y is in $\mathscr{S}^\perp$.

8. Let the element $x^l \ne \theta$ be removed from the given basis. If the modified set were still a basis, by definition of a basis, the element could be represented in the form $x^l = \sum_{k=1}^{\infty} (x^l, x^k)x^k$. By the orthogonality of $\{x^k\}$, however, then $x^l = 0 + 0 + \cdots$, so that $x^l = \theta$, contrary to the assumption.

9. Select an arbitrary vector x^1 and find a vector x^2 independent of x^1. Such a vector exists, for otherwise the space would be of finite dimension. The process can be repeated indefinitely with a similar argument.

Chapter 7

1. To $(x^k, y^k) - (x, y)$ we add and subtract (x, y), (x^k, y), and (x, y^k). This gives

$$(x^k, y^k) - (x, y) = (x^k - x, y^k - y) + (x, y^k - y) + (y, x^k - x).$$

Using the Cauchy–Schwarz and triangle inequalities gives

$$|(x^k, y^k) - (x, y)| \le \|x^k - x\| \|y^k - y\| + \|x\| \|y^k - y\| + \|y\| \|x^k - x\| \to 0$$

as $k \to \infty$ by hypothesis.

2. (a) If $x \ne \theta$, then $\|x\| > 0$; if $x = \theta = (0, 0, \ldots, 0)$, then $\|x\| = 0$.
   (b) $\|\alpha x\| = [\sum_{i=1}^{n} (\alpha x_i)^2]^{1/2} = |\alpha| \sum_{i=1}^{n} (x_i)^2]^{1/2} = |\alpha| \|x\|$.
   (c) $\|x + y\|^2 = [\sum_{i=1}^{n} (x_i + y_i)^2] = \sum_{i=1}^{n} (x_i^2 + 2x_i y_i + y_i^2) \le \|x\|^2 + \|y\|^2$
   $+ 2|(x, y)| \le \|x\|^2 + \|y\|^2 + 2\|x\| \|y\| = (\|x\| + \|y\|)^2.$

3. By definition of a spanning set, $\mathscr{S}$ consists of all vectors of the form $x = \alpha_1 x^1 + \cdots + \alpha_m x^m$. If the set $\{x^i\}$ is linearly independent, then $n = m$ by definition [see Group D, equation (7.3)] of the dimension of a subspace. If the set is linearly dependent, then at least one of the x^i's, say x^m for simplicity, can be represented as a linear combination of the remaining vectors; hence, $x^1, x^2, \ldots, x^{m-1}$ span $\mathscr{S}$. Thus, n is certainly less than m, since there are at most $m - 1$ linearly independent vectors in $\mathscr{S}$ in this case.

4. From (5.27), we first find $P_1(t) = 1/(\int_{-1}^{1} dt)^{-1/2} = 1/2^{1/2}$; (5.30) and (5.31) then give $c_{22} = (3/2)^{1/2}$, $P_2(t) = (3/2)^{1/2}t$; thus, $c_{33} = \frac{3}{2}(5/2)^{1/2}$, $P_3(t) = (1/2)(5/2)^{1/2} \times (3t^2 - 1)$.

5. From the triangle inequality [7.12(c)], $d(x, z) - d(y, z) \leq d(x, y)$; interchanging the roles of x and y, $d(y, z) - d(x, z) \leq d(y, x)$, or $-d(y, x) \leq d(x, z) - d(y, z)$. Since $d(y, x) = d(x, y)$, by [7.12(b)], we find $|d(x, z) - d(y, z)| \leq d(x, y)$, by combining the preceding inequalities.

6. Let a sequence of functions $\{x^n\}$ belonging to $C_{a \leq t \leq b}$ be Cauchy, i.e., $d(x^n, x^m) = \max_{a \leq t \leq b} |x^n(t) - x^m(t)| \to 0$ as $n, m \to \infty$. Since this says that the sequence $\{x^n(t)\}$ is Cauchy for each t, there exists $x(t) = \lim_{n \to \infty} x^n(t)$ for each t by the completeness of the real numbers. The function x so defined will be shown to be in $C_{a \leq t \leq b}$, with $x^n \to x$ in $C_{a \leq t \leq b}$. We have $|x^n(t) - x(t)| \leq |x^n(t) - x^m(t)| + |x^m(t) - x(t)| \leq \max_{a \leq \tau \leq b} |x^n(\tau) - x^m(\tau)| + |x^m(t) - x(t)|$ for any t in $[a, b]$ and for any m, n. Given $\varepsilon > 0$, choose m so large (depending on t) that the second term is less than $\varepsilon/4$, and the first term is less than $\varepsilon/4$ for all sufficiently large n (independent of t). Thus, $|x^n(t) - x(t)| \leq \varepsilon/2$ for all sufficiently large n, independent of t, whence $\max_{a \leq t \leq b} |x^n(t) - x(t)| < \varepsilon$ for all sufficiently large n. This shows that $x^n \to x$ as $n \to \infty$ in the norm of $C_{a \leq t \leq b}$. Finally, for t and τ in $[a, b]$ and any n, write

$$|x(t) - x(\tau)| \leq |x(t) - x^n(t)| + |x^n(t) - x^n(\tau)| + |x^n(\tau) - x(\tau)|$$

$$\leq 2 \max_{a \leq z \leq b} |x(z) - x^n(z)| + |x^n(t) - x^n(\tau)|.$$

Given $\varepsilon > 0$, choose n so that the first term is $\varepsilon/2$. Since x^n is uniformly continuous, the second term is less than $\varepsilon/2$ provided that $|t - \tau|$ is sufficiently small. Hence, $|x(t) - x(\tau)| < \varepsilon$ if $|t - \tau|$ is sufficiently small, showing that x is (uniformly) continuous on $[a, b]$. Thus, $C_{a \leq t \leq b}$ is complete.

7. For any points x^1 and x^2 in the closed ball, $\|x^1 - c\| \leq R$, $\|x^2 - c\| \leq R$. Take any point y on the line segment through x^1 and x^2: $y = \beta_1 x^1 + (1 - \beta_1)x^2$ where $0 < \beta_1 < 1$. Then

$$\|y - c\| = \|\beta_1 x^1 + (1 - \beta_1)x^2 - c\| = \|\beta_1 x^1 + (1 - \beta_1)x^2 - \beta_1 c - (1 - \beta_1)c\|$$

$$\leq \beta_1 \|x^1 - c\| + (1 - \beta_1)\|x^2 - c\|,$$

by Axiom N (b), (c), $\leq \beta_1 R + (1 - \beta_1)R = R$. This proves the assertion.

8. For $k = 1$, inequality (7.21) becomes $\|x\| \geq |(x, x^1)|$ where $\|x^1\| = 1$. Taking $x^1 = y/\|y\|$, where $y \neq \theta$, we obtain $\|x\| \geq (x, y/\|y\|)$, whence $|(x, y)| \leq \|x\| \|y\|$, just the Cauchy–Schwarz inequality.

9. Take an arbitrary function $x(t)$ in $C_{0 \leq t \leq 1}$. Let S_0 denote the set in $C_{0 \leq t \leq 1}$ of all polynomials $P_0(t)$ with *rational* coefficients. By the Weierstrass approximation theorem [cf. the text preceding equation (13.34)], there exists a polynomial $P(t)$ such that [see (7.12a)] $\max_{0 \leq t \leq 1} |x(t) - P(t)| < \varepsilon/2$, $\varepsilon > 0$. Likewise, there exists a polynomial $P_0(t) \in S_0$ such that $\max_{0 \leq t \leq 1} |P(t) - P_0(t)| < \varepsilon/2$. Hence, $d(x, P_0) = \max_{0 \leq t \leq 1} |x(t) - P_0(t)| < \varepsilon$ by the triangle inequality. This proves the assertion [cf. the text following (7.5)], since the set S_0 is countable.

Chapter 8

1. By the linearity of $f[x]$ and the equality $0 = 0 + 0$, there is $f[0] = f[0 + 0] = f[0] + f[0]$ or $f[0] - f[0] = 0 = f[0]$. Alternately, observe that $f[0] = f[0x] = 0f[x] = 0$.

2. For an arbitrary vector $x = (x_1, x_2, \ldots, x_n)$ and an orthonormal basis $e_1, e_2, \ldots, e_n$, there is $f[x] = f[\sum_{i=1}^{n} x_i e_i] = \sum_{i=1}^{n} x_i f[e_i] = \sum_{i=1}^{n} x_i a_i = (x, \ a)$, where $a = (a_1, a_2, \ldots, a_n)$ is the vector given by $a_i = f[e_i]$.

3. Since x^1 and x^2 are linearly independent vectors in $\mathscr{E}_2$, we can write $x^3 = \alpha_1(12, 5) + \alpha_2(2, 2) = (12\alpha_1 + 2\alpha_2, \ 5\alpha_1 + 2\alpha_2)$. By hypothesis, $12\alpha_1 + 2\alpha_2 = 14$, $5\alpha_1 + 2\alpha_2 = 8$, so $\alpha_1 = 6/7$, $\alpha_2 = 13/7$. Thus (by linearity of L), $y^3 = 6/7y_1 + 13/7y_2 = (122/7, 15/7)$.

4. $L_1 L_2 \bar{x} = L_1 \bar{y} = L_1(\bar{x}_1, 0) = (0, \bar{x}_1)$ for any $\bar{x} = (\bar{x}_1, \bar{x}_2)$. $L_2 L_1 x = L_2 y = L_2(0, x_1) = (0, 0)$ for any $x = (x_1, x_2)$. Note that $L_2 L_1 \neq L_1 L_2$; in fact, $L_1 L_2 = L_1$, while $L_2 L_1 = 0$.

5. Properties (b) and (c) are clearly satisfied, e.g., $(x + y, \ z) = \int_0^t [x(\tau) + y(\tau)]z(t - \tau) \, d\tau = (x, z) + (y, z)$. Property (a) is satisfied by the known convolution relation $\int_0^t x(\tau)y(t - \tau) \, d\tau = \int_0^t x(t - \tau)y(\tau) \, d\tau$ (obtained by a simple change of variable). Property (d) generally is not satisfied, as shown by the following example. Let $x(\tau) = \cos \tau$, so that $(x, x) = \int_0^T \cos \tau \cos(T - \tau) \, d\tau = \frac{1}{2}(\sin T + T \cos T)$. For $T = \pi$, say, we have $(x, x) = -\pi/2 < 0$.

6. Hint: apply the divergence theorem to (u, v).

7. Note that $(e_i, e_j) = (f_i, f_j) = \delta_{ij}$. Using the summation convention, let $x \equiv \alpha_i e_i$, $y = \beta_j e_j$. Then $(Lx, Ly) = \alpha_i \beta_j(Le_i, \ Le_j) = \alpha_i \beta_j(f_i, f_j) = \alpha_i \beta_j \delta_{ij} = (\alpha_i e_i, \ \beta_j e_j) = (x, y)$.

8. There is $(Lx^m, y^n) - (Lx, y) = (L(x^m - x), \ y^n) + (Lx, \ y^n - y)$. By the Cauchy–Schwarz inequality, $|(L(x^m - x), \ y^n)| \leq \|L(x^m - x)\| \, \|y^n\|$. Now, the sequence $\{\|y^n\|\}$ is bounded since it converges (to $\|y\|$). Also $\|L(x^m - x)\| = \|Lx^m - Lx\| \to 0$ for $m \to \infty$ by hypothesis. Thus, $\lim_{m, n \to \infty} (L(x^m - x), y^n) = 0$. Similarly, the Cauchy–Schwarz inequality shows that $\lim_{n \to \infty} (Lx, y^n - y) = 0$ because $\|y^n - y\| \to 0$ for $n \to \infty$. These facts give the desired result.

Chapter 9

1. Simply observe that the given sequence is an infinite linearly independent set. For, suppose $\sum_{i=0}^{n} \alpha_i t^i = 0$, $-1 \leq t \leq 1$. Setting $t = 0$ gives $\alpha_0 = 0$. Differentiating the relation and setting $t = 0$ gives $\alpha_1 = 0$. The process can be continued indefinitely, so that $C_{-1 \leq t \leq 1}$ is infinite dimensional, since no finite set of vectors can span $C_{-1 \leq t \leq 1}$.

2. Let $i^1, i^2, \ldots, i^n$ be an orthonormal basis for $\mathscr{S}$ and let x be any vector in $\mathscr{H}$. Construct the orthogonal projection $\bar{x}$ of x on $\mathscr{S}$. Clearly, $\bar{x} = \sum_{k=1}^{n} (x, i^k)i^k$.

Next, consider $y = x - \bar{x}$. We have $(y, i^k) = (x, i^k) - (\bar{x}, i^k) = 0$, and so $y \perp \mathscr{S}$. Consequently, $x = \bar{x} + y$ with $\bar{x}$ in $\mathscr{S}$ and y in $\mathscr{S}^{\perp}$, i.e., $\mathscr{H} = \mathscr{S} \oplus \mathscr{S}^{\perp}$.

3. Let $(x, x) = (y, y)$. Then $0 = (x, x) + (x, y) - (x, y) - (y, y) = (x, x + y) - (y, x + y) = (x + y, x - y)$; the sum and difference represent the diagonals of a rhombus.

4. Let $\{f^1, f^2, \ldots, f^m\}$ and $\{g^1, g^2, \ldots, g^n\}$ be bases for the subspaces $\mathscr{S}'$ and $\mathscr{S}''$, respectively, and let $\mathscr{V}$ denote the space $\mathscr{S}' \oplus \mathscr{S}''$. By definition of a direct sum, every vector z in $\mathscr{V}$ may be represented in the form $z = x + y$, where $x \in \mathscr{S}'$ and $y \in \mathscr{S}''$. Clearly, we can write $x = \sum_{i=1}^{m} \alpha_i' f^i$ and $y = \sum_{i=1}^{n} \alpha_i'' g^i$, whence it follows that the set $\{f^1, f^2, \ldots, f^m, g^1, g^2, \ldots, g^n\}$ spans $\mathscr{V}$. Suppose now that this set is linearly dependent: $\theta = \sum_{i=1}^{m} \bar{\alpha}_i' f^i + \sum_{i=1}^{n} \bar{\alpha}_i'' g^i = \theta + \theta$ for some scalars $\bar{\alpha}_1'$, $\bar{\alpha}_2', \ldots, \bar{\alpha}_m', \bar{\alpha}_1'', \bar{\alpha}_2'', \ldots, \bar{\alpha}_n''$, at least one of which is nonzero. The uniqueness of representation of θ as the sum $\theta + \theta$, however, implies that separately $\sum_{i=1}^{m} \bar{\alpha}_i' f^i = \theta$ and $\sum_{i=1}^{m} \bar{\alpha}_i'' g^i = \theta$. But $\{f^i\}$ and $\{g^i\}$ are sets of linearly independent vectors; hence, $\bar{\alpha}_i' = 0$, $i = 1, \ldots, m$, $\bar{\alpha}_i'' = 0$, $i = 1, \ldots, n$, and our assumption is false. Consequently, the set $\{f_1', f^1, \ldots, f^m, g^1, g^2, \ldots, g^n\}$ is a basis for $\mathscr{V}$ and the dimension of $\mathscr{V}$ is $m + n$.

5. Let $\mathscr{S}'$ and $\mathscr{S}''$ represent the xy-plane and z-axis, respectively, both subspaces of $\mathscr{E}_3$. Any $u = (u_1, u_2, u_3)$ in $\mathscr{E}_3$ can be written in a unique way as $u = u' + u''$, where $u' = u_1 e_1 + u_2 e_2$ and $u'' = u_3 e_3$. Thus, $\mathscr{E}_3 = \mathscr{S}' \oplus \mathscr{S}''$.

6. Every $u' = u_1 e_1 + u_2 e_2 + 0 e_3$ and every $u'' = 0 e_1 + 0 e_2 + u_3 e_3$. Now, $(u', u'') = ([u_1 e_1 + u_2 e_2] + 0 e_3, 0[e_1 + e_2] + u_3 e_3) = ([u_1 e_1 + u_2 e_2], 0[e_1 + e_2]) + ([u_1 e_1 + u_2 e_2], u_3 e_3) + (0 e_3, 0[e_1 + e_2]) + (0 e_3, u_3 e_3) = 0$, since $(e_1, e_3) = (e_2, e_3) = 0$ by definition.

7. We are looking for any two linearly independent vectors of the form $x = (x_1, x_2, x_3, x_4)$ such that $(x, f) = \sum_{i=1}^{4} x_i f_i = 0$ and $(x, g) = \sum_{i=1}^{4} x_i g_i = 0$ hold for each of them. We are thus free to choose two of the coordinates x_i at our convenience. If we take, say, $x_2 = 0$, $x_4 = 1$ and then $x_2 = -1$, $x_4 = 0$, we find $x_1 = -17$, $x_3 = 5$ and $x_1 = 2$, $x_3 = 0$, respectively. Thus, an acceptable basis for $\mathscr{S}^{\perp}$ is $\bar{f} = (-17, 0, 5, 1)$ and $\bar{g} = (2, -1, 0, 0)$.

8. Let $x^1 = \bar{x}^1 + \bar{\bar{x}}^1$ and $x^2 = \bar{x}^2 + \bar{\bar{x}}^2$, where $\bar{x}^1, \bar{x}^2 \in \mathscr{S}$ and $\bar{\bar{x}}^1, \bar{\bar{x}}^2 \in \mathscr{S}^{\perp}$, so $Pr_{\mathscr{S}} x^1 = \bar{x}^1$ and $Pr_{\mathscr{S}} x^2 = \bar{x}^2$. Then $(Pr_{\mathscr{S}} x^1, x^2) = (\bar{x}^1, \bar{x}^2 + \bar{\bar{x}}^2) = (\bar{x}^1, \bar{x}^2)$ and $(x^1, Pr_{\mathscr{S}} x^2) = (\bar{x}^1 + \bar{\bar{x}}^1, \bar{x}^2) = (\bar{x}^1, \bar{x}^2)$. This proves the assertion.

Chapter 10

1. Clearly, the limit function is given by $f(t) = 0$ for $0 \leq t \leq 1$ and $f(t) = 1$ for $t = 1$. The convergence is not uniform because $\sup_{0 \leq t \leq 1} |f_n(t) - f(t)| = 1$ cannot be made arbitrarily small for all sufficiently large n.

2. For any x in $[0, 1]$ there is $\lim_{n \to \infty} f_n(x) = 0$, but $\int_0^1 f_n^2(x)\, dx = \int_0^{1/n} n^2\, dx = n \to \infty$.

3. By the Cauchy–Schwarz inequality, $|(f_n - f, g)| \leq \|f_n - f\|\,\|g\|$. Since $\|f_n - f\| \to 0$ for $n \to \infty$, then, for any fixed g, also $(f_n - f, g) \to 0$ for $n \to \infty$.

4. The arithmetic mean distance is $\int_{-1}^{1} (x^2 - x^3)\,dx = 2/3 \approx 0.667$; the mean-square distance is $[\int_{-1}^{1} (x^2 - x^3)^2\,dx]^{1/2} \approx 0.828$.

5. **Necessity:** (1) Construct the Fourier series (7.19a) for an arbitrary element x in the space in terms of a *complete* orthonormal set $\{\phi^i\}$: $s \equiv \sum_{i=1}^{\infty} \alpha_i \phi^i$, where $\alpha_i = (x, \phi^i)$. (2) Bessel's inequality (7.21) implies that $\sum_{i=1}^{n} \alpha_i{}^2 \leq \|x\|^2$ for each n, so that the α-series converges (as a series of positive terms with bounded partial sums). (3) By the orthonormality of $\{\phi^i\}$:

$$\|s_n - s_m\|^2 = \left\| \sum_{i=m+1}^{n} \alpha_i \phi^i \right\|^2 = \sum_{m+1}^{n} \alpha_i{}^2,$$

and by the convergence of the α-series, $\|s_n - s_m\| \to 0$ for $m, n \to \infty$. But the space is *complete* so that $\lim_{n \to \infty} \sum_{i=1}^{n} \alpha_i \phi^i$ exists and equals $\sum_{i=1}^{\infty} \alpha_i \phi_i = s$. (4) We find, if $n \geq i$, $(x - s, \phi^i) = \alpha_i - (s, \phi^i) = \alpha_i - (s_n, \phi^i) + (s_n - s, \phi^i) = (s_n - s, \phi_i)$. The Cauchy–Schwarz inequality gives $|(x - s, \phi^i)| \leq \|s_n - s\|\,\|\phi^i\| = \|s_n - s\| \to 0$ for $n \to \infty$. The result holding for each i, we have $(x - s, \phi^i) = 0$, $i = 1, 2, \ldots$. By the completeness of $\{\phi^i\}$, $x - s = 0$ or $x = s = \sum_{i=1}^{\infty} \alpha_i \phi_i$. Consequently, $\|x\|^2 = \sum_{i=1}^{\infty} \alpha_i{}^2$, the Parseval equality.

Sufficiency: Let the Parseval equality be satisfied. We take the subspace $\mathscr{S}$ generated by $\{\phi_n\}$, and represent any element x in the Hilbert space in the form $x = y + z$, where $y = Pr_{\mathscr{S}}\, x$ and $z \perp \mathscr{S}$ (i.e., $z \perp$ all ϕ_n in $\mathscr{S}$). Now $(x, \phi_n) = (y, \phi_n)$, so that x and y have the same Fourier coefficients α_n. By the Pythagorean theorem, $\|x\|^2 = \|y\|^2 + \|z\|^2$ and, by Bessel's inequality (7.21), $\|x\|^2 = \|y\|^2 + \|z\|^2 \geq \|y\|^2 \geq \sum \alpha_n{}^2 = \|x\|^2$, the last equality since x satisfies Parseval's equality. Then each inequality is actually an equality, and $\|y\|^2 + \|z\|^2 = \|y\|^2$, so that $z = \theta$ and $x = y$, i.e., x belongs to $\mathscr{S}$.

A simple intuitive illustration of the correspondence existing between a complete orthonormal system in a complete space and the associated Parseval equality is provided in Chapter 7, in which a connection between an orthonormal triad of coordinate vectors in $\mathscr{E}_3$ and the Pythagorean theorem is examined (cf. the text describing Figure 7.2).

6. The dimension of $\mathscr{S}$ is $2n + 1$. The set is orthonormal. By (7.19), e.g., $\alpha_k = (x, \cos kt) = 1/\pi \int_{-\pi}^{\pi} x \cos kt\,dt$. Bessel's inequality:

$$\|x\|^2 = 1/\pi \int_{-\pi}^{\pi} x^2\,dt \geq \sum_{i=0}^{n} \alpha_i{}^2 + \sum_{i=1}^{n} \beta_i{}^2.$$

7. For two eigenvectors $x, y \in \mathscr{V}$ belonging to λ, we have $Lx = \lambda x$, $Ly = \lambda y$. By the linearity of L, $L(\alpha x + \beta y) = \alpha Lx + \beta Ly = \lambda(\alpha x + \beta y)$ for any two scalars α and β. Thus, the vector $\alpha x + \beta y$ belongs to λ, i.e., $\alpha x + \beta y$ is in $\mathscr{S}$, whence $\mathscr{S}$ is a subspace.

8. By hypothesis, $Lx = \lambda x$, $L^*y = \lambda^* y$, where x and y are eigenvectors associated with λ for L and λ^* for L^*, respectively. Now $\lambda(x, y) = (\lambda x, y) = (Lx, y) = (x, L^*y) = (x, \lambda^* y) = \lambda^*(x, y)$ or $(\lambda^* - \lambda)(x, y) = 0$. Thus, $(x, y) = 0$ if $\lambda^* \neq \lambda$.

Chapter 11

1. For the first term of the series, inequalities (11.1) and (11.3) reduce to $\frac{2}{3} > \frac{1}{4}$.

2. We find $t = (\pi/2) - (4/\pi) \sum_{n=1,3,5,\dots}^{\infty} (\cos nt/n^2)$ or, after normalization, $t = (\pi^{3/2}/2)(1/\pi^{1/2}) - 2(2/\pi)^{1/2} \sum_{n=1,3,5,\dots}^{\infty} 1/n^2 (2/\pi)^{1/2} \cos nt$. Parseval's equality gives $\pi^4/96 = 1 + 1/3^4 + 1/5^4 + \cdots$, which is exactly the expansion (342) of $\pi^4/96$ on p. 64 of L. B. Jolley, *Summation of Series*, Dover, 1961.

3. $\int f^2(x)\, dx \int g^2(x)\, dx - \left(\int f(x)g(x)\, dx \right)^2$

$$= \frac{1}{2} \int f^2(x) \left[\int g^2(y)\, dy \right] dx + \frac{1}{2} \int g^2(x) \left[\int f^2(y)\, dy \right] dx$$

$$- \int f(x)g(x) \left[\int f(y)g(y)\, dy \right] dx = \frac{1}{2} \int \left\{ \int [f(x)g(y) - g(x)f(y)]^2\, dx \right\} dy.$$

6. Set, for example, $i^1 = \cos 2t/\pi^{1/2}$ and $g = \alpha t$, $\alpha \geq 1$. Upon using (11.4), we find $(f, f) \leq 2\alpha^2 \pi^3/3$ (note that in the present case, $(f, f) = \pi^3/3$). Verify that i^1 is normalized, that $f \perp i^1$, and that the condition (11.10) is satisfied.

7. For $f = t$ and $c = \alpha t$, $-\pi \leq t \leq \pi$, equation (11.25) gives $r = (1 - \alpha)(2\pi^3/3)^{1/2}$; moreover, $\|c\| = \alpha(2\pi^3/3)^{1/2}$. Thus, $(2\alpha - 1)^2(2\pi^3/3) \leq \|f\|^2 \leq 2\pi^3/3$; $\alpha = 1$ implies equality, while the hypersphere reduces to a point. Note that, in the present case, $\|f\|^2 = 2\pi^3/3$.

8. Condition (11.35a) is obviously satisfied. We find $\alpha^2(2\pi^3/3) \leq \|f\|^2 \leq 2\pi^3/3$; $\alpha = 1$ implies equality, while there must be $|\alpha| \leq 1$, in general. Note that, in the present case, $\|f\|^2 = 2\pi^3/3$.

9. $(Lu, v) = \int_0^1 e^{\alpha x}(u'' + \alpha u')v\, dx = \int_0^1 (e^{\alpha x}u')'v\, dx$

$$= e^{\alpha x}u'v \Big|_0^1 - e^{\alpha x}uv' \Big|_0^1 + \int_0^1 (e^{\alpha x}v')'u\, dx$$

$$= e^{\alpha}u'(1)v(1) + u(0)v'(0) + \int_0^1 (e^{\alpha x}v')'u\, dx.$$

Thus, $L^*v = (e^{\alpha x}v')' = e^{\alpha x}(v'' + \alpha v')$, and the required boundary conditions are $v(1) = 0$, $v'(0) = 0$; see equation (11.82). Since $L^* = L$, the operator L is called *self-adjoint*.

Chapter 12

2. (a) four; base vectors 1, sin t, cos t, cos $2t$; (b) three; base vectors sin t, cos t, sin $2t$.

4. By equations (9.3), $X = \alpha_1 S_1 + \alpha_2 S_2$, $Y = \beta_1 S_1 + \beta_2 S_2$, where $\alpha_1 + \alpha_2 = \beta_1 + \beta_2 = 1$. Thus, $X - Y = (\alpha_1 - \beta_1)S_1 + (\alpha_2 - \beta_2)S_2$ and

$$\gamma_1 + \gamma_2 = (\alpha_1 - \beta_1) + (\alpha_2 - \beta_2) = 0.$$

5. By the Cauchy–Schwarz inequality, $-1 \le (I_1{}^\varepsilon, \ I_2{}^\varepsilon) \le 1$. Also $|(I_1{}^\varepsilon, \ I_2{}^\varepsilon + I_3{}^\varepsilon)| \le \|I_1{}^\varepsilon\| \, \|I_2{}^\varepsilon + I_3{}^\varepsilon\| = (I_2{}^\varepsilon + I_3{}^\varepsilon, \ I_2{}^\varepsilon + I_3{}^\varepsilon)^{1/2} = [(I_2{}^\varepsilon, I_2{}^\varepsilon) + (I_3{}^\varepsilon, I_3{}^\varepsilon)]^{1/2} = 2^{1/2}$.

6. Let $S = S_1{}' + S_2{}' = S_1{}'' + S_2{}''$, where $S_1{}'$, $S_1{}''$ are in $\mathscr{S}_1$ and $S_2{}'$, $S_2{}''$ are in $\mathscr{S}_2$. Set $T = S_1{}' - S_1{}'' = S_2{}'' - S_2{}'$. Since the intrinsic vector T is in both $\mathscr{S}_1$ and $\mathscr{S}_2$ while $\mathscr{S}_1 \perp \mathscr{S}_2$, then T is perpendicular to itself and must be the zero vector. Thus, $S_1{}'' = S_1{}''$, $S_2{}' = S_2{}''$.

7. By (12.25) extended to $H_{(2)}$, we have $N = (S, I^1)I^1 + (S, I^2)I^2$ [see also (9.14)]. Therefore, $\|N\| = (a^2 + b^2)^{1/2}$.

8. We establish the correspondences $X \leftrightarrow f(t)$ and $S \leftrightarrow g(t) = t$ and find $\|S\|^2 = \int_a^b t^2 \, dt = (b^3 - a^3)/3$. Upon normalizing, $I = S/\|S\| = 3^{1/2}/(b^3 - a^3)^{1/2}t$. We now recognize in the equalities $\int_a^b [f(t)]^2 \, dt = \bar{\alpha}^2$, $\int_a^b tf(t) \, dt = \bar{\beta}$ the equations (12.111) and (12.112), respectively, in which $C_0 = 0$, $R_0{}^2 = \bar{\alpha}^2$, $\alpha = \bar{\beta}[3/(b^3 - a^3)]^{1/2}$. It follows that the tip of the vector X is on a hypercircle of class one with a parametric representation (12.113). Equations (12.115) and (12.117) imply that

$$C = \alpha I \leftrightarrow \frac{3}{b^3 - a^3}\, \bar{\beta}t,$$

$$R^2 = \bar{\alpha}^2 - \bar{\beta}^2 \frac{3}{b^3 - a^3}.$$

Thus, "approximately," $f(t) \approx [3/(b^3 - a^3)]\bar{\beta}t$.

Chapter 13

1. Take, for example, the subspace spanned by $\{\cos kt\}$ or by any subset thereof.

2. (1) The plane perpendicular to the given line and passing through the origin.
   (2) The line perpendicular to the given plane and passing through the origin.
   (3) The line perpendicular to the given line and passing through the origin.

3. Consider the orthogonal decompositions $f_1 = f_1{}^* + f_1{}^\perp$ and $f_2 = f_2{}^* + f_2{}^\perp$, where $\mathrm{Pr}_{\mathscr{S}} f_i = f_i{}^*$, $\mathrm{Pr}_{\mathscr{S}\perp} f_i = f_i{}^\perp$, $i = 1, 2$. Adding: $f_1 + f_2 = (f_1{}^* + f_2{}^*) + (f_1{}^\perp + f_2{}^\perp)$. But $(f_1{}^* + f_2{}^*) \in \mathscr{S}$ and $(f_1{}^\perp + f_2{}^\perp) \in \mathscr{S}^\perp$, whence $\mathrm{Pr}_{\mathscr{S}}(f_1 + f_2) = f_1{}^* + f_2{}^* = \mathrm{Pr}_{\mathscr{S}} f_1 + \mathrm{Pr}_{\mathscr{S}} f_2$, by the uniqueness of the orthogonal decomposition.

4. Let $f = f^* + f^\perp$ be the orthogonal decomposition of f. Then, by multiplying by c, we infer directly that $cf = cf^* + cf^\perp$ is the orthogonal decomposition of cf. Thus, $\mathrm{Pr}_{\mathscr{S}}(cf) = cf^* = c\mathrm{Pr}_{\mathscr{S}} f_1$ reasoning as in the preceding solution.

5. (a) The subspaces $\mathscr{E}_x$ and $\mathscr{E}_{yz}$ include all vectors $v_x = (\alpha, 0, 0)$ and $v_{yz} = (0, \beta, \gamma)$, respectively, where α, β, γ are scalars, By (5.13), therefore, for (1) and (2) in Problem 2, we have, for every v_x in $\mathscr{E}_x$ and every v_{yz} in $\mathscr{E}_{yz}$, $(v_x, v_{yz}) = \alpha \cdot 0 + 0 \cdot \beta + 0 \cdot \gamma = 0$, so that $\mathscr{E}_x \subset \mathscr{E}_{yz}^{\perp}$, $\mathscr{E}_{yz} \subset \mathscr{E}_x^{\perp}$. If $v = (\bar{\alpha}, \bar{\beta}, \bar{\gamma})$ is in $\mathscr{E}_{yz}^{\perp}$, then $0 = (v, v_{yz}) = \bar{\beta}\beta + \bar{\gamma}\gamma$ for every β and γ. Thus, $\bar{\beta} = \bar{\gamma} = 0$, so $v = (\bar{\alpha}, 0, 0)$. Hence, $\mathscr{E}_{yz}^{\perp} \subset \mathscr{E}_x$. Similarly, $\mathscr{E}_x^{\perp} \subset \mathscr{E}_{yz}$. We conclude: $\mathscr{E}_x = \mathscr{E}_{yz}^{\perp}$, $\mathscr{E}_{yz} = \mathscr{E}_x^{\perp}$.
   (b) $\mathrm{Pr}_{\mathscr{E}_{yz}} v = v_{yz}$; $\mathrm{Pr}_{\mathscr{E}_x} v = v_x$, where $v = (\alpha, \beta, \gamma)$.

6. By (13.1): $u' = \alpha' g_1$, $u'' = \alpha'' g_2$, $u''' = \beta_1 g_1 + \beta_2 g_2$. By (13.9): $(u', g_1) = (u, g_1)$, $(u'', g_2) = (u, g_2)$, $(u''', g_1) = (u, g_1)$, $(u''', g_2) = (u, g_2)$. Upon substituting the first three of the preceding equations into the remaining four, we find $\alpha' = (u, g_1)/(g_1, g_1)$, $\alpha'' = (u, g_2)/(g_2, g_2)$, and likewise for β_1 and β_2. Thus,

$$u''' = \frac{(u, g_1)g_1}{(g_1, g_1)} + \frac{(u, g_2)g_2}{(g_2, g_2)} = u' + u''.$$

7. Let x be any vector in $\mathscr{S}$. Then for any $y \in \mathscr{S}^{\perp}$, we have $y \perp x$. Hence, $x \perp \mathscr{S}^{\perp}$, which means that x is in $\mathscr{S}^{\perp\perp}$. Since $x \in \mathscr{S}$ implies that $x \in \mathscr{S}^{\perp\perp}$, then $\mathscr{S} \subset \mathscr{S}^{\perp\perp}$. Conversely, assume that $x \in \mathscr{S}^{\perp\perp}$. By (9.29), we can set $x = y + z$, where $y \in \mathscr{S}$ and $z \in \mathscr{S}^{\perp}$. But if $y \in \mathscr{S}$, then also $y \in \mathscr{S}^{\perp\perp}$, so that $z = x - y \in \mathscr{S}^{\perp\perp}$ or $z \perp \mathscr{S}^{\perp}$. Since $z \in \mathscr{S}^{\perp}$ and $z \perp \mathscr{S}^{\perp}$, then $z = \theta$ and $x = y \in \mathscr{S}$. Consequently, $\mathscr{S}^{\perp\perp} \subset \mathscr{S}$. Since $\mathscr{S} \subset \mathscr{S}^{\perp\perp}$ and $\mathscr{S}^{\perp\perp} \subset \mathscr{S}$, then $\mathscr{S} = \mathscr{S}^{\perp\perp}$.

8. The function $f(x)$, being continuous, has its maximum at a point x_0, $a \le x_0 \le b$. But if $|f(x_0)| \ge |f(x)|$, $a \le x \le b$, then also $|cf(x_0)| \ge |cf(x)|$. Thus,

$$\|cf\| = \max_{a \le x \le b} |cf(x)| = |cf(x_0)| = |c||f(x_0)| = |c| \max_{a \le x \le b} |f(x)| = |c|\|f\|.$$

9. We have, for each x, $a \le x \le b$, $|f(x) + g(x)| \le |f(x)| + |g(x)| \le \max_{a \le y \le b} |f(y)| + \max_{a \le y \le b} |g(y)| = \|f\| + \|g\|$. Thus, also

$$\max_{a \le x \le b} |f(x) + g(x)| \le \|f\| + \|g\|.$$

Consequently, $\|f + g\| = \max_{a \le x \le b} |f(x) + g(x)| \le \|f\| + \|g\|$. Note that $\max_{a \le x \le b} |f(x) + g(x)|$ does not generally occur at the same value of x as the maximum of either $|f(x)|$ or $|g(x)|$.

Chapter 14

1. Since u_0 is the solution of (1), for any $u_n = u_0 + h$, $h \in \mathscr{H}$, we have $F[u_n] = (L(u_0 + h), u_0 + h) - 2(u_0 + h, v) = F[u_0] + (L(u_n - u_0), u_n - u_0)$ by symmetry of L. But $F[u_0] = \min F[u]$, and since $F[u_n] \to \min F[u]$, we must have $(L(u_n - u_0), u_n - u_0) \to 0$. Now, by coerciveness, $\|u_n - u_0\|^2 \le (1/c)(L(u_n - u_0), u_n - u_0)$, so that $u_n \to u_0$ strongly.

2. We represent the solution u_0 in the form $u_0 = \sum_{i=1}^{\infty} \alpha_i \phi_i$ and let the element $v_n \in \mathcal{H}_n$ be given by $v_n = \sum_{i=1}^{n} \alpha_i \phi_i$. Then for $n \to \infty$, we have $v_n \to u_0$ and $F[v_n] \to F[u_0]$. Let $u_n \in \mathcal{H}_n$ be the element of $\mathcal{H}_n$ for which $F[u_n] = \min_{u \in \mathcal{H}} F[u]$. Keeping in mind that $F[u_0] = \min_{u \in \mathcal{H}} F[u]$, we have evidently $F[u_0] \le F[u_n] \le F[v_n]$. Since $F[v_n] \to F[u_0]$, also $F[u_n] \to F[u_0]$, and by the conclusion of the previous problem, $u_n \to u_0$.

3. The error is $\|u_n - u_0\|$. We have $L(u_n - u_0) = Lu_n - v$, and $L^{-1}L(u_n - u_0) = L^{-1}(Lu_n - v)$ or $u_n - u_0 = L^{-1}(Lu_n - v)$, whence $\|u_n - u_0\| = \|L^{-1}(Lu_n - v)\| \le c\|Lu_n - v\|$.

4. Any $u_n \in \mathcal{H}_n$ can be represented as $u_n = \sum_{i=1}^{n} \beta_i \phi_i$. Thus,

$$F[u_n] = (Lu_n, u_n) - 2(u_n, v)$$

$$= \left(\sum_{i=1}^{n} \beta_i(L\phi_i), \sum_{i=1}^{n} \beta_i \phi_i \right) - 2\left(\sum_{i=1}^{n} \beta_i \phi_i, v \right)$$

$$= \sum_{i,j=1}^{n} \beta_i \beta_j(L\phi_i, \phi_j) - 2\sum_{i=1}^{n} \beta_i(\phi_i, v).$$

Setting $\partial F[u_n]/\partial \beta_i = 0$, $i = 1, 2, \ldots, n$, we arrive at the equations

$$\sum_{j=1}^{n} \beta_j(L\phi_j, \phi_i) - (v, \phi_i) = \left(\sum_{j=1}^{n} \beta_j L\phi_j - v, \phi_i \right)$$

$$= \left(L\left(\sum_{j=1}^{n} \beta_j \phi_j \right) - v, \phi_i \right) = (Lu_n - v, \phi_i) = 0, \quad i = 1, 2, \ldots, n.$$

This proves our assertion. The latter equations are known as Galerkin's equations. It is often advantageous to select coordinate functions $\{\phi_i\}$ satisfying the boundary conditions if the problem of interest is a boundary value problem.

5. In Green's third identity $\int_{\Omega} (v\nabla^2 u - u\nabla^2 v)\, d\Omega = \int_{\partial\Omega} [v(\partial u/\partial n) - u(\partial v/\partial n)]\, ds$, we replace u by $\sum_{i=1}^{n} \beta_i \nabla^2 \phi_i$ and v by ϕ_j, $j = 1, 2, \ldots, n$. Bearing in mind the boundary value problem under discussion, viz., $\nabla^4 w = q/D$ in Ω and $w = \partial w/\partial n = 0$ on $\partial\Omega$, we select ϕ_i's satisfying the boundary conditions and find the Galerkin equations $\int_{\Omega} \phi_j \sum_{i=1}^{n} \beta_i \nabla^4 \phi_i\, d\Omega = \sum_{i=1} \beta_i \int_{\Omega} \nabla^2 \phi_i \nabla^2 \phi_j\, d\Omega = \int_{\Omega} (q/D)\phi_j\, d\Omega$, $j = 1, 2, \ldots, n$. By inspection, the symmetry is proved.

6. We have, using the third Green identity, $\int_{\Omega} qv\, d\Omega = \int_{\Omega} v\nabla^2 u\, d\Omega = \int_{\Omega} u\nabla^2 v\, d\Omega + \int_{\partial\Omega} (v(\partial u/\partial n) - u(\partial v/\partial n))\, ds = \int_{\Omega} u\nabla^2 v\, d\Omega$. Hence,

$$F[v] + \int_{\Omega} u^2\, d\Omega = \int_{\Omega} [v\nabla^4 v - 2vq + u^2]\, d\Omega = \int_{\Omega} [(\nabla^2 v)^2 - 2vq + u^2]\, d\Omega$$

$$= \int_{\Omega} [(\nabla^2 v)^2 - 2u\nabla^2 v + u^2]\, d\Omega = \int_{\Omega} (\nabla^2 v - u)^2\, d\Omega \ge 0,$$

and our assertion is proved.

7. Set $\varepsilon_i = u_i - u_0$, $i = 1, 2, \ldots, n$. Then $\sum_{i=1}^{n} L_i \varepsilon_i = \sum_{i=1}^{n} L_i u_i - \sum_{i=1}^{n} L_i u_0 = f - Lu_0 = 0$, (a). Now,

$$F[u_i] = F[u_0 + \varepsilon_i] = \sum_{i=1}^{n} (L_i(u_0 + \varepsilon_i), (u_0 + \varepsilon_i))$$

$$= \sum_{i=1}^{n} [(L_i u_0, u_0) + (L_i u_0, \varepsilon_i) + (L_i \varepsilon_i, u_0) + (L_i \varepsilon_i, \varepsilon_i)]$$

$$= \left(\sum_{i=1}^{n} L_i u_0, u_0 \right) + \sum_{i=1}^{n} (L_i u_0, \varepsilon_0) + \left(\sum_{i=1}^{n} L_i \varepsilon_i, u_0 \right) + \sum_{i=1}^{n} (L_i \varepsilon_i, \varepsilon_i)$$

$$= (Lu_0, u_0) + \sum_{i=1}^{n} (L_i \varepsilon_i, \varepsilon_i),$$

where we have used equation (a) and the symmetry of the operators L_i. Thus, $F[u_0 + \varepsilon_i] = (Lu_0, u_0) + \sum_{i=1}^{n} (L_i \varepsilon_i, \varepsilon_i) \geq (Lu_0, u_0)$ on account of definiteness of the L_i's. Consequently, $F[u_i] \geq (Lu_0, u_0)$, with equality for $\varepsilon_i = 0$ or $u_i = u_0$, $i = 1, 2, \ldots, n$. We note that the inner product (Lu_0, u_0) represents the square of the norm earlier denoted by $\|u_0\|_H$ [cf. equation (10.10)].

8. The bilinear form associated with $F[u]$ is $F[u, v] = \int_\Omega \nabla u \nabla v \, d\Omega = F[u] + 2F[u, v] + F[v]$, (3). Now let u satisfy the boundary condition (2) and let v be a harmonic function. Then, by the first Green identity: $F[u, v] = \int_\Omega \nabla u \nabla v \, d\Omega = - \int_\Omega u \nabla^2 v \, d\Omega + \int_{\partial\Omega} u(\partial v/\partial n) \, ds = 0$, (4). Now if u is a solution of Poisson's equation and u_0 is the solution to the boundary value problem under scrutiny, then $v = u - u_0$ is a harmonic function, and, on account of (4), equation (3) yields $F[u] = F[u_0 + v] = F[u_0] + F[v] > F[u_0]$. Thus, assuming that $v \neq 0$, $F[u]$ becomes minimum for $u = u_0$.

Chapter 15

1. The symmetry of the alternative form is easily shown by interchanging the order of integration. Furthermore,

$$\langle Lu, u \rangle = \int_0^1 [du(t)/dt] \left[\int_0^{1-t} u(\tau) \, d\tau \right] dt$$

$$= u(t) \int_0^{1-t} u(\tau) \, d\tau \Big|_0^1 + \int_0^1 u(t)u(1-t) \, dt = \int_0^1 u(t)u(1-t) \, dt,$$

where the integral was differentiated with respect to the parameter t and account was taken of the condition $u(0) = 0$.

2. We have $\langle Lu, u \rangle = (Lu, Lu) = \int_0^1 (du/dt)^2 \, dt$ and $\langle f, u \rangle = (f, Lu) = \int_0^1 f(du/dt) \, dt$.

3. By equation (12.8), $(S_1, S_2) = \int_V \tau_{ij}^1 e_{ij}^2 \, dV = 1/2 \int_V (u_{i,j}^1 + u_{j,i}^1)\tau_{ij}^2 \, dV = \int_V [(u_i^1 \tau_{ih}^2)_{,j} - u_i^1 \tau_{ij,j}^2] \, dV$, where the symmetry of τ_{ij} and the arbitrariness of dummy indices were considered. Use of the Gauss–Green theorem and the rela-

tions (15.12) and (15.13) now gives: $(S_1, \ S_2) = \int_S u_i{}^1 \tau_{ij}^2 n_j \, dS - \int_V u_i{}^1 \tau_{ij,j}^2 \, dV = \int_S u_i{}^1 f_i{}^2 \, dS - \int_V u_i{}^1 F_i \, dV = 0.$

4. On account of the orthotropy, we have $e_{11}^{(1)} = \tau_{11}^{(1)}/E_1$, $e_{22}^{(1)} = -v_{12}\tau_{11}^{(1)}/E_1$, $e_{22}^{(2)} = \tau_{22}^{(2)}/E_2$, $e_{11}^{(2)} = -v_{21}\tau_{22}^{(2)}/E_2$, where v_{12} is associated with a contraction in the 2-direction under tension in the 1-direction. By (12.8), we have

$$(S_1, S_1) = (\tau_{11}^{(1)})^2/E_1, \ (S_2, S_2) = (\tau_{22}^{(2)})^2/E_2,$$
$$(S_1, S_2) = -v_{21}\tau_{11}^{(1)}\tau_{22}^{(2)}/E_2 = -v_{12}\tau_{11}^{(1)}\tau_{22}^{(2)}/E_1.$$

Now

$$\cos \measuredangle \, [S_1, S_2] = (S_1, S_2)/\|S_1\| \, \|S_2\| = -v_{21}(E_1/E_2)^{1/2} = -v_{12}(E_2/E_1)^{1/2}.$$

Thus, the states fail to be orthogonal, since Poisson's constants are different from zero. Incidentally, we arrive here at the well-known relation $E_1 v_{21} = E_2 v_{12}$.

5. Take $u(x) = u_n(x) + \varepsilon v(x)$, where ε is a real number. We have $\rho(u) = [\lambda_n(u_n, u_n) + 2\varepsilon\lambda_n(u_n, v) + \varepsilon^2(v, Lv)]/[(u_n, u_n) + 2\varepsilon(u_n, v) + \varepsilon^2(v, v)] = \lambda_n + \varepsilon^2[(v, Lv) - \lambda_n(v, v)]/[(u_n, u_n) + 2\varepsilon(u_n, v) + \varepsilon^2(v, v)]$, after utilizing the symmetry of L. Hence, the functional $\rho(u)$ has a stationary value for $\varepsilon = 0$, inasmuch as $d(\varepsilon^2)/d\varepsilon = 0$ for $\varepsilon = 0$, and out claim is proved.

6. (1) Since $e^{\alpha t} - c$ is a continuous monotonic function defined in a closed interval, it takes its extreme values at the ends of the interval, where the approximation error also has its extreme values. We make the error at the ends equal if $|e^{\alpha t} - c|_{t=0} = |e^{\alpha t} - c|_{t=1}$. This gives $c = \frac{1}{2}(1 + e^\alpha)$ and the error norm $\frac{1}{2}(e^\alpha - 1)$.

(2) In this case, the best approximation corresponds to the value of c that minimizes $\{\int_0^1 [e^{\alpha t} - c]^2 \, dt\}^{1/2}$. This gives $c = (e^\alpha - 1)/\alpha$ and the error norm $[(e^{2\alpha} - 1)/2\alpha - (e^\alpha - 1)^2/\alpha^2]^{1/2}$.

If $\alpha = 1$, the error norms in the cases in question are about 0.86 and 0.49, respectively.

7. The symmetry of L on the domain D is easily shown by repeated integration by parts of the integral $(Lv, w) = -S \int_0^l v''w \, dx$, where $v, w \in D$. Now let v be a virtual displacement, $v \in D$; then $\Pi(v) = (S/2) \int_0^l (v')^2 \, dx - \int_0^l pv \, dx$. But $(Lv, v) = -Svv'|_0^l + S \int_0^l (v')^2 \, dx \geq 0$ and $\Pi(v) = \frac{1}{2}(Lv, v) - (p, v)$. Take $v = w + f$, where w and $f \in D$. Then $\Pi(v) - \Pi(w) = \frac{1}{2}(L(w + f), w + f) - \frac{1}{2}(Lw, w) - (p, f) = (Lw - p, f) + (Lf, f)$. Setting $w = u$, the solution of the Problems (1) and (2), and considering that $(Lf, f) \geq 0$ for any $f \in D$, we prove our assertion.

8. (a) If $F[v]$ is stationary for $v = u$, then, setting $v = u + \varepsilon f$ for any $f \in S$, we have

$$0 = (d/d\varepsilon)\{F[u + \varepsilon f]\}_{\varepsilon = 0}$$
$$= (d/d\varepsilon)[b(u, u) + 2\varepsilon(u, f) + \varepsilon^2 b(f, f) - 2l(u) - 2\varepsilon l(f)]_{\varepsilon = 0} = 2b(u, f) - 2l(f),$$

as claimed.

(b) If u is the solution defined above, then by simply setting $f = u$ in the preceding equation, we obtain the desired result.

9. We first recall that a set S is convex [compare the footnote preceding equation (6.18)], if whenever v and w are in S, so is $\alpha v + \beta w$, where $\alpha, \beta \geq 0, \alpha + \beta = 1$. Now, if $v, w \in S, \alpha, \beta \geq 0$, and $\alpha + \beta = 1$, then $\alpha v(1) + \beta w(1) \geq 0$, so that S is convex, and if $u, v \in S$, then $u + \gamma(v - u) \in S$ for each $\gamma, 0 \leq \gamma \leq 1$. Consequently, if u satisfies (1) and $v \in S$, then $F[v] = F[u + v - u] = b(u + v - u, u + v - u) - 2l(u + v - u) = F[u] + b(v - u, v - u) + 2[b(u, v - u) - l(v - u)] \geq F[u]$, and our assertion is proved.

10. The left-hand bound is gained by observing that $b(v - w, v - w) \geq 0$ and replacing w by $wb(v, w)/b(w, w)$; note that $b(v, w)$ is here a number. In order to obtain the right-hand bound, we write $b(w, w) = b(v + w - v, v + w - v) = b(v, v) + 2b(w - v, v) + b(w - v, w - v) \geq b(v, v)$.

References

1. J. Céa, *Optimisation: Théorie et Algorithmes*, Dunod, Paris (1971).
2. G. B. Edelen, in: *Continuum Physics* (A. C. Eringen, ed.), Vol. 4, pp. 76–204, Academic Press, New York (1976).
3. A. C. Eringen, in: *Continuum Physics* (A. C. Eringen, ed.), Vol. 4, pp. 205–267, Academic Press, New York (1976).
4. R. M. Christensen, *Theory of Viscoelasticity*, Academic Press, New York (1971).
5. Z. Nehari, *Conformal Mapping*, Dover Publications, New York (1975).
6. *The Random House Dictionary of the English Language* (L. Urdang, ed.), Random House, New York (1968).
7. S. K. Berberian, *Introduction to Hilbert Space*, Oxford University Press, New York (1961).
8. S. G. Mikhlin, *Variational Methods in Mathematical Physics*, Macmillan Company, New York (1964).
9. A. W. Naylor and G. R. Sell, *Linear Operator Theory in Engineering and Science*, Holt, Rinehart and Winston, New York (1971).
10. I. N. Sneddon, in: *Continuum Physics* (A. C. Eringen, ed.), Vol. 1, pp. 356–490, Academic Press, New York (1971).
11. A. E. Taylor, *Introduction to Functional Analysis*, J. Wiley and Sons, New York (1958).
12. T. J. Fletcher, *Linear Algebra Through Its Applications*, Van Nostrand Reinhold Company, London (1972).
13. M. Hausner, *A Vector Space Approach to Geometry*, Prentice-Hall, Englewood Cliffs (1965).
14. J. T. Oden, *Applied Functional Analysis, A First Course for Students of Mechanics and Engineering Science*, Prentice-Hall, Englewood Cliffs (1979).
15. E. Kreyszig, *Introductory Functional Analysis with Applications*, J. Wiley and Sons, New York (1978).
16. S. Lipschutz, *Set Theory and Related Topics*, Schaum Publishing Company, New York (1964).
17. P. K. Rashevski, *Geometry of Riemann and Tensor Analysis* (Polish translation) Government Scientific Publication (P.W.N.), Warsaw (1958).
18. K. O. Friedrichs, *From Pythagoras to Einstein*, Random House, New York (1965).
19. C. Lanczos, *The Variational Principles of Mechanics*, University Press, Toronto (1960).
20. P. R. Halmos, *Finite-Dimensional Vector Spaces*, D. Van Nostrand Company, Princeton (1958).

21. I. S. Sokolnikoff, *Tensor Analysis*, J. Wiley and Sons, New York (1962).
22. B. Friedman, *Principles and Techniques of Applied Mathematics*, J. Wiley and Sons, New York (1956).
23. B. Z. Vulikh, *Introduction to Functional Analysis for Scientists and Technologists*, Pergamon Press, Oxford (1963).
24. I. Stakgold, *Green's Functions and Boundary Value Problems*, J. Wiley and Sons, New York (1979).
25. S. G. Gould, *Variational Methods for Eigenvalue Problems*, University of Toronto Press, Toronto (1957).
26. A. N. Kolmogorov and S. V. Fomin, *Introductory Real Analysis* (freely revised English version by R. A. Silverman), Dover Publications, New York (1970).
27. I. S. Sokolnikoff and R. M. Redheffer, *Mathematics of Physics and Modern Engineering*, McGraw-Hill Book Company, New York (1958).
28. L. Collatz, *Functional Analysis and Numerical Mathematics* (in German), Springer-Verlag, Berlin (1968).
29. J. B. Diaz, Upper and lower bounds for quadratic functionals, *Collectanea Mathematica*, Vol. 4, pp. 1–50 (1951).
30. K. Yosida, *Functional Analysis*, Springer-Verlag, Berlin (1968).
31. I. S. Sokolnikoff, *Mathematical Theory of Elasticity*, McGraw-Hill Book Company, New York (1956).
32. S. Timoshenko, *Theory of Plates and Shells*, McGraw-Hill Book Company, New York (1940).
33. P. Rafalski, Orthogonal projection method, I, II, and III, *Bull. Pol. Acad. Sci., Ser. Sci. Techn.*, Vol. 17, pp. 63–67, 69–73, and 167–171 (1969).
34. F. Mandl, *Quantum Mechanics*, Butterworths Publications, London (1957).
35. A. Messiah, *Quantum Mechanics* (2 vols.), North-Holland Publishing Co., Amsterdam (1962).
36. E. C. Kemble, *The Fundamental Principles of Quantum Mechanics with Elementary Applications*, Dover Publications, New York (1958).
37. J. L. Powell and B. Crasemann, *Quantum Mechanics*, Addison-Wesley Publishing Company, Reading (1961).
38. D. T. Gillespie, *A Quantum Mechanics Primer*, J. Wiley and Sons, New York (1974).
39. J. von Neumann, *Mathematical Foundations of Quantum Mechanics* (in German), J. Springer, Berlin (1932); English translation, Princeton University Press (1955).
40. J. L. Synge, *The Hypercircle in Mathematical Physics*, University Press, Cambridge (1957).
41. G. E. Shilov, *An Introduction to the Theory of Linear Spaces*, Dover Publications (1974).
42. G. Rieder, in: *Applications of Methods of Functional Analysis to Problems of Mechanics* (P. Germain and B. Nayroles, eds.), pp. 450–461, Springer-Verlag, Berlin (1976).
43. S. Lipschutz, *General Topology*, Schaum's Outline Series, McGraw-Hill Book Company, New York (1965).
44. J. W. Dettman, *Mathematical Methods in Physics and Engineering*, McGraw-Hill Book Company, New York (1962).
45. J. B. Diaz and A. Weinstein, The torsional rigidity and variational methods, *Am. J. Math.*, Vol. 70, pp. 107–116 (1948).
46. R. Weinstock, *Calculus of Variations*, McGraw-Hill Book Company, New York (1952).
47. S. G. Lekhnitskii, *Theory of Elasticity of an Anisotropic Body*, Holden-Day Inc., San Francisco (1963).
48. J. L. Nowinski, Cauchy-Schwarz inequality and the evaluation of torsional rigidity of anisotropic bars, *SIAM J. Appl. Math.*, Vol. 24, pp. 324–331 (1973).

49. J. C. S. YANG, in: *Developments in Mechanics* (T. C. Huang and M. Johnson Jr., eds.), Vol. 3, pp. 377–391, J. Wiley and Sons, New York (1965).

50. A. WEINSTEIN, in: *Proceedings of Symposia in Applied Mathematics* (R. V. Churchill, E. Reissner, and A. H. Taub, eds.), Vol. 3, pp. 141–162, McGraw-Hill Book Co., New York (1950).

51. Y. A. PRATUSEVITCH, *Variational Methods in Structural Mechanics* (in Russian), Government Publication of Technical Theoretical Literature (OGIZ), Moscow (1958).

52. J. L. NOWINSKI, Bilateral bounds for the solution of a generalized biharmonic boundary value problem, *SIAM J. Appl. Math., Part A*, Vol. 39, p. 193 (1980).

53. S. BERGMAN and M. SCHIFFER, *Kernel Functions and Elliptic Differential Equations in Mathematical Physics*, Academic Press, New York (1953).

54. J. B. DIAZ and H. J. GREENBERG, Upper and lower bounds for the solution of the first biharmonic boundary value problem, *J. Math. Phys.*, Vol. 27, pp. 193–201 (1948).

55. G. FICHERA, On some general integration methods employed in connection with linear differential equations, *J. Math. Phys.*, Vol. 29, pp. 59–68 (1950).

56. L. E. PAYNE and H. F. WEINBERGER, New bounds in harmonic and biharmonic problems, *J. Math. Phys.*, Vol. 33, pp. 291–307 (1958).

57. J. H. BRAMBLE, Continuation of biharmonic functions across circular arcs, *J. Math. Mech.*, Vol. 7, pp. 905–924 (1958).

58. I. N. SNEDDON and D. S. BERRY, in: *Handbuch der Physik* (S. Flugge, ed.), Vol. 6, pp. 1–126, Springer-Verlag, Berlin (1958).

59. N. I. MUSKHELISHVILI, *Some Basic Problems of Mathematical Theory of Elasticity* (transl. by J. R. M. Radok), P. Noordhoff Ltd., Groningen (1963).

60. S. TIMOSHENKO and S. WOINOWSKY-KRIEGER, *Theory of Plates and Shells*, McGraw-Hill Book Company, New York (1959).

61. M. T. HUBER, Theory of plates rectilinearly directional (in Polish), *Archive of the Scientific Society*, Lvov (1921).

62. M. T. HUBER, *Problems of Statics of Technically Important Orthotropic Plates* (in German), Academy of Technical Sciences, Warsaw (1929).

63. W. NOWACKI, Contributions to the theory of orthotropic plates (in German), *Acta Techn. Acad. Sci. Hung.*, Vol. 8, pp. 109–128 (1954).

64. J. MOSSAKOWSKI, Singular solutions in the theory of orthotropic plates (in Polish), *Arch. Appl. Mech.*, (a) Vol. 6, pp. 413–432 (1954), (b) Vol. 7, pp. 97–110 (1955).

65. R. T. S. HEARMON, *An Introduction to Applied Anisotropic Elasticity*, Oxford University Press, Oxford (1961).

66. J. R. VINSON and T. W. CHOU, *Composite Materials and Their Use in Structures*, Applied Science Publishers, London (1975).

67. K. BRODOVITSKII, On the integral $\int_0^\pi \sin^m x/(p + q \cos x)\, dx$ (in Russian), *Dokl. Acad. Sci. USSR*, Vol. 120, No. 6 (1958).

68. T. KATO, On some approximate methods concerning the operator T^*T, *Math. Ann.*, Vol. 126, pp. 253–262 (1953).

69. H. FUJITA, Contributions to the theory of upper and lower bounds in boundary value problems, *J. Phys. Soc. Japan*, Vol. 10, pp. 1–8 (1955).

70. Y. NAKATA and H. FUJITA, On upper and lower bounds of the eigenvalues of a free plate, *J. Phys. Soc. Japan*, Vol. 10, pp. 823–824 (1955).

71. T. KATO, H. FUJITA, Y. NAKATA, and M. NEWMAN, Estimation of the frequencies of thin elastic plates with free edges, *J. Nat. Bur. Stand.*, Vol. 59, pp. 169–186 (1957).

72. M. D. GREENBERG, *Application of Green's Functions in Science and Engineering*, Prentice-Hall, Englewood Cliffs (1971).

73. F. J. Murray, Linear transformation between Hilbert spaces, *Trans. Am. Math. Soc.*, Vol. 37, pp. 301–338 (1935).

74. S. Timoshenko and J. N. Goodier, *Theory of Elasticity*, McGraw-Hill Book Company, New York (1951).

75. B. Diaz and H. J. Greenberg, Upper and lower bounds for the solution of the first boundary value problem of elasticity, *Q. Appl. Math.*, Vol. 6, pp. 326–331 (1948).

76. K. Washizu, Bounds for solutions of boundary value problems in elasticity, *J. Math. Phys.*, Vol. 32, pp. 117–128 (1953).

77. W. Prager and J. L. Synge, Approximations in elasticity based on the concept of function space, *Q. Appl. Math.*, Vol. 5, pp. 241–269 (1947).

78. C. E. Pearson, *Theoretical Elasticity*, Harvard University Press, Cambridge (1959).

79. J. L. Nowinski and K. H. Cho, The hypercircle method in the problem of a case bounded cylinder in the gravity field, *Iran. J. Sci. Tech.*, Vol. 6, pp. 171–176 (1977).

80. B. S. Seidel, J. F. Sontowski, and R. D. Swope, in: *Proc. Fifth Int. Symp. Space Techn.*, pp. 73–86, Tokyo (1963).

81. C. G. Maple, The Dirichlet problem: bounds at a point for the solution and its derivatives, *Q. Appl. Math.*, Vol. 8, pp. 213–228 (1950).

82. S. Lipschutz, *Linear Algebra*, McGraw-Hill Book Company, New York (1968).

83. J. L. Nowinski, *Theory of Thermoelasticity With Applications*, Sijthoff and Noordhoff, Alphen a.d. Rijn (1978).

84. A. P. Wills, *Vector Analysis with an Introduction to Tensor Analysis*, Dover Publications, New York (1958).

85. J. L. Nowinski and K. H. Cho, On the application of the method of orthogonal projection in heat conduction, *J. Therm. Stresses*, Vol. 1, pp. 63–74 (1978).

86. P. J. Davis, *Interpolation and Approximation*, Dover Publications, New York (1975).

87. N. I. Achieser, *Theory of Approximations*, F. Ungar, New York (1956).

88. S. Buck, in: *Numerical Analysis* (R. E. Langer, ed.), Vol. 1, pp. 11–23, University of Wisconsin Press, Madison (1959).

89. S. Timoshenko, *Vibration Problems in Engineering*, D. Van Nostrand Company, Princeton (1955).

90. J. L. Oden, *Finite Elements of Nonlinear Continua*, McGraw-Hill Book Company, New York (1972).

91. R. Courant and D. Hilbert, *Methods of Mathematical Physics* (in German), Vol. 1, Springer, Berlin (1931).

92. K. Girkmann, *Flaechentragwerke*, Springer-Verlag, Vienna (1959).

93. M. E. Gurtin, Variational principles for linear initial value problems, *Q. Appl. Math.*, Vol. 22, pp. 252–256 (1964).

94. N. Laws, The use of energy theorems to obtain upper and lower bounds, *J. Inst. Math. Appl.*, Vol. 15, pp. 109–119 (1975).

95. A. M. Arthurs, *Complementary Variational Principles*, Oxford University Press, Oxford (1970).

96. B. Noble and M. J. Sewell, On dual extremum principles in applied mathematics, *J. Inst. Math. Appl.*, Vol. 9, pp. 123–193 (1972).

97. W. Prager, *The Extremum Principles of the Mathematical Theory of Elasticity and Their Use in Stress Analysis*, University of Washington, Engineering Experimental Station, Bulletin No. 119 (1951).

98. G. Duvant and J. L. Lions, *Inequalities in Mechanics and Physics*, Springer-Verlag, Berlin (1976).

99. S. L. Sobolev, *Applications of Functional Analysis in Mathematical Physics*, American Mathematical Society, Providence (1950).

100. L. Schwartz, *Theory of Distributions* (in French), Vols. 1 and 2, Hermann et Cie, Paris (1950–1951).

101. J. Mikusinski and R. Sikorski, The elementary theory of distributions, *Rozpr. Mat.*, No. 25, Warsaw (1961).

102. G. Temple, Theories and applications of generalized functions, *J. London Math. Soc.*, Vol. 28, pp. 134–148 (1953).

103. M. J. Lighthill, *Introduction to Fourier Analysis and Generalized Functions*, Cambridge University Press, Cambridge (1970).

104. P. M. Morse and H. Fesbach, *Methods of Theoretical Physics*, Vols. 1 and 2, McGraw-Hill Book Company, New York (1953).

105. A. N. Kolmogorov and S. V. Fomin, *Elements of the Theory of Functions and Functional Analysis* (in Russian), Nauka, Moscow (1968).

106. H. S. Carslaw and J. C. Jaeger, *Conduction of Heat in Solids*, Clarendon Press, Oxford (1959).

107. A. Kufner, J. Oldrich, and S. Fucik, *Function Spaces*, Noordhoff International Publishing, Leyden (1977).

108. R. J. Weinacht, Weak solutions of partial differential equations, *Math. Mag.*, Rensselear Polytechnic Institute, No. 3, New York (1970).

109. G. Szefer and L. Demkowicz, *Application of Sobolev Space Approximation Method to the Solution of Elastic Plates*, Report 10.3-05.12, coordinated with Institute of Technological Research, Cracow (private communication in 1978).

100. A. E. H. Love, *A Treatise on the Mathematical Theory of Elasticity*, Dover Publications, New York (1944).

111. R. J. Knops and L. E. Payne, *Uniqueness Theorems in Linear Elasticity*, Springer-Verlag, New York (1971).

112. C. C. Wang and C. Truesdell, *Introduction to Rational Elasticity*, Noordhoff International Publishing Company, Leyden (1973).

113. G. Fichera, in: *Handbuch der Physik* (C. Truesdell, ed.), Vol. 6a/2, pp. 347–424, Springer-Verlag, Berlin (1972).

114. M. E. Gurtin, in: *Handbuch der Physik* (C. Truesdell, ed.), Vol. 6a/2, pp. 1–295, Springer-Verlag, Berlin (1972).

115. F. Magri, Variational formulation for every linear equation, *Int. J. Eng. Sci.*, Vol. 12, pp. 537–549 (1974).

116. E. Tonti, On the variational formulation for linear initial value problems, *Ann. Mat. Pura Appl.*, Vol. 95, pp. 331–359 (1973).

117. Z. K. Rafalson, A problem arising in the solution of the biharmonic equation (in Russian), *Dokl. Acad. Sci. USSR*, Vol. 64, p. 779 (1949).

118. M. G. Slobodianskii, The transformation of minimum functional problems into problems of mechanics (in Russian), *Dokl. Acad. Sci. USSR*, Vol. 91, No. 4 (1953).

119. S. G. Lekhnitskii, *Anisotropic Plates* (in Russian), G.I.T.T.L., Moscow (1957); Eng. translation: Holden-Day, San Francisco (1963).

120. K. Rektorys, ed., *Survey of Applicable Mathematics*, M.I.T. Press, Cambridge (1969).

Index